BARRON'S

FCAT

HIGH SCHOOL MATH

FLORIDA COMPREHENSIVE ASSESSMENT TEST

2ND EDITION

Pamela Windspirit, M.Ed.
High School Alternative Education Teacher
The Academy at Charlotte Technical Center
Port Charlotte, Florida

About the Author: Pamela Windspirit has a master's degree in education and is an alternative high school math teacher in Florida. She is National Board Certified and is the National Board Coordinator for Charlotte County, Florida.

All inquiries should be addressed to:
Barron's Educational Series, Inc.
250 Wireless Boulevard
Hauppauge, New York 11788
www.barronseduc.com

Library of Congress Catalog Card No. 2008036385
ISBN-13: 978-0-7641-4015-0
ISBN-10: 0-7641-4015-9

Library of Congress Cataloging-in-Publication Data

Windspirit, Pamela J.
Barron's FCAT high school math : Florida Comprehensive Assessment Test / Pamela J. Windspirit.—2nd ed.
p. cm.
Includes index.
ISBN-13: 978-0-7641-4015-0
ISBN-10: 0-7641-4015-9
1. Mathematics—Examinations, questions, etc. 2. Florida Comprehensive Assessment Test—Study guides. 3. Mathematical ability—Testing. I. Windspirit, Pamela J. Barron's how to prepare for the FCAT. II. Title. III. Title: FCAT high school math.

QA43.W57 2009
510.76—dc22

2008036385

PRINTED IN THE UNITED STATES OF AMERICA
9 8 7 6 5 4 3 2 1

Contents

Introduction to the FCAT

WHAT IS THE FCAT?

FCAT stands for Florida Comprehensive Assessment Test. The FCAT is based on a set of standards called the Sunshine State Standards—a list of things students should know and be able to do in a given subject area. The relevant Sunshine State Standards are listed at the beginning of each chapter. The FCAT was designed to test what *you* know and are able to do at the end of the tenth grade.

HOW TO USE THIS BOOK

FCAT High School Math will help you do your best on the mathematics portion of the FCAT. This book reviews what you have learned in math and helps you become familiar not only with the types of questions on the FCAT but also with the way these questions are asked.

In this chapter we will look at some tips on reading FCAT questions; review some basics about FCAT questions, including examples of each type of question; and review calculator use.

Chapters 1–3 each deal with a particular topic covered on the FCAT, and each is divided into sections. Each section gives a review of problems and sample problems to work. Immediately following the sample questions are the answers and an explanation of how the problems should have been worked. At the end of each chapter are mixed practice problems and their answers. Try to work the sample problems and practice problems without looking at the answers. If you get stuck or miss a problem, read the explanation carefully, returning to the chapter explanation for similar problems if necessary.

Identifying Your Strengths and Weaknesses

The next section is a diagnostic test. You should take this test first to identify your strongest and weakest areas.

At the end of this book are two practice tests. These tests were designed to be similar to the actual FCAT. Answers and explanations are provided at the end of each test so that you can check to see if you are prepared.

Test Tips

When taking the FCAT, you should keep the following test tips in mind.

- **Read the directions.** Reading the directions carefully is one of the keys to doing well on any test, especially the FCAT. You've probably heard this before. If you don't understand the directions given for the test, you should ask the teacher for an explanation immediately.

- **Read the problem carefully.** If you are having trouble doing a problem or getting an answer that makes sense, it may not be because you don't know how to do the problem. You may have missed something simple in the original question. Read the problem *at least twice* before beginning to work on it.

- **Estimate before calculating.** Use estimation and common sense to get a rough idea of what your answer should be before beginning to work. Then compare your answer with your estimate. This allows you to spot calculation errors quickly.

- **Check your work.** You've probably heard this too. The FCAT is required for high school graduation in Florida. It's worth the extra time to make certain you've worked every problem correctly. Use all the time you are given to check your answers.

Types of Questions

The FCAT has four types of questions.

- Multiple-choice
- Gridded response
- Short response (think, solve, explain)
- Extended response (think, solve, explain)

Multiple-Choice

- Ⓐ Ⓑ Ⓒ Ⓓ or Ⓕ Ⓖ Ⓗ Ⓘ
- Multiple-choice problems are worth 1 point each.
- A multiple-choice problem should take you about 1 minute to solve.
- Each problem has four answer choices (A, B, C, D or F, G, H, I).
- Read the question carefully and try to estimate what your answer should be before making a selection. You can often select the best answer based on your estimate without making any calculations.
- Know that there are always extra "attractive" answers in multiple-choice questions and that finding your answer among the choices does not guarantee that it is correct. These extra attractive answers are called distracters and can mislead you if you haven't read the question carefully.

- Make your choice by filling in the bubble completely. If you have to go back and erase, make certain you have erased everything. Tests are machine-scored, and the machine may pick up stray marks.

Sample Multiple-Choice Problems

Problem 1: The product of a number and 2.5 is less than 2.5. Which of the following could be the number?

A. 2.5
B. 1.25
C. 2
D. .25

Answer: Ⓐ Ⓑ Ⓒ ● (D)

Problem 2: Which of the following is closest to $2\sqrt{3}$?

F. 2.3
G. 4.2
H. 3.5
I. 6.0

Answer: Ⓕ Ⓖ ● (H) Ⓘ

Gridded Response

- Gridded response items use a bubble format.
- These problems should take approximately 1.5 minutes each to complete.
- Your answer should be written directly in the answer boxes in the grid and then filled in below.
- Align your answer all the way to the left or all the way to the right. Do *not* leave a blank answer box in the middle of your answer (see later examples).
- Use only one blank box per number or symbol. Do not add boxes. (If your answer is longer, you may need to put it into another form or round.)
- Be certain to completely fill one and *only* one bubble for each number or symbol that you write in a box.
- Do not add bubbles or fill in bubbles that are left over underneath blank answer boxes.
- Do not add symbols to the box or outside the box. Use only what you are given.
- Decimal points take up an entire box and should be filled in.

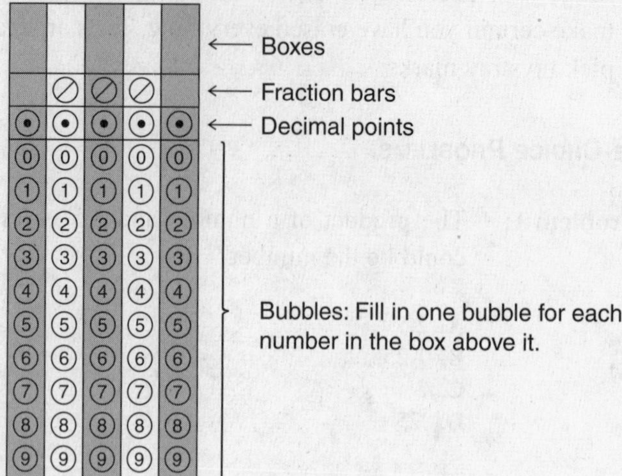

- Mixed numbers *cannot* be filled in. The machine reads an empty space as a "stop reading," so if you try to grid $3\frac{1}{4}$, the machine will read it as either $\frac{31}{4}$ or 3. Instead, change the mixed number to an improper fraction $\left(\frac{13}{4}\right)$ or the decimal equivalent (3.25) and grid it.
- Answers containing zeros such as .01 and 100 should be written in exactly and all numbers (including the zeros) filled in.
- You *must* fill in the bubbles correctly to receive credit for your answer.

Sample Gridded-Response Problems

Whole Numbers

Problem 1: Add 100 + 25.

Answer:

Numbers should be all the way left or right.

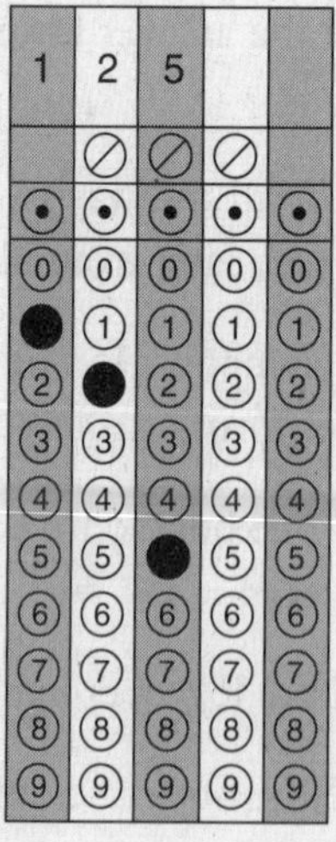

or

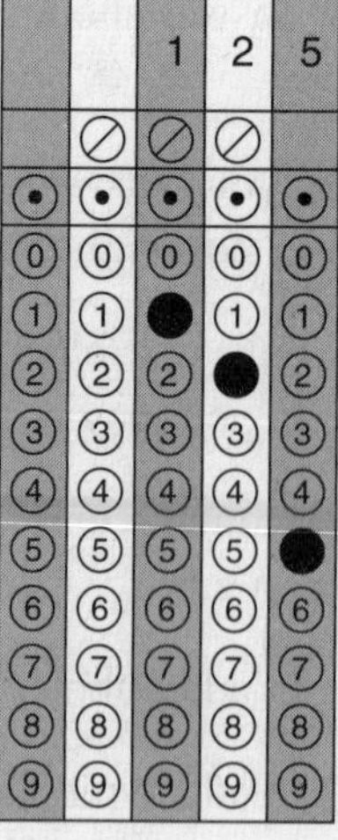

Decimals

Problem 2: Show the decimal equivalent of $2\frac{3}{100}$.

Answer:

2 • 0 3

or

2 • 0 3

Read as "two and three hundredths."

Fractions

Important: You *cannot* write a mixed number such as $2\frac{3}{8}$ in the answer grid. You must convert your answer to an improper fraction $\left(\frac{19}{8}\right)$ or to a decimal number (such as 2.375). Gridding $2\frac{3}{8}$ as a mixed number would cause the answer to be read as $\frac{23}{8}$, which is incorrect.

Problem 3: Add $1\frac{1}{4}+4\frac{3}{8}$.

Answer: $1\frac{1}{4}+4\frac{3}{8}=5\frac{5}{8}=\frac{45}{8}$ or 5.625

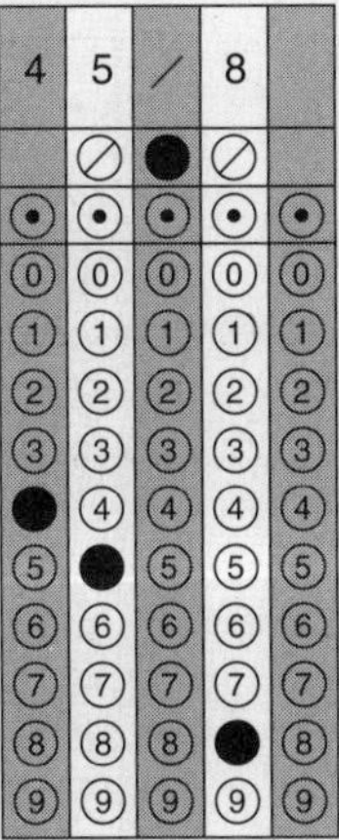

or

Decimals or Fractions

Some answers can be shown in either decimal form or fraction form and still be counted as correct.

Problem 4: Fran misses a basket 5 out of 25 times. Based on her record, what is the probability that she will make the basket the next time she shoots the ball?

Answer:

All are correct.

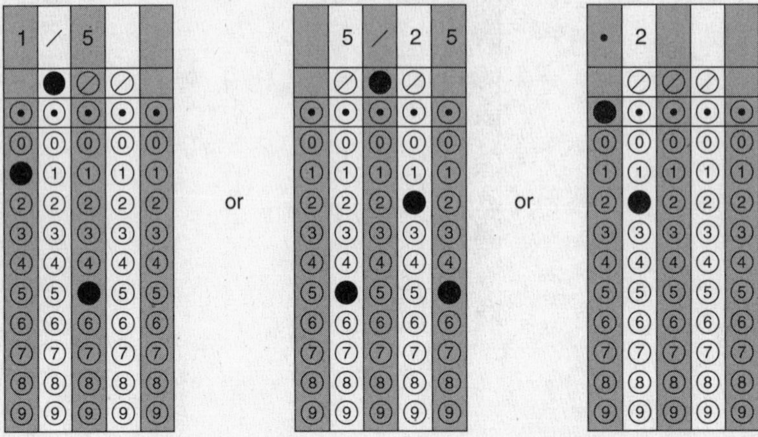

Range of Answers

If your answer is within a range of values, several values may be correct.

Problem 5: Give a possible value for x if $1.2 < x < 1.25$.

Answer: Values of x could be shown as

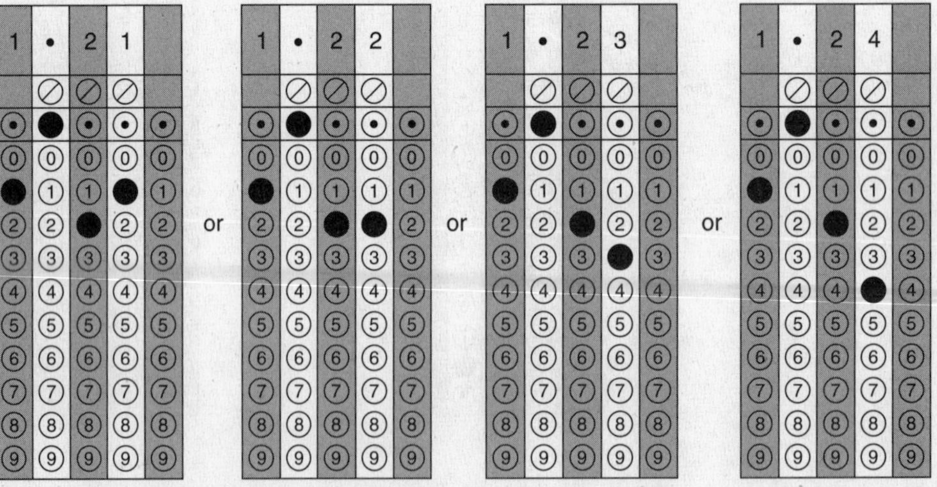

There are also many other correct responses for this problem.

Short-Response Problems

Short-response problems have the think-solve-explain symbol next to the problem.

THINK
SOLVE
EXPLAIN

- These problems are worth 2 points each. A partial answer may be worth 1 point.
- You should allow approximately 5 minutes for solving each problem.
- Read the question carefully.
- Use the numbers and information you are given in the problem to find the answer.
- If you cannot work the entire problem, work the parts that you can. You may receive partial credit.
- If you are asked to explain something, make your explanation clear and keep it short.
- *Show your work.* To receive partial credit it is *important* to show the steps you used to arrive at your answer even if you are using a calculator. For example, if you need to take the square root of 2, write $\sqrt{2}$ on your paper before finding the answer on your calculator.
- Check your work!

SAMPLE SHORT-RESPONSE PROBLEM

Slither's House Moving Service has moved a house from one block to another as shown here. Explain in words the two transformations that were used to move the house from block A to block B. Use mathematical terms such as *slide*, *flip*, or *rotate*.

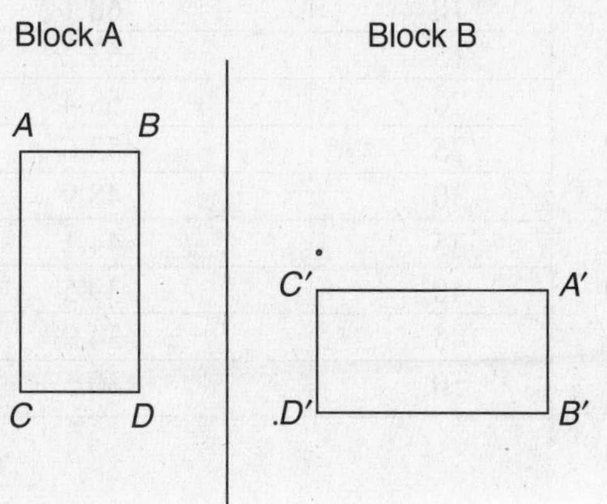

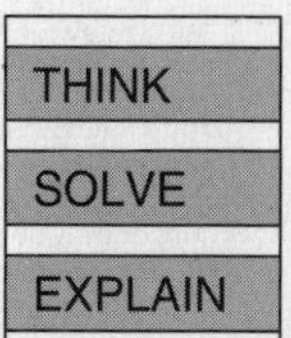

Answer: First, the house was translated (slid) over the line. Then it was rotated (turned) 90 degrees.

Extended Response

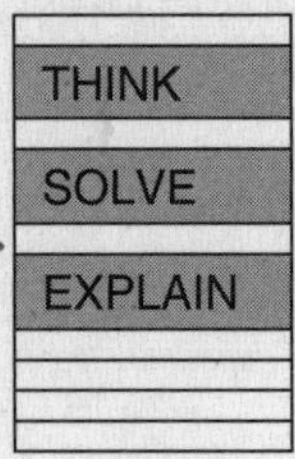

Extended-response problems have a larger think-solve-explain symbol next to the problem.

- These problems are worth 4 points each. Partial answers may be worth 1, 2, or 3 points.
- You should allow approximately 10 minutes for solving each problem.
- Read the question carefully.
- Use the numbers and information you are given in the problem to find the answer.
- If you cannot work the entire problem, work the parts that you can. You may receive partial credit.
- If you are asked to explain something, make your explanation clear and keep it short enough to fit in the space provided for the answer.
- *Show your work.* In order to receive full credit for the problem, you must show all the steps you used to arrive at the answer. Even if you make errors, showing correct steps can allow you to receive partial credit.
- Check your work!

SAMPLE EXTENDED-RESPONSE PROBLEM

The accompanying table shows the combined life expectancy in 2003 for all races.

Age in 2003	Life Expectancy (years)
5	73.1
10	68.1
15	63.2
20	58.4
25	53.6
30	48.9
35	44.1
40	39.5
45	34.9
50	30.5

(*a*) Use the grid to construct a scatterplot showing the relationships between a person's age and his or her life expectancy.

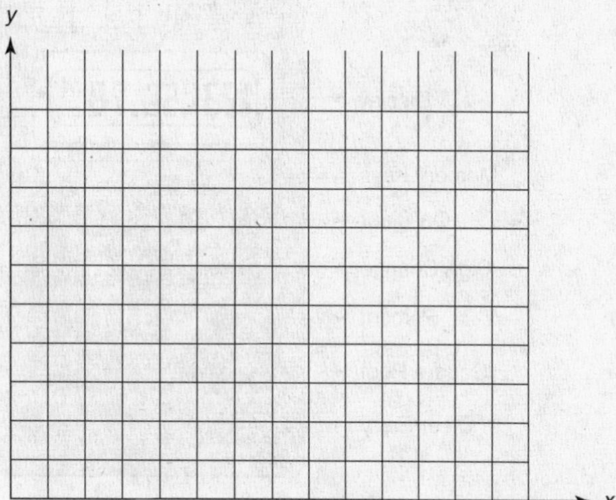

(*b*) Describe the relationship between age and life expectancy shown in the scatterplot.

(*c*) Use the scatterplot to predict the life expectancy of a person aged 65 years.

Answer: The scatterplot should look like the one shown here. The trend line has already been added to the scatterplot and extended to show that a 65-year-old person can expect to live approximately 17.5 more years. There is a *negative correlation* between age and life expectancy (the older you are, the shorter your life expectancy).

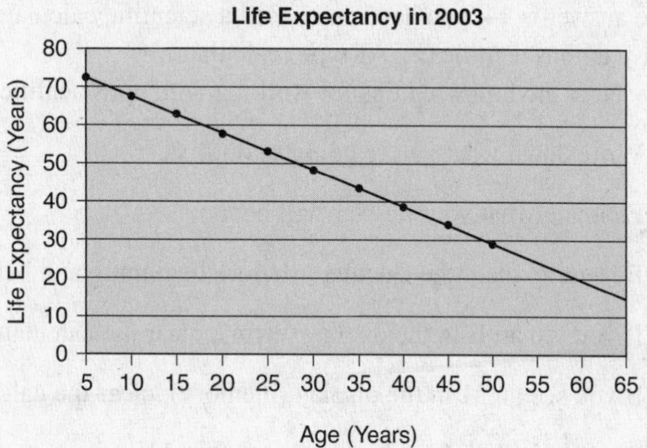

The FCAT Calculator

The FCAT calculator is a small nonscientific calculator. You will not be allowed to use any calculator other than the one provided for you. It is important that you become familiar with how a nonscientific calculator works and how it is different from scientific calculators, which are used in many classrooms.

A Generic Calculator and Its Functions

Solar Cell
Memory Indicator
Display
Negative Sign
Error Indicator
Memory Keys
On/Clear
Off
Sign Change
Division
Percent
Multiplication
Square Root
Subtraction
Clear Entry
Addition
Equals Sign
Decimal Point

Note: An FCAT calculator (nonscientific) is much simpler than a scientific calculator. Scientific calculators are able to do operations in the correct order but FCAT calculators are not. You must break the problems down into simpler parts before entering into the calculator. You can purchase a nonscientific calculator almost anywhere. They are quite inexpensive and well worth the money for practicing for the FCAT. If you want to be certain that you have purchased a nonscientific calculator, enter the following keystrokes in order: 2 + 4 × 3 =. If the answer is 14, you have purchased a scientific calculator. If you get an answer of 18, you have a nonscientific (FCAT-type) calculator.

Note the following tips for working with nonscientific calculators.

- Write down what you're going to do first.
- Estimate what your answer will be.
- Be sure to clear the calculator before beginning and between steps.
- If you see an **E** in the display (error), clear the calculator before beginning.
- If you see an **M** in the display (memory), clear the calculator before beginning.
- Do all calculations at least twice to avoid key-in errors.
- Know the order of operations (your calculator does not).
- Some answers (particularly those involving fractions) may have to be rounded.

Error Messages

Error messages (signified by an **E** in the display) can occur if you are working too fast for the calculator to keep up. They can also occur if you attempt to work with numbers that are too big for the display. An FCAT calculator holds only eight digits.

Example: If you multiply 10,250 by 56,198, you might get an answer that appears to be smaller than the two numbers you started with, that is, 5.7602950 **E**. This means you have gone over the display's maximum number of digits (eight). If this happens, you may need to rethink the way you are working the problem. Read the problem over again and use smaller numbers.

Calculator Memory

Because an FCAT calculator is not programmed to do strings of calculations, you can use the memory to hold a number while you perform other calculations. Here's an example of how you can use the memory to hold onto numbers while you continue to work. Calculate the following.

$$4(3 + 5 \times 9) + 5(24 \div 3) =$$

Keystrokes	Display	Memory
5 × 9 =	45	0
+ 3 =	48	0
× 4 =	192	0
M+	M 192	192
2 4 ÷ 3 =	M 8	192
× 5 =	M 40	192
M+	M 40	232
MRC	M 232	232

In addition to the calculator tips mentioned in this chapter, many more tips are included in Chapter 1.

Rounding Answers

Sometimes you will get an answer that doesn't match those given in a multiple-choice problem. When this happens, it isn't always because you've done the problem incorrectly. You may simply need to round your answer.

Example: A shirt is on sale at $\frac{2}{3}$ of the original price of $15. How much does it cost?

Find $\frac{2}{3}$ of $15, or $\frac{2}{3} \times 15$.

Method 1: $2 \times 15 \div 3 = 10$
Method 2: $2 \div 3 \times 15 = 9.999999$

Both answers are correct, but with method 2, you need to round your answer to $10.

Finding Percents

The FCAT calculator has a percentage key. You can use this key to find percents.

Example 1: Florida charges 7% tax. How much tax would there be on a purchase of $525?

5 2 5 × 7 % (don't hit =). Answer: $36.75

Example 2: Maria bought new tires for her car. She paid $265 for four tires plus 7% tax. What was her total bill?

2 6 5 + 7 % (don't hit =). Answer: $283.55

Example 3: Patrick bought a new car. The sticker price was $24,500, but the dealer gave him a 12% discount. What was the price of the new car?

2 4 5 0 0 − 1 2 % (don't hit =). Answer: $21,560

Example 4: If 6% of a number is 15, find the number.

1 5 ÷ 6 % (don't hit =). Answer: 250

Now might be a good time to practice your calculator skills. Try the following sample questions.

Samples

1. Find 6% of $250.
2. How much will a $17.96 CD cost if 7% tax is added?
3. How much will a $17.96 CD cost if there is a 25% discount?
4. How much will a $17.96 CD cost if there is a 25% discount and then 7% tax is added?
5. Find 5.5% of 200.
6. If tax is 6.5%, what will the tax on a $30 meal be?
7. How much will a $40 meal cost if the tax is 7%?
8. Find the discount and the price: 60% off $500.
9. Find the discount and the price: 25% off $15.50.
10. Find $4.50 plus $4\frac{1}{2}$% tax.
11. What number is 15% of 12?
12. 4.5 is what percent of 15?
13. Ann left a 15% tip for a $35 meal. What was the total cost of the meal?
14. Brett earns an 11% commission on his sales. If he earned $3500, what were his total sales?

Answers

1. 2 5 0 × 6 % (don't hit =). Answer: \$15
2. 1 7 . 9 6 + 7 % = (don't hit =). Answer: \$19.22
3. 1 7 . 9 6 − 2 5 % (don't hit =). Answer: \$13.47
4. 1 7 . 9 6 − 2 5 % + 7 % (don't hit =). Answer: \$14.41
5. 2 0 0 × 5 . 5 % (don't hit =). Answer: 11
6. 3 0 × 6 . 5 % (don't hit =). Answer: \$1.95
7. 4 0 + 7 % (don't hit =). Answer: \$42.80
8. To find the discount: 5 0 0 × 6 0 % (don't hit =). Answer: \$300 (discount)

   To find the price (after the discount): 5 0 0 − 6 0 % (don't hit =). Answer: \$200 (price)
9. To find the discount: 1 5 . 5 0 × 2 5 % (don't hit =). Answer: \$3.88 (discount)

   To find the price: 1 5 . 5 0 − 2 5 % (don't hit =). Answer: \$11.63 (price)
10. 4 . 5 0 + 4 . 5 % (don't hit =). Answer: \$4.70
11. 1 2 × 1 5 % (don't hit =). Answer: 1.8
12. 4.5 is 30% of 15. The easiest way to do this problem is to use the proportion method: $\frac{part}{whole} = \frac{percent}{100}$. 4.5 is part of 15 (the whole). You are solving for the percent.

    $\frac{4.5}{15} = \frac{percent}{100}$

    Multiply diagonally (4.5 × 100) and divide by 15.
13. The total cost of the meal is \$35 + 15%. 3 5 + 1 5 % (don't hit =). Answer: \$40.25
14. What the problem is saying is that 11% of sales is equal to \$3500.

    Use the ratio and proportion method: $\frac{part}{whole} = \frac{percent}{100}$. The *whole* represents total sales. The *part* is Brett's commission. As before, 11 goes over the 100 to represent 11% as 11 out of 100. $\frac{3500}{whole} = \frac{11}{100}$. Multiply 3500 × 100 and divide by 11. Answer: \$31,818.18

The FCAT Mathematics Reference Sheet

You will not have to memorize a lot of formulas to take the FCAT. The following formulas will be provided for you.

- Area
- Surface area
- Volume
- The Pythagorean Theorem
- Distance, rate, time
- Distance between two points
- Midpoint between two points
- Simple interest
- Slope-intercept form of an equation

Test Tip: Be sure you take advantage of the reference sheet. When taking the FCAT, you can tear the reference sheet out so that you can use it more easily.

Many conversions are also provided on the reference sheet, both in US customary and metric. Even unusual conversions such as 1 acre = 43,560 square feet are provided. So you won't have to memorize these either. Just concentrate on *how* to make a conversion if you have the conversion unit.

DIAGNOSTIC TEST

Directions: Answer every question on this test. Place your answers in either the answer spaces or, where provided, in the "gridded-response" boxes. (Longer answers should be written on a separate sheet of paper.) Do all of your calculations on separate sheets of paper. Once you have answered every question, be sure to check your answers and to review the answer explanations at the end of the test.

1. Which of the following is *not* equivalent to (-2^3)?

A. $-\sqrt{64}$
B. -8×10^0
C. $|-8|$
D. $-\frac{32}{4}$

1. ________

2. A beekeeper measures four of his bees. Their lengths are 1.905 cm, 1.9 cm, 1.89 cm, and 1.91 cm. What is the length of the longest bee measured?

F. 1.89 cm
G. 1.9 cm
H. 1.905 cm
I. 1.91 cm

2. ________

3. A salesclerk earns a weekly salary of $250 plus $3 for every pair of shoes he sells. Last week he earned $400. In which of the following equations does n represent the number of pairs of shoes he sold last week?

A. $3n + 250 = 400$
B. $3n + 250 + 3 = 400$
C. $n + 250 = 400$
D. $3n = 400$

3. ________

4. An after-holiday sale offers an additional 25% off already discounted merchandise. If the original discount was 50%, how much would a $225 stereo system cost before tax using both discounts?

4.

	/	/	/	
.	.	.	.	.
0	0	0	0	0
1	1	1	1	1
2	2	2	2	2
3	3	3	3	3
4	4	4	4	4
5	5	5	5	5
6	6	6	6	6
7	7	7	7	7
8	8	8	8	8
9	9	9	9	9

GO ON →

5. A bacterium measures approximately 1.25 microns or 1.25×10^{-6} meters. How many meters, written in standard notation, equal 100,000 bacteria placed end to end?

F. 0.125
G. 1.25
H. 12.5
I. 125

5.

---

6. Eva paid \$22.40 for a pair of shoes that were marked down from \$28.00. What percent of the original price was the sale price?

6.

---

7. Which point on this number line represents a number that when squared will result in the largest number?

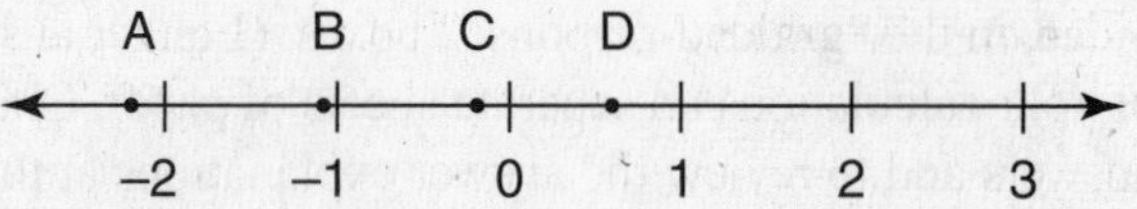

A. A
B. B
C. C
D. D

7. ______

---

8. Find the value of the expression $(-4.5)^2 - 4^2$.

8.

---

9. Which of the following shows the correct application of the distributive property when simplifying the expression $4(2x + y) - 2(x + 2y)$?

F. $8x + 4y - 2x - 4y$
G. $8x + 4y - 2x + 4y$
H. $8x + y - 2x + 2y$
I. $8x + 4y - 2x + 2y$

9. ______

10. If $2a = b$, $c = b$, and $d = 3c$, find d when $a = -2$. Show your work and write the answer in simplified form.

THINK

SOLVE

EXPLAIN

10. ________

11. A giant sequoia in Sequoia National Park, California, is the largest tree in the United States. It is 105% the size of a coast redwood located in Jedidiah Smith State Park, California. If the giant sequoia is 998 feet tall, how tall is the coast redwood?

A. 59 feet
B. 950 feet
C. 992 feet
D. 1048 feet

11. ________

12. The senior class sold candy to raise money for their prom. They bought 50 pounds of candy and divided it into $\frac{1}{4}$-pound bags to be sold at $4.50 per bag. What was their profit if the 50 pounds of candy cost $125.00?

12.

	⊘	⊘	⊘	
⊙	⊙	⊙	⊙	⊙
0	0	0	0	0
1	1	1	1	1
2	2	2	2	2
3	3	3	3	3
4	4	4	4	4
5	5	5	5	5
6	6	6	6	6
7	7	7	7	7
8	8	8	8	8
9	9	9	9	9

13. This diagram shows a city park. The shaded square measures 6 feet on a side. *Estimate* the area of the sidewalk surrounding the park.

F. 384 square feet
G. 576 square feet
H. 720 square feet
I. 1536 square feet

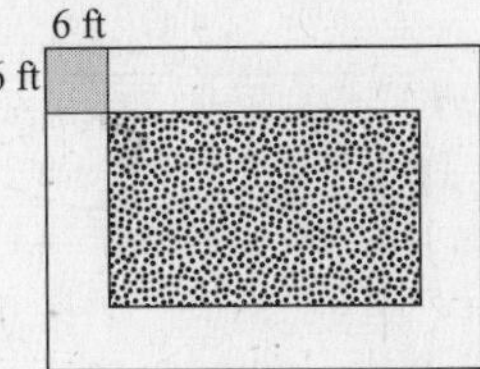

13. ________

14. Shalon is having a barbeque dinner for 20 people. The list of ingredients she needs to make dinner is shown here. *Estimate* the bill to the nearest dollar if she purchases 3 packages of steak, 4 chickens, 2 bags of corn on the cob, 3 loaves of bread, 4 heads of lettuce, and 2 packs of tomatoes. Show your work or explain in words how you arrived at your answer.

Item	Price ($)
Package of steak	18.25
Whole chicken	4.40
Bag of corn on the cob	2.79
Loaf of bread	1.89
Head of lettuce	1.29
Pack of tomatoes	4.29

THINK

SOLVE

EXPLAIN

14. ________

GO ON →

15. A right circular cylinder has a diameter of 4 feet and a height of 6 feet. Find the surface area. Use 3.14 for π.

4 ft

6 ft

15.

16. Shannon and Mark have given up on their lawn. They've decided to cover the entire front yard in oyster shell. Their lawn measures 80 feet long by 40 feet wide. Oyster shell costs $1.50 per cubic foot.

(*a*) What else do you need to know to find out how much covering the front yard with oyster shell will cost?

(*b*) Use the value you selected in part (*a*) to *estimate* the total cost of the oyster shell. Show your work.

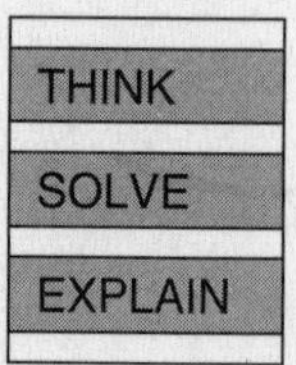

16. (*a*) ________
(*b*) ________

17. Tile manufacturers know that only certain tiles can fit together with no gaps between them (tessellate). Which of the following regularly shaped tiles *cannot* tessellate?

A. triangle
B. square
C. pentagon
D. hexagon

17. ________

18. Raven and Mike took a 4-hour canoe trip down the Peace River. While traveling with the current (downstream), they were able to go 12 miles. The next day it took 6 hours to paddle back against the current (upstream). The water was moving at a speed of 0.5 mile per hour.

(*a*) Write an equation you could use to find their paddling speed (p) in miles per hour when the canoe traveled with the current. Write another equation you could use to find their paddling speed (p) in miles per hour when the boat traveled in the direction opposite the current.

(*b*) Solve one of the equations from part (*a*) to determine their paddling speed.

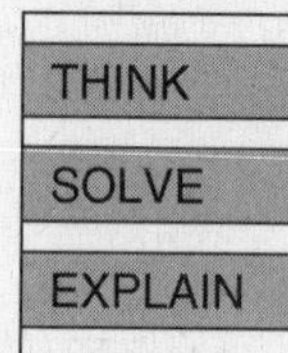

18. (*a*) ________
(*b*) ________

19. Marcus is painting his house. Before beginning, he measured to see how many square feet of walls he would have to paint and found he needed to paint a total of 3760 square feet of wall space. He has bought 3 gallons of paint so far and has painted only 1120 square feet. Which of the following comes closest to the amount of paint Marcus still needs to buy?

F. 7 gallons
G. 10 gallons
H. 15 gallons
I. 20 gallons

19. ________

20. A diagonal brace is used to stabilize the frame of a home being constructed. If the frame boards are 8 feet long and spaced 3 feet apart, about how long does the brace need to be?

A. 73 feet
B. 12 feet
C. 11 feet
D. 8.5 feet

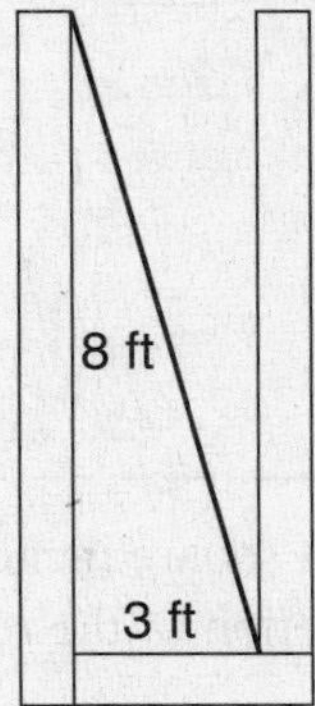

20. ________

21. Cliff weighed 7 pounds at his birth. He gained $\frac{1}{2}$ pound per week for the first 12 weeks. If he continues to gain weight at this rate, how much will he weigh when he is 16 years old?

F. 103 pounds
G. 416 pounds
H. 423 pounds
I. 445 pounds

21. ________

22. How many hours longer will it take a car averaging 60 miles per hour than a car traveling 75 miles per hour to travel a distance of 600 miles?

22.

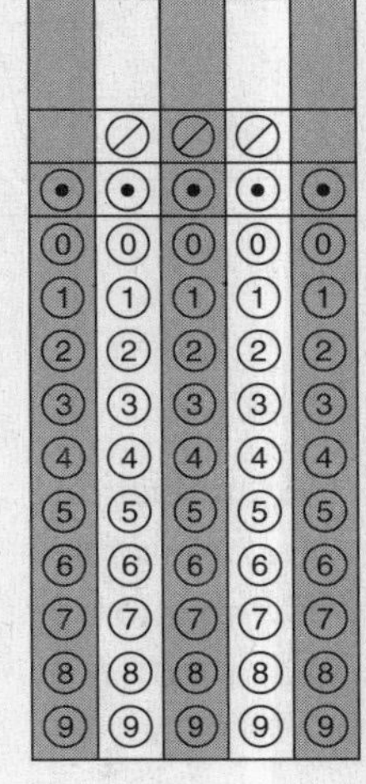

23. The figure shown here is a parallelogram. What is the measure, in degrees, of $\angle ABC$ if $m\angle CDE = 75°$?

A. 285°
B. 180°
C. 105°
D. 75°

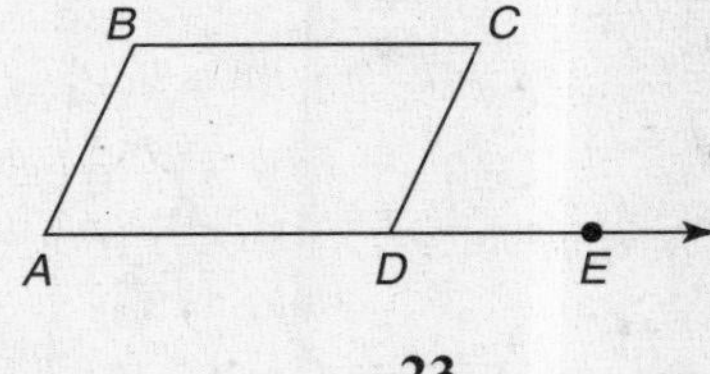

23. ________

24. Use the drawing of a regular hexagon shown here to find $m\angle FEG$. Show your work or explain in writing how you arrived at your answer.

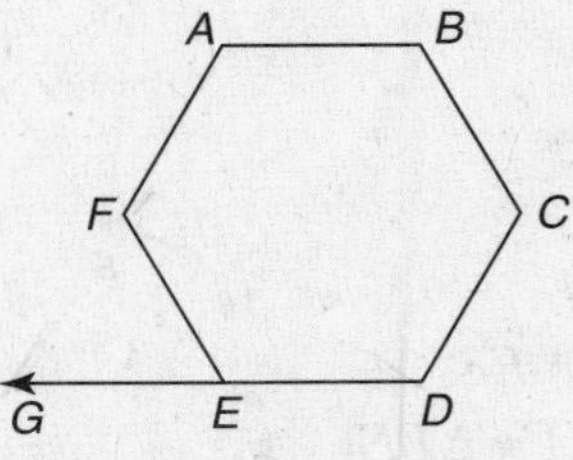

THINK

SOLVE

EXPLAIN

24. ________

GO ON

25. The figure shown here is an example of what type of transformation?

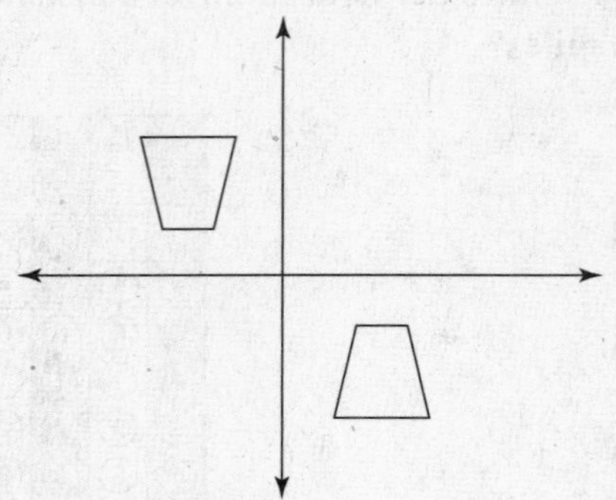

F. reflection
G. translation
H. dilation
I. rotation

25. ________

26. Based on the information in the diagram shown here, which of the following statements is *not* true?

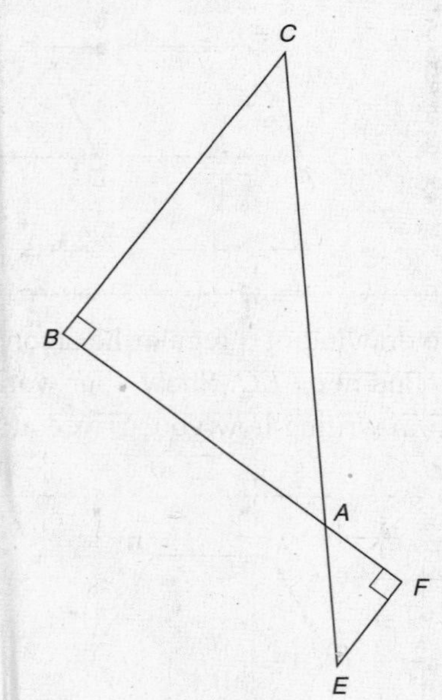

A. $\overline{BC} \parallel \overline{FE}$
B. $\Delta CBA \cong \Delta EFA$
C. $\frac{\overline{BA}}{\overline{FA}} = \frac{\overline{CB}}{\overline{EF}}$
D. $\angle BCA \cong \angle FAE$

26. ________

27. Which shapes *cannot* result from the intersection of a plane and a right circular cone?

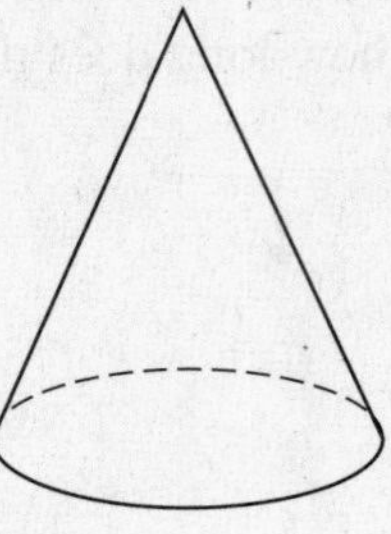

F. circle
G. triangle
H. oval
I. rhombus

27. ________

28. Find the area in square inches of the shaded planar cross section of the rectangular solid shown here.

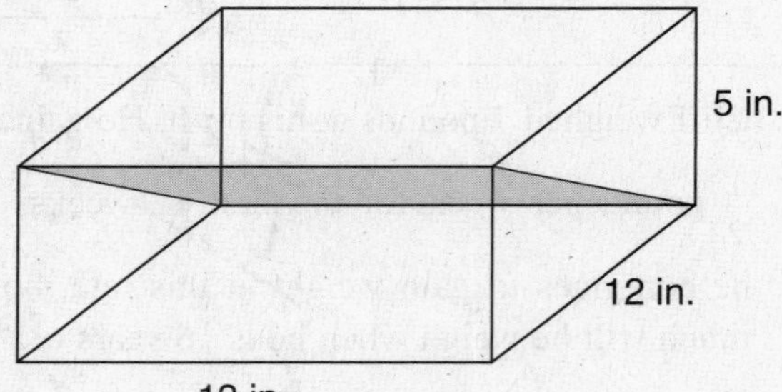

A. 169 square inches
B. 156 square inches
C. 144 square inches
D. 52 square inches

28. ________

29. A recipe for banana bread calls for 2 teaspoons of baking soda and 3 cups of flour. Kerry accidentally put in 6 teaspoons of baking soda. If Kerry decides to add enough flour to make the recipe proportional, which proportion below will show her how much flour to add?

F. $\frac{2}{3} = \frac{6}{x}$

G. $\frac{2}{2} = \frac{3}{x}$

H. $\frac{x}{6} = \frac{2}{3}$

I. $\frac{2}{3} = \frac{3}{x}$

29. ________

30. The two triangles shown here are similar. Find the length of side x.

10 cm
15 in.
8 cm
x

30.

31. In the accompanying diagram, $\overline{LP}$ is parallel to $\overline{MO}$.

(*a*) Write a proportion that can be used to find x.

(*b*) Solve the proportion. Show your work.

(*c*) Explain in geometric terms why ΔMNO is similar to ΔLNP.

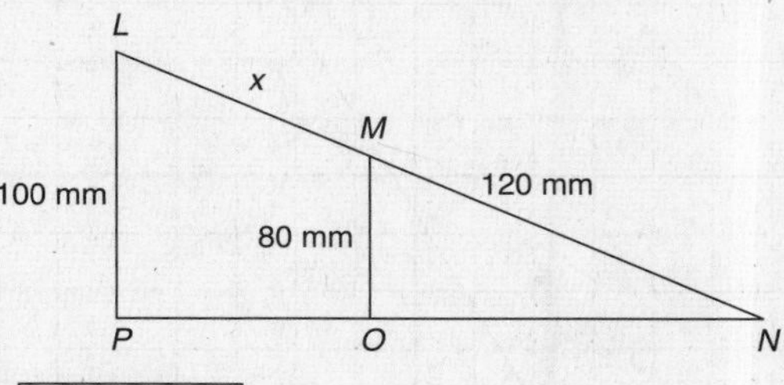

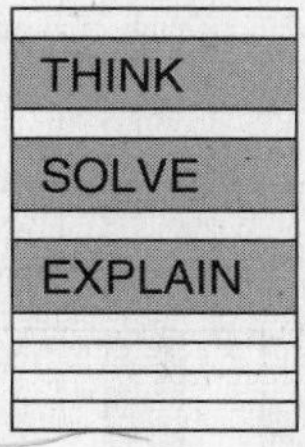

31. (*a*) ________
(*b*) ________
(*c*) ________

32. What is the slope of the line parallel to the line graphed here?

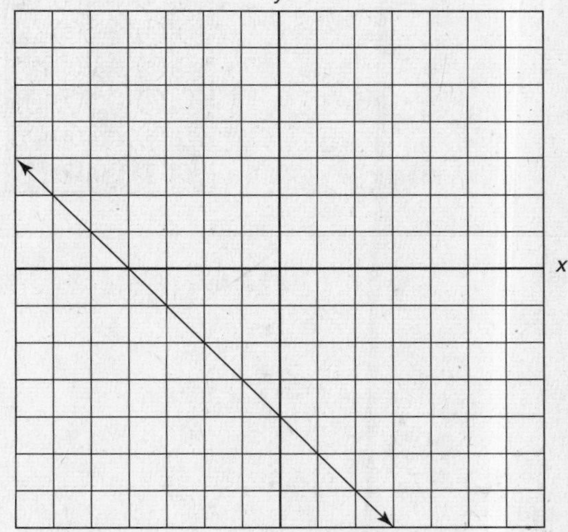

A. −2
B. −1
C. 1
D. 2

32. ________

GO ON

33. Find the sum of the x- and y-coordinates of the midpoint of line segment $\overline{AB}$ shown on the accompanying graph.

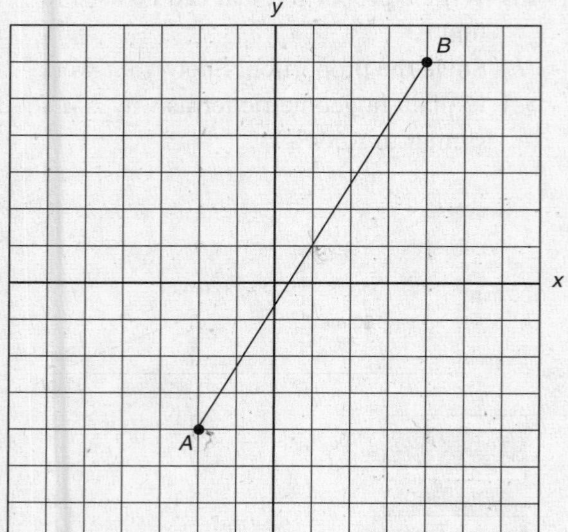

33.

34. (*a*) Find the slope of each side of the triangle.

(*b*) Use the slopes to demonstrate that ΔCBE is a right triangle. Explain your reasoning.

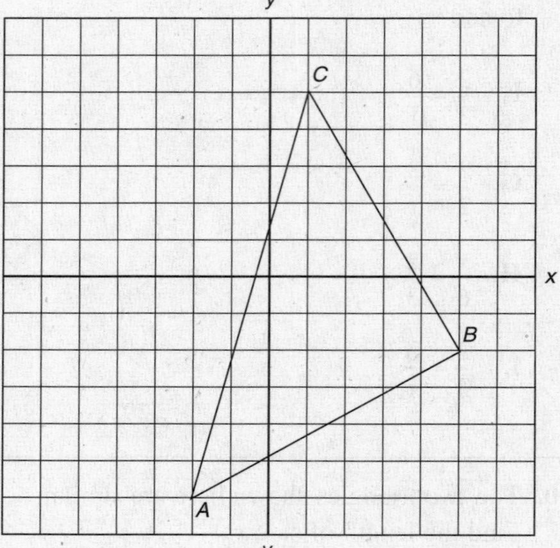

THINK

SOLVE

EXPLAIN

34. (*a*) ________
(*b*) ________

35. Juan has a balance of \$2650 in his savings account. If he saves \$70 in change each month to increase his balance, which equation describes a balance of \$3500 after n months?

F. $3500 = 2650 + n$
G. $3500 = 70n + 2650$
H. $3500 = 70n$
I. $3500 = 70(n + 2650)$

35. ________

36. What is the next number in the pattern $\frac{7}{81}, \frac{7}{27}, \frac{7}{9}, \ldots$?

36.

---

37. Tint Tile is having a sale on tile. The 12-inch × 12-inch tiles cost \$1.25 each, and the 17-inch × 17-inch tiles cost \$2.25 each. If you need to tile the rectangular room shown here, about how much cheaper would it be to tile it with the 17-inch × 17-inch tiles? Show or explain the steps you used to arrive at your answer.

THINK

SOLVE

EXPLAIN

20 ft

10 ft

37. ______

---

38. Pam set a schedule for the book she is writing. She figured she had about 50 pages left to write over a period of 4 weeks. She decided to play it safe and write 20 pages the first week, and 10 pages each of the next 3 weeks. Something unexpected came up the first week, and she wrote only $5\frac{1}{2}$ pages. Then, the second week she was able to write 13 pages. How many pages will she have to write in *each* day of the last 2 weeks to meet her deadline?

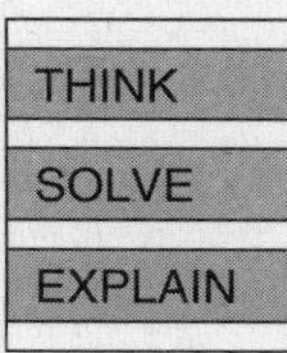

38. ______

---

39. Which of the following expressions is equivalent to $(a^4b^3c)^2$?

A. $a^8b^6c^2$
B. $a^2b^1c^0$
C. $a^6b^5c^2$
D. $a^8b^5c^2$

39. ______

---

40. Julio currently earns \$40,000 per year. He wants to take a new job paying \$25,000 in salary and 25% sales commission. How much will he need to sell to make the same salary as before?

40.

GO ON →

41. A local computer shop charges \$50 per hour to repair desktop computers and \$75 per hour to repair laptop/notebook computers. Last year they made a total of \$75,000 and repaired 1100 computers.

(*a*) Write a system of equations that can be used to find how many of each type of computer they repaired.

(*b*) Solve the system of equations to find how many desktops and how many laptops were repaired.

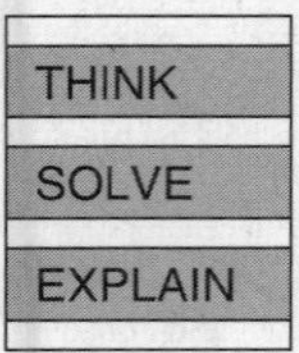

41. (*a*) ________
(*b*) ________

42. The accompanying box-and-whisker plot shows the average number of vacation days per year in nine countries, including the United States.

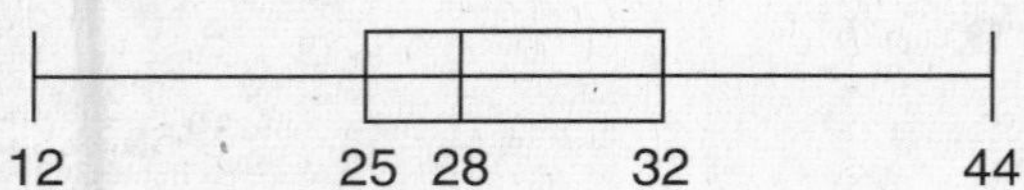

How many countries have more than 28 vacation days per year?

F. 8
G. 5
H. 4
I. 3

42. ________

43. The top US states by total traveler spending for 2007 are shown in the table.

State	US Travel Dollars (%)
California	23.8
Florida	19.0
New York	12.1
Texas	11.1
Illinois	7.5
Nevada	6.6
New Jersey	5.2
Pennsylvania	5.0
Hawaii	4.9
Georgia	4.8

(*a*) Make a bar graph on the accompanying grid that shows the percentage of traveler spending for each state.

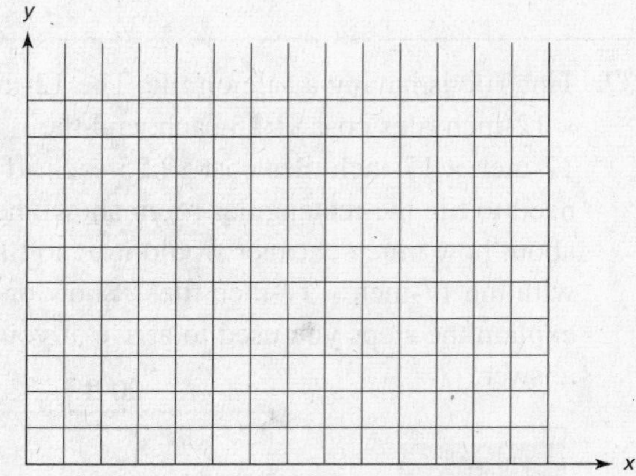

(*b*) Which four states receive nearly two-thirds of the total traveler spending?

(*c*) If the total traveler spending in 2007 was \$282 billion, predict how many traveler dollars were spent in Florida that year.

THINK
SOLVE
EXPLAIN

43. (*a*) ________
(*b*) ________
(*c*) ________

44. The average golf scores for ten players are 120, 82, 72, 88, 85, 78, 72, 77, 88, 86. Which of the following measures of central tendency best represent this data?

A. mean
B. median
C. mode
D. range

44. ________

45. Frank was timing his brother Andy's running time. He clocked Andy's times as 3 minutes, 45 seconds; 3 minutes 30 seconds; and 3 minutes, 15 seconds. What is Andy's mean time in minutes?

45.

	⊘	⊘	⊘	
⊙	⊙	⊙	⊙	⊙
0	0	0	0	0
1	1	1	1	1
2	2	2	2	2
3	3	3	3	3
4	4	4	4	4
5	5	5	5	5
6	6	6	6	6
7	7	7	7	7
8	8	8	8	8
9	9	9	9	9

46. Cindy stuffs all her shorts and tee shirts in one drawer. She has 4 pairs of shorts: red, brown, blue, and green. In the same drawer are 5 tee shirts: white, blue, green, tan, and yellow. What is the probability that she will pull out a green pair of shorts first and a green tee shirt second?

F. $\frac{1}{72}$

G. $\frac{2}{17}$

H. $\frac{2}{9}$

I. $\frac{1}{4}$

46. ________

47. The Weather Channel has predicted a 60% chance of rain for each of the next 5 days. What is the probability it will *not* rain tomorrow?

47.

	⊘	⊘	⊘	
⊙	⊙	⊙	⊙	⊙
0	0	0	0	0
1	1	1	1	1
2	2	2	2	2
3	3	3	3	3
4	4	4	4	4
5	5	5	5	5
6	6	6	6	6
7	7	7	7	7
8	8	8	8	8
9	9	9	9	9

GO ON →

48. The accompanying table gives the average US television viewing time.

Group	Age	Hours and Minutes of TV Per Week
Women	18+	32:46
	18–24	21:30
	25–54	30:35
	55+	41:20
Working women		27:43
Men	18+	28:54
	18–24	20:10
	25–54	27:33
	55+	36:28
Teens	12–17	19:40
Children	2–11	19:40
All viewers		28:13

Based on the table, which of the following is *not* true?

A. Women watch more television than men.
B. Men watch more television than the average viewer.
C. Men spend most of their television time watching sports.
D. Working women watch about 5 hours less television a week than other women.

48. ________

49. During the last year, the number of children living with their grandparents increased from 73,500 to 74,602. What is the percent of increase?

F. 1.5%
G. 5%
H. 15%
I. 22.5%

49. ________

50. Angelina conducted a survey at her school to see how many students liked their new principal. She decided to survey only the freshmen because they had not known the old principal and would be unbiased. Was her survey valid? Write your answer and explain your reasoning.

THINK

SOLVE

EXPLAIN

50. ________

STOP

SOLUTIONS: DIAGNOSTIC TEST

Answer Key

1. **C**
2. **I**
3. **A**
4. **$84.38**
5. **F**
6. **80%**
7. **A**
8. **4.25**
9. **F**
10. **–12**
11. **B**
12. **775**
13. **H**
14. **$100**
15. **100.5**
16. (*a*): See answer explanation
    (*b*): **$2400**
17. **C**
18. **2.5 mph:** See answer explanation
19. **G**
20. **D**
21. **H**
22. **2**
23. **C**
24. **60°**
25. **I**
26. **D**
27. **I**
28. **A**
29. **F**
30. **27**
31. (*a*): See answer explanation
    (*b*): See answer explanation
    (*c*): $x = 30$
32. **B**
33. **2**
34. (*a*): See answer explanation
    (*b*): See answer explanation
35. **G**
36. **7/3**
37. **$25**
38. **2.25**
39. **A**
40. **$60,000**
41. (*a*): **$75,000.** See answer explanation
    (*b*): $x = 300$
42. **H**
43. (*a*): See answer explanation
    (*b*): **California, Florida, New York, Texas**
    (*c*): **$53.4 billion**
44. **B**
45. **3.5**
46. **F**
47. **40**
48. **C**
49. **F**
50. **No**

Answers Explained

1. C. |–8| is read as "the absolute value of negative eight." The absolute value of a number is the distance of the number from zero on the number line. Since negative eight is exactly eight spaces away from zero on the number line, |–8| = 8.

2. I. Add zeros to the end of each number until they all have the same number of decimal places: {1.890, 1.900, 1.905, 1.910}. Line them up vertically, ignoring the leading 1 (because all the numbers begin with 1) and compare 890, 900, 905, and 910.

3. A. He earns $3 for every pair of shoes sold which is represented by *3n*. Add to that his base weekly salary of $250. Then, $3n + 250 = 400$.

4. $84.38. Use your calculator:

225 – 50 % = 112.50

and

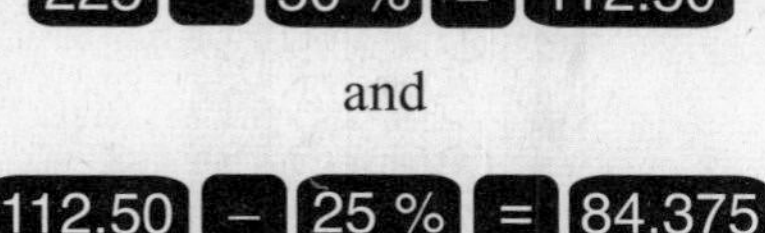

,

which rounds to $84.38.

5. F. $1.25 \times 10^{-6} = 0.00000125$. Multiply by 100,000 = 0.125. Multiplying by 10 moves the decimal one place to the right. Multiplying by 100 moves the decimal two places to the right . . . Multiplying by 100,000 will move the decimal in 0.00000125 five places to the right.

6. Write a ratio comparing the discounted cost to the original cost: 22.40/28. Then, put it in lowest terms and convert to a percent.

$$\frac{22.4}{28} = \frac{224}{280} = \frac{56}{70} = \frac{28}{35} = \frac{4}{5} = \frac{8}{10} = \frac{80}{100} = 80\%$$

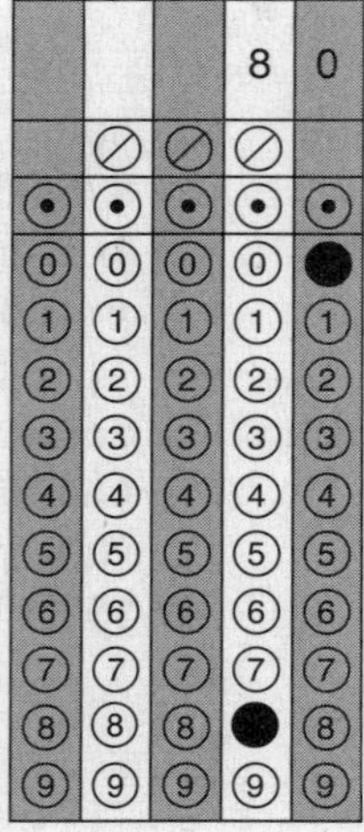

7. A. $(-2.2)^2 = 4.84$. One strategy would be to take each number and multiply it by itself. Even if you are not certain what number the point represents, you can approximate.

8.

4 . 2 5

Work the problem out step by step:

$(-4.5)^2 - 4^2$	Deal with powers first
$(-4.5)(-4.5) - 4 \cdot 4$	by multiplying.
$20.25 - 16 = 4.25$	Subtract.

9. F.

$4(2x + y) - 2(x + 2y)$	
$4(2x + y) + 2(-x - 2y)$	Use "add the opposite."
$4 \cdot 2x + 4 \cdot y + (2 \cdot -x + 2 \cdot -2y)$	Distribute the 4 and the 2.
$8x + 4y + (-2x - 4y)$	Remove the parentheses.
$8x + 4y - 2x - 4y$	

10. If $2a = b$ and $a = -2$, substitute -2 for a. $2(-2) = b$, and $-4 = b$. Since $c = b$, this means that c also equals -4. $d = 3c$; substitute -4 for c; $d = 3(-4) = -12$.

11. B. *Method 1*: 105% of a number equals 998. Write an equation: $1.05n = 998$, and so $n = 998 \div 1.05 \approx 950$.

Method 2: Use ratio and proportion: $\frac{part}{whole} = \frac{percent}{100}$. Substitute 105 for *percent.* Since 998 equals the 105%, it should go on top, next to 105. Substitute it for *part*: $\frac{998}{n} = \frac{105}{100}$.

Solve for n by multiplying $998 \cdot 100$ and then dividing by 105. The answer is approximately ($\approx$) 950.

12.

7 7 5

The profit (the amount they will make) will equal the amount of money raised minus what they paid for the candy. The amount raised equals the number of bags of candy multiplied by \$4.50. They sold $50 \div \frac{1}{4}$ or $50 \div .25$ or 200 bags of candy.

Multiply 200 by the cost per bag of \$4.50 to find that they raised \$900 worth of candy. From this amount subtract the original cost of \$125. Their profit was \$775.

13. H. By copying the shaded square as shown in the diagram, you can see there are about 20 squares in all. Since the area of the original shaded square is 36 square feet (6 feet × 6 feet), multiply 36 by 20 to get the total square feet. There are about 720 square feet in all.

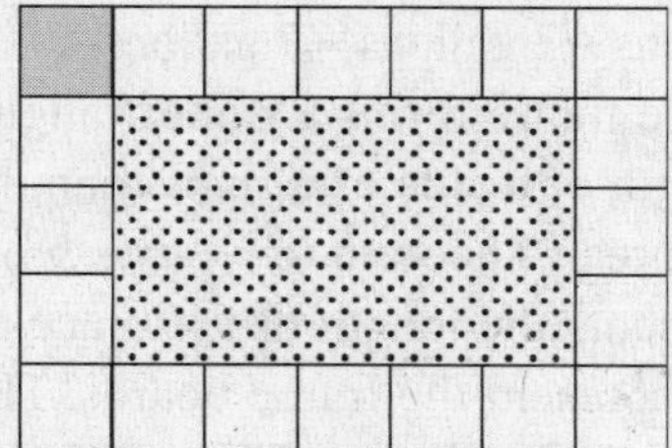

14. To *estimate*, round each number before beginning calculations. The rounded numbers are shown in the table. Multiply by the number of items (third column) to find the total estimated cost for each item (fourth column). Add to find the total cost.

Item	**Price (\$)**	**Item**	**Cost (\$)**
Package of steak	20	3	60
Whole chicken	4	4	16
Bag of corn on the cob	3	2	6
Loaf of bread	2	3	6
Head of lettuce	1	4	4
Pack of tomatoes	4	2	8
The total cost is about \$100			

15.

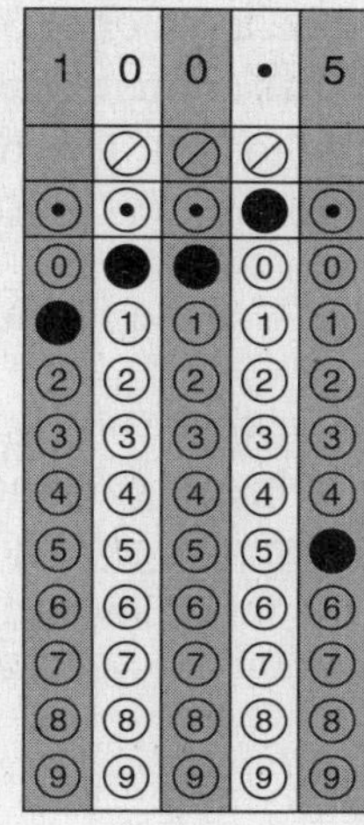

Use the formula for the surface area of a cylinder: $SA_{cyl} = 2\pi r^2 + 2\pi rh$ (where r = radius and h = height). On the FCAT, you will be given this formula (and many others) on the FCAT reference sheet. Substitute the given numbers. Remember that the radius is half the diameter.

$SA_{cyl} = 2\pi r^2 + 2\pi rh$

$2 \times \pi \times r \times r + 2 \times \pi \times r \times h$

$2 \times 3.14 \times 2 \times 2 + 2 \times 3.14 \times 2 \times 6$

Do the multiplication first.

$25.12 + 75.36 = 100.48$ square feet

Round to 100.5 to fit the grid.

16. (*a*) You need to know how deep they want to make the shell. Since shell is sold in cubic feet, you must calculate the volume. To calculate the volume you need the length, width, and height (or, in this case, depth).

(*b*) **\$2400** If (for example) the shell was supposed to be 6 inches deep [pretty deep, but an easy number to use because it's half (0.5) a foot], then to find the number of cubic feet, multiply $80 \times 40 \times 0.5 = 1600$ cubic feet. Multiply $1600 \times \$1.50 = \$2400$.

17. C. For a regular polygon to tessellate, a single angle must be a factor of 360 degrees so that when the tiles are put together, all their corners will add to 360. This means that the number of degrees in one angle must divide evenly into 360. Use the formula: $\frac{180(n-2)}{n}$ to calculate the number of degrees in a single angle of each polygon. (n stands for the number of sides in the polygon.) You will not have to memorize the formulas, as it will be provided on the FCAT reference sheet.

Triangle: $\frac{180(3-2)}{3} = \frac{180}{3} = 60$
60 divides evenly into 360.

Square: $\frac{180(4-2)}{4} = \frac{180(2)}{4} = \frac{360}{4} = 90$
90 divides evenly into 360.

Pentagon: $\frac{180(5-2)}{5} = \frac{180(3)}{5} = \frac{540}{5} = 108$
108 does *not* divide evenly into 360 (will *not* tessellate).

Hexagon: $\frac{180(6-2)}{6} = \frac{180(4)}{6} = \frac{720}{6} = 120$
120 divides evenly into 360.

18. (*a*) Use $d = rt$. The distance (d) is 12 miles, the downstream time (r) is 4 hours, and the upstream time (r) is 6 hours. Finding the numbers to substitute for rate (r) takes a little more time. Let p stand for their paddling speed. Going downstream, their boat's speed (rate) is $p + 0.5$ mph. And going upstream the boat's rate is $p - 0.5$ mph. Substitute $p + 0.5$ for rate (r) in the downstream formula, and $p - 0.5$ for rate (r) in the upstream formula. Substitute hours traveled for time (t). Remember, it took 4 hours to go downstream and 6 hours to go upstream. The equations are as follows.

Downstream: $d = rt = 12 = 4(p + 0.5)$
Upstream: $d = rt = 12 = 6(p - 0.5)$

(*b*) Downstream solution:
$12 = 4(p + 0.5)$
$12 = 4p + 2$
$10 = 4p$
$2.5 = p$

Upstream solution:
$12 = 6(p - 0.5)$ $12 = 6p - 3$
$15 = 6p$
$2.5 = p$

They were paddling at 2.5 miles per hour.

19. G. You can use a proportion to solve this problem. Three gallons paints 1120 square feet $\left(\frac{3}{1120}\right)$, and x gallons paints 3760 feet $\left(\frac{x}{3760}\right)$.

Set the two ratios equal, $\frac{3}{1120} = \frac{x}{3760}$, and solve. Multiply 3760 by 3 and divide by 1120. 10 gallons – 3 gallons = 7 gallons.

20. D. The figure forms a right triangle. To find the length of a side of a right triangle, use the Pythagorean Theorem ($c^2 = a^2 + b^2$), where c represents the length of the brace and a and b represent the frame boards. Then, $c^2 = 8^2 + 3^2 = 64 + 9 = 73$. $c = \sqrt{73} = 8.54$.

21. H. $\frac{1}{2}$ pound × 52 weeks × 16 years = 416. Then add the original 7 pounds for a total of 423 pounds.

22.

2				
	⊘	⊘	⊘	
⊙	⊙	⊙	⊙	⊙
0	0	0	0	0
1	1	1	1	1
●	2	2	2	2
3	3	3	3	3
4	4	4	4	4
5	5	5	5	5
6	6	6	6	6
7	7	7	7	7
8	8	8	8	8
9	9	9	9	9

600 miles ÷ 60 miles per hour = 10 hours. 600 miles ÷ 75 miles per hour = 8 hours. It will take the car averaging 60 miles per hour 2 hours longer.

23. C. If m$\angle CDE$ = 75°, then m$\angle CDA$ is supplementary and equals 105°. In a parallelogram, angles that are diagonal to each other are congruent; therefore, m$\angle ABC$ = 105°.

24. m$\angle FEG$ = 60°. Use the formula $\frac{180(n-2)}{n}$ to find the measure of one interior angle of $\frac{180(6-2)}{6} = \frac{180(4)}{6} = 120°$. Each interior angle of the hexagon (including $\angle ABC$) measures 120°. Because interior and exterior angles are supplementary, m$\angle FEG$ = 60° (180 − 120 = 60).

25. I. The figure was rotated 180 degrees about the origin. One reflection would not have accomplished the move.

26. D. Because $\overline{BF}$ is perpendicular to both $\overline{BC}$ and $\overline{FE}$, $\overline{BC}$ and $\overline{FE}$ are parallel. $\triangle CBA$ is similar to $\triangle EFA$ because the angles are congruent (angle-angle). Both have a right angle, and $\triangle BAC$ is congruent to $\triangle FAE$ because they are vertical angles. If both triangles are similar, then their corresponding sides are proportional; that is, $\frac{\overline{BA}}{\overline{FA}} = \frac{\overline{CB}}{\overline{EF}}$. $\angle BCA$ is not congruent with $\angle FAE$.

27. I. A plane intersecting a cone does not produce a four-sided figure.

28. A. The length of the planar cross section is 13. The width can be found by using the Pythagorean Theorem: $c^2 = a^2 + b^2$. Substitute 5 and 12 for a and b, respectively. Then $c^2 = 5^2 + 12^2 = 25 + 144$. $c^2 = 169$, and so $c = 13$. Since both the length and width of the cross section are 13, find the area by multiplying: 13 × 13 = 169 square inches.

29. F. The proportion is arranged comparing two ratios, one for baking soda and another for flour. The first ratio compares 2 teaspoons of baking soda to 3 cups of flour (the original recipe). The second ratio compares 6 teaspoons of baking soda to x cups of flour (the adjusted recipe).

Ratio 1 Ratio 2

$$\frac{2 \text{ teaspoons}}{3 \text{ cups}} = \frac{6 \text{ teaspoons}}{x \text{ cups}}$$

Set the ratios equal to make the recipes proportional.

30.

2	7			
	⊘	⊘	⊘	
⊙	⊙	⊙	⊙	⊙
0	0	0	0	0
1	1	1	1	1
●	2	2	2	2
3	3	3	3	3
4	4	4	4	4
5	5	5	5	5
6	6	6	6	6
7	●	7	7	7
8	8	8	8	8
9	9	9	9	9

If the triangles are similar, then corresponding sides are proportional. Compare the smaller triangle to the larger triangle using a proportion: 10/18 = 15/x. Solve by cross-multiplying: 18 × 15 ÷ 10 = x; x = 27.

31. (*a*) Corresponding sides in similar triangles are proportional. $\overline{MO}$ corresponds to $\overline{LP}$, and $\overline{NM}$ corresponds to $\overline{NL}$. Set the proportions up accordingly: $\frac{\overline{MO}}{\overline{NM}} = \frac{\overline{LP}}{\overline{NL}}$.

Then substitute values: $\frac{80}{120} = \frac{100}{120+x}$.

Notice the expression for $\overline{NL}$ is $120 + x$.

(*b*) Cross-multiply to solve:

$80(120 + x) = 100(120)$ Distribute the 80.

$80 \times 120 + 80 \times x = 100 \times 120$ Multiply first.

$9600 + 80x = 12000$ Subtract 9600 from both sides.

$80x = 2400$ Divide by 80.

$x = 30$

(*c*) These triangles share $\angle N$. In addition, because $\overline{LP}$ is parallel to $\overline{MO}$, $\angle O$ is congruent to $\angle P$, and $\angle NMO$ is congruent to $\angle NLP$. Since all the angles are congruent, the two triangles are similar.

32. B. The slope equals the vertical change divided by the horizontal change. This is also called "rise over run." If you select two lattice points on the line and count the rise and divide by the run, you will have the slope. The rise is −2, and the run is 2. −2/2 = −1. Parallel lines have the same slope.

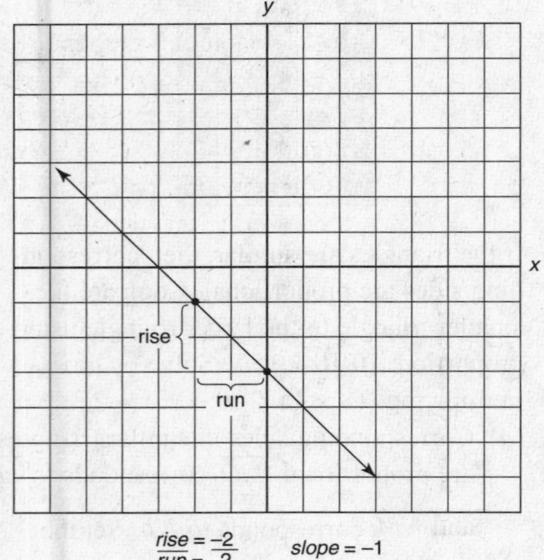

Be careful! The term *rise* can be deceptive. It really means how many places you had to move vertically to get from one point to the next, so if you had to move down, as in this case, the rise can be negative. When calculating the slope, it is easier to always work from left to right. The run refers to the horizontal change, or how many places you had to move horizontally to get from one point to the next.

33.

2				
	⊘	⊘	⊘	
⊙	⊙	⊙	⊙	⊙
0	0	0	0	0
1	1	1	1	1
●	2	2	2	2
3	3	3	3	3
4	4	4	4	4
5	5	5	5	5
6	6	6	6	6
7	7	7	7	7
8	8	8	8	8
9	9	9	9	9

To find the midpoint, average the *x*-coordinates and then average the *y*-coordinates. This average gives you the location of the midpoint. The formula for midpoint is $\left(\frac{x_1 + x_2}{2}, \frac{y_1 + y_2}{2}\right)$. The endpoints of the graphed line segment are at (−2, −4) and (4, 6). After substituting the *x*- and *y*-coordinates, we find the midpoint is located at $\frac{-2+4}{2}, \frac{-4+6}{2} = \left(\frac{2}{2}, \frac{2}{2}\right) = (1, 1)$. The sum of the *x*- and *y*-coordinates is 1 + 1 = 2.

34. (*a*) The slope of $\overline{AB} = \left(\frac{4}{7}\right.$; the slope of $\overline{BC} = -\left.\frac{7}{4}\right)$; the slope of $\overline{CA} = 7$.

(*b*) Because the slopes of $\overline{AB}$ and $\overline{BC}$ are opposite reciprocals $\left(\frac{4}{7} \text{ and } -\frac{7}{4}\right)$ of each other, the two line segments are perpendicular. This is a right triangle.

35. G. Saving \$70 per month for *n* months is represented by the expression 70*n*. Juan began with \$2650, so the expression for his entire bank balance is 70*n* + 2650. The final balance of \$3500 equals 70*n* + 2650.

36.

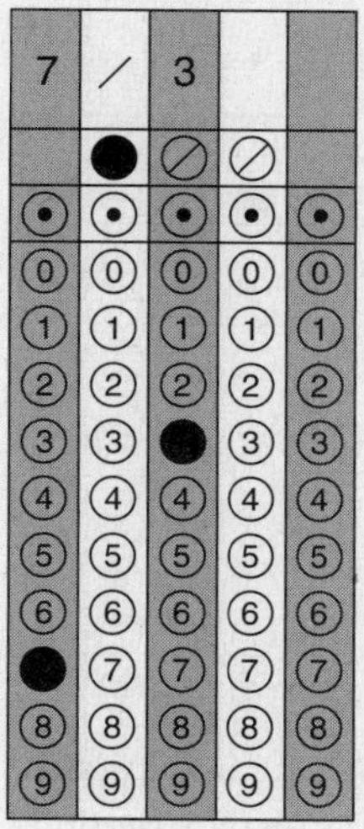

The pattern stays the same in the numerator, so you need to deal only with changes in the denominator. The denominator is divided by 3 with each new term. The next number will be $\frac{7}{3}$.

37. The total square footage to be tiled is 200 square feet. Each 12-inch × 12-inch tile represents one square foot. The cost of one tile is \$1.25, so multiply 200 by \$1.25 to get the cost of tiling with the 12-inch tiles (\$250). Each 17-inch × 17-inch tile represents $(17 \times 17)/144 \approx 2$ square feet. (There are 144 square inches in one square foot.) 200 square feet divided by 2 square feet = 100 tiles. $100 \bullet 2.25 = \$225$. It is \$25 cheaper to use the larger tiles.

38. 2 • 25 pages per day. If she had 50 pages to write and has written 5.5 and 13 pages, then she has $50 - 5.5 - 13 = 31.5$ pages left to write. There are 7 days in each week and 14 days in the 2 remaining weeks. 31.5 divided by 14 is 2.25.

39. **A.** $(a^4b^3c)^2 = (a^4b^3c) \times (a^4b^3c) = a^{4+4}b^{3+3}c^{1+1} = a^8b^6c^2$. When multiplying variables with like bases, *add* the exponents.

40.

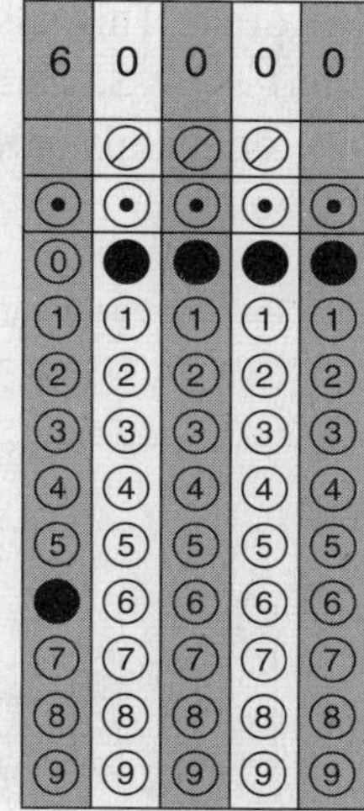

He will need to have \$60,000 in sales and to make \$15,000 in commissions. Subtract the new salary from the old salary: \$40,000 – 25,000. 25% of his sales must equal \$15,000. Translate into an algebraic equation: $0.25s = 15{,}000$. Solve: $15{,}000 \div .25 = 60{,}000$.

41. 300 desktops and 800 laptops were repaired.

(*a*) Let x = the number of desktop computers repaired, and y = the number of laptops repaired. Then, $x + y = 1100$ (representing the number of computers), and $50x + 75y = 75{,}000$ (representing the charges for repairs).

(*b*) Solve the system of equations: Substitute $x = 1100 - y$ for x in the second equation and solve for y. (If $x + y = 1100$, then $x = 1100 - y$.)

$50(1100 - y) + 75y = 75{,}000$
Substitute $x = 1100 - y$ for x.
$50 \bullet 1100 - 50 \bullet y + 75y = 75{,}000$
Do multiplication first.
$55{,}000 - 50y + 75y = 75{,}000$
Add like terms.
$55{,}000 + 25y = 75{,}000$
Subtract 55,000 from both sides.
$25y = 20{,}000$
Divide by 25.
$y = 800$

Substitute $y = 800$ in the first equation, $x + y = 1100$: $x + 800 = 1100$; $x = 300$.

42. H. Nine numbers arranged in order have a median number of 28. This means the middle number (median) is 28. Since there are 9 countries shown on the histogram, 4 must be above 28 and 4 below.

43. (*a*)

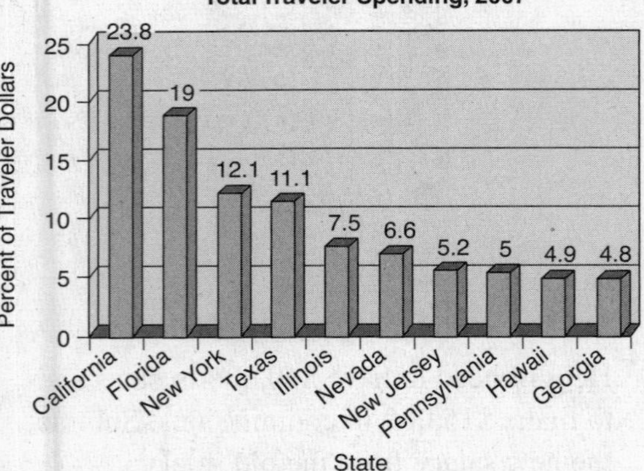

(*b*) California, Florida, New York, Texas—they add to about 67% (23.8 + 19 + 12.1 + 11.1 = 66).

(*c*) $53.4 billion. Florida spent 19% of 282 billion. Multiply 282 billion by 0.19. Since your calculator will not accept more than eight digits, multiply 282 by 0.19 and add the word *billion*.

44. B. The median best represents the data set because the score of 120 is an outlier; it would pull the mean up too far to represent the rest of the data. The data is bimodal (72 and 88), so there is no one number to represent all the data.

45.

3 • 5

Add: 3 minutes, 45 seconds
3 minutes, 30 seconds
3 minutes, 15 seconds
9 minutes, 90 seconds

Divide by 3: 3 minutes 30 seconds, or 3.5 minutes.

46. F. The probability that she will pull out a pair of green shorts first is $\frac{1}{9}$. If she does not replace the shorts, then the probability that she will pull out the green tee shirt next is $\frac{1}{8}$. The probability of green shorts and then a green tee shirt is $\frac{1}{9} \times \frac{1}{8} = \frac{1}{72}$.

47.

4 0

If there is a 60% chance of rain, then there is a 40% chance of it not raining. The probabilities must add to 100%.

48. C. There is nothing in the table to indicate *what* men watch.

49. F. Percent of increase is found by:

$$\frac{\textit{amount of change}}{\textit{original amount}} = \frac{\textit{percent}}{100}.$$

$$\frac{74{,}602 - 73{,}500}{73{,}500} = \frac{\textit{percent}}{100}$$

$$= \frac{1102}{73{,}500} = \frac{\textit{percent}}{100}$$

(To find *percent*, multiply 1102 by 100 and divide by 73,500) $= \frac{1102}{73{,}500} = \frac{1.5}{100}$.

50. No. Angelina wanted to find out whether *all* the students at her high school liked the new principal. By asking only freshmen, she received no information about how sophomores, juniors, and seniors felt about the principal.

Chapter 1 | Algebra

KEY TERMS/CONCEPTS

- Real, rational, and irrational numbers
- Forms of numbers: integers, fractions, decimals, absolute value, powers, scientific notation, square roots
- Operations on numbers and order of operations
- Tables, charts, and graphs
- Algebraic expressions, equations, and inequalities
- Slope, rate, and ratio
- Monomials, polynomials, and quadratics
- Relations and functions

SECTION 1: REAL NUMBERS

In this section we will look at the Real Number System, its properties, and how numbers are defined and used. For the FCAT, you must be able to recognize types of numbers by name, put them into order by size, and understand how their properties allow you to add, subtract, multiply, and divide them following certain rules. To get the most out of this section, read the definitions and look at the examples. Then try the sample questions. If you miss a sample question, read the answer carefully and compare it with what you learned about that particular type of number, rereading if necessary. At the end of the section are some problems for additional practice.

REAL NUMBERS

Real numbers are divided into rational and irrational numbers.

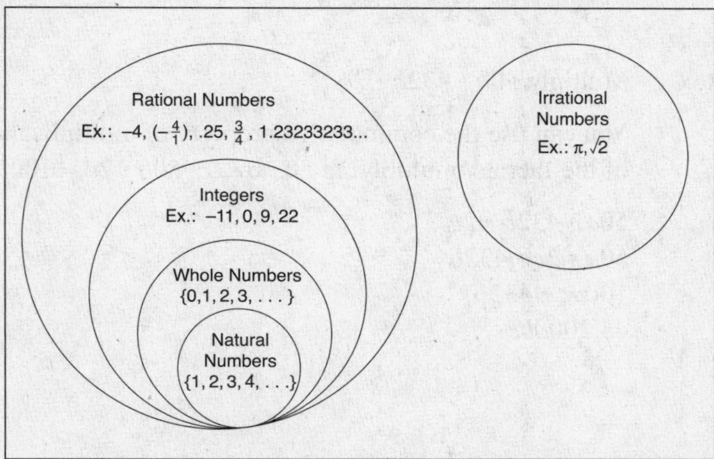

A real number can be placed on a number line.

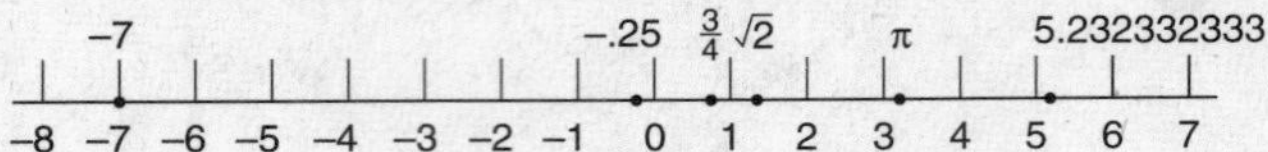

PROPERTIES OF REAL NUMBERS

Commutative Properties

If you think of the word *commute*, and its meaning, "switch places," or "go back and forth," you will have a pretty good idea of the *commutative property*. This property means that you can add or multiply a group of numbers in any order.

Commutative Property of Addition

The *commutative property of addition* states that if you change the order of the numbers you are adding (addends), the sum will not change.

Examples: $1 + 2 + 3 = 3 + 1 + 2$
$a + b = b + a$

You can save time working addition problems if you make use of this property.

Example: $2x + 19 + 6 + 4 + 1 + 6x$

By changing the order of addition, you can add $2x + 6x$, $19 + 1$, and $6 + 4$, arriving at a sum of $8x + 30$.

Commutative Property of Multiplication

The *commutative property of multiplication* states that if you change the order of the factors, the product will not change.

Examples: $2 \cdot 3 \cdot 4 = 4 \cdot 2 \cdot 3$
$ab = ba$

Problems that might seem complex and difficult can be made considerably easier using this property.

Example: Multiply $50a \cdot 432b \cdot 2c$

You can use the commutative property of multiplication to change the order of the factors, multiplying $50a$ by $2c$: $50a \cdot 2c = 100\ ac$.

$50a \cdot 432b \cdot 2c$
$50a \cdot 2c \cdot 432b$
$100ac \cdot 432b$
$43{,}200abc$

Associative Properties

If you think of the terms *associate* and *associate with* or their meaning "to socialize with," you already understand something about the *associative property*. This property means that you can regroup addends or factors without changing the sum or product.

Note: Subtraction and division are not commutative. $2 - 5$ is not the same as $5 - 2$, just as $6 \div 2$ is not the same as $2 \div 6$.

Associative Property of Addition

The *associative property of addition* states that you can regroup addends and still get the same sum.

Examples: $(9 + 1) + 5 = 9 + (1 + 5)$
$a + b + c = a + (b + c) = (a + b) + c$

Associative Property of Multiplication

The *associative property of multiplication* states that you can regroup factors and still get the same product.

Examples: $4 \cdot 12 \cdot 10 = 4(12 \cdot 10)$
$abc = a(bc) = (ab)c$

Identity Properties

Identity Property of Addition

If you add zero to a number, you will get the same number. This commonsense idea is called the *identity property of addition*. Zero is the *additive identity*. This name comes from the word *identical*, which is what you wind up with if you add zero to a number.

Note: There are no identity properties for subtraction or division.

Examples: $-4 + 4 = 0$ — -4 and 4 are additive inverses.
$-4 + 0 = -4$ — 0 is called the additive identity.

Identity Property of Multiplication

According to the *identity property of multiplication*, if you multiply a number by 1, the product will be the original number.

Examples: $4 \times 1 = 4$ — 1 is called the *multiplicative identity*.
$a \times 1 = a$
$4 \times \frac{1}{4} = 1$ — 4 and $\frac{1}{4}$ are *reciprocals*. When you multiply them, you get 1 as a product.

Distributive Property

The *distributive property* allows you to distribute or share a number outside the parentheses with each term inside the parentheses. When multiplying a number by a sum or difference, you can use the distributive property.

Example 1: $5(6 + 1)$
$5(6) + 5(1)$
$30 + 5 = 35$

Example 2: $4(5 - 2)$
$4(5) - 4(2)$
$20 - 8 = 12$

Example 3: $a(b + c) = ab + ac$

Example 4: $a(b - c) = ab - ac$

At first, this might seem to be a waste of time. You might think that when using the distributive property to simplify 5(6 + 1), it would be easier to go ahead and find the sum of 6 and 1 and then multiply by 5. The distributive property's real value comes into play when you have to distribute with variables. The idea of this property is to distribute the number outside the parentheses to each term inside the parentheses.

Example 1: $5(2x + 4)$
$5(2x) + 5(4)$
$10x + 20$

Example 2: $(2a - 5b)2$
$2a(2) - 5b(2)$
$4a - 10b$

Sample Questions

Use the distributive property to simplify the following expressions.

1. $4(9 - 2)$
2. $3x(x + 9)$
3. $-5(3 - x)$
4. $2(-1 + 7)$
5. $(2x + 3)4$

Answers

1. $4(9 - 2) = 4(9) - 4(2) = 36 - 8 = 28$
2. $3x(x + 9) = 3x(x) + 3x(9) = 3x^2 + 27x$
3. $-5(3 - x) = -5(3) - (-5)(x) = -15 + 5x$
4. $2(-1 + 7) = 2(-1) + 2(7) = -2 + 14 = 12$
5. $(2x + 3)4 = 2x(4) + 3(4) = 8x + 12$

The Distributive Property and Division

The distributive property can also be used with division. You can distribute the divisor over each term in the dividend. In plain English, it means you are "sharing" the bottom number with each number or term on the top.

Example 1: Simplify $\frac{32-24}{8}$.

$\frac{32-24}{8}$ Distribute the 8.

$\frac{32}{8}-\frac{24}{8}$ Complete the division.

$4-3=1$

Example 2: Simplify $\frac{9x^2+6x}{3x}$.

$\frac{9x^2+6x}{3x}$ Distribute the $3x$ under each term in the dividend.

$\frac{9x^2}{3x}+\frac{6x}{3x}$ Simplify: $9x^2 \div 3x = 3x$ and $6x \div 3x = 2$.

$3x+2$

Zero Property of Multiplication

The *zero property of multiplication* states that if you multiply any number by zero, the product will be zero.

Example: $4 \cdot 0 = 0$

This concept is used frequently in algebra when attempting to factor quadratic equations. If $ab = 0$, then one of the two factors, either a or b must be zero; otherwise the product could not be zero.

Transitive Property

The *transitive property* states that if $a = b$ and $b = c$, then $a = c$. A more commonsense example is 12 + 8 = 20 and 4 + 16 = 20, and so 12 + 8 = 4 + 16.

Conversions

Metric	US Customary
1 kilometer = 1000 meters	1 mile = 1760 yards = 5280 feet
1 meter = 100 centimeters = 1000 milliliters	1 yard = 3 feet = 36 inches
1 liter = 1000 milliliters = 1000 cubic centimeters	1 acre = 43,560 square feet
1 gram = 1000 milligrams	1 gallon = 4 quarts = 8 pints
1 kilogram = 1000 grams	1 quart = 2 pints = 4 cups
	1 pint = 2 cups = 16 ounces
Time	1 cup = 8 fluid ounces
1 hour = 60 minutes	1 ton = 2000 pounds
1 minute = 60 seconds	1 pound = 16 ounces

CONVERSIONS

It is helpful to have an approximate idea of the size of each measurement unit. For example, 1 inch is approximately equal to 2.5 centimeters. One mile is farther than one kilometer. In fact, a kilometer is only about six-tenths of a mile. A yardstick is 36 inches long (a yard). That's only about 3 inches shorter than a meter stick which is exactly 1 meter long. The average male adult is about 2 meters tall or close to 6 feet. A car can be about 18 feet long or 6 yards (which is close to 6 meters). A penny is only about 2 millimeters thick.

Test Tip: When converting from larger to smaller units, *multiply* by the conversion unit.

Example 1: Convert 5 kilometers to meters.
5 kilometers = 5 • 1000 meters = 5000 meters

Example 2: Convert 2.75 miles to yards.
2.75 miles = 2.75 • 1760 yards = 4840 yards

Test Tip: To convert from a smaller unit to a larger unit, *divide* by the conversion unit.

Example 3: Convert 620 millimeters to centimeters.
620 ÷ 10 = 62 centimeters (Since 100 centimeters = 1000 millimeters, there are 10 millimeters in 1 centimeter; therefore, the conversion unit is 10.)

Example 4: Convert 120,600 inches to yards.
120,600 inches ÷ 36 inches = 3350 yards
(36 inches = 1 yard)

Example 5: Convert 2 miles to feet.
2 • 5280 = 10,560 feet

Sample Questions

1. Convert 17 meters (m) to kilometers (km).
2. Convert 450 centimeters (cm) to meters.
3. Convert 812 millimeters (mm) to meters.
4. Convert 104 kilometers to centimeters.
5. Convert 22 meters to millimeters.
6. A typical model building is $\frac{1}{144}$ the size of the original. If the original building is 1080 feet tall, how tall should the model be?
7. The distance from Jerry's house to school is 448 yards. At the end of 10 days of school, to the nearest tenth of a mile, how many miles has Jerry walked?
8. Convert 12 feet, 9 inches to inches.
9. Convert 4 yards, 2 feet to feet.
10. Convert 30 inches to feet.
11. Convert $1\frac{1}{2}$ miles to yards.

Answers

1. $17 \div 1000 = .017$ km
2. $450 \div 100 = 4.5$ m
3. $812 \div 1000 = .812$ m
4. $104 \bullet 1000 = 104{,}000$ m; $104{,}000 \bullet 100 = 10{,}400{,}000$ cm (100 cm/m)
5. $22 \bullet 1000 = 22{,}000$ mm
6. $1080 \bullet \frac{1}{144} = 7.5$ feet tall. (Remember that multiplying by the fraction $\frac{1}{144}$ is the same as dividing by 144.)
7. 448 yards • 2 (walking to and from school) = 896 yards per day
   896 • 10 = 8960 yards in 10 days
   8960 ÷ 1760 yards ≈ 5.09 miles
8. $12 \bullet 12 + 9 = 153$ inches. (First convert 12 feet to inches and then add 9 inches.)
9. $4 \bullet 3 + 2 = 14$ feet. (First convert yards to feet and then add 2 feet.)
10. $30 \div 12 = 2\frac{1}{2}$ feet.
11. $1.5 \bullet 1760 = 2640$ yards. (For ease of calculation, convert from $1\frac{1}{2}$ to 1.5.)

IRRATIONAL NUMBERS

Irrational numbers are decimal numbers that do not terminate (stop) or repeat in a regular pattern. Although the actual number itself is nonrepeating and nonterminating, pi is usually rounded to 3.14 or the fraction $\frac{22}{7}$. These are called rational representations of π.

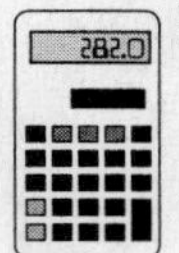

Examples: π

$\sqrt{2}$

$3\sqrt{10}$

.01001000100001

On the FCAT, students usually use 3.14 as a stand-in for π in calculations.

Frequently, when you take a square root of a number, the answer is irrational. Take the square root of 2 on your calculator. Enter the number first and then hit the $\sqrt{\ }$ key. Notice that the numbers to the right of the decimal point do not repeat in any particular pattern. Again, it is important to look closely at any answer you find by taking the square root to see if it terminates or repeats (which means it is rational).

RATIONAL NUMBERS

Rational numbers include all integers, whole numbers, natural numbers, fractions, and decimals (repeating and terminating).

A rational number is any number that can be written as a fraction a/b, where b is not equal to zero. This does not necessarily mean that the number *must* be a fraction, only that it can be written as one.

The denominator b cannot be zero because it is not possible to divide by zero.

Examples: -4 $\left(\text{because it can be written as } -\dfrac{4}{1}\right)$

$4\dfrac{1}{4}$ or 4.25 $\left(\text{because they can be written in the form } \dfrac{17}{4}\right)$

$4\dfrac{1}{4}$ can also be written as $4\dfrac{25}{100}$.

4.25 and $4\dfrac{25}{100}$ are called equivalent forms of $4\dfrac{1}{4}$.

INTEGERS

Integers represent zero, all whole numbers and their opposites: {. . . , –4, –3, –2, –1, 0, 1, 2, 3, . . .}. Natural numbers, whole numbers, and integers can be odd or even (divisible by 2). We say the opposite of 4 is –4, and the opposite of –9 is 9.

Positive integers are all integers that are greater than zero: {1, 2, 3, 4, . . .}. Negative integers are less than zero: {. . . , –4, –3, –2, –1}. Zero is neither positive nor negative.

Sample Questions

1. Which of the following numbers is rational?

A. $2\sqrt{2}$
B. 3.14
C. π
D. $\sqrt{10}$

2. Identify each of the following numbers as rational or irrational.

$\sqrt{5}$	______	$.\overline{125}$	______
$\dfrac{22}{7}$	______	$\sqrt{36}$	______
3.14	______	$\dfrac{3}{7}$	______
π	______	.135	______
–9	______	$\dfrac{1}{9}$	______

3. The group of numbers {–7, –6, 0, 9} can be best classified as

A. whole numbers
B. natural numbers
C. integers
D. irrational Numbers

4. Which integer is larger, –5 or –6?

5. Which integer represents the smallest number?
   A. 9
   B. −9.5
   C. $-8\frac{3}{4}$
   D. −9

Answers

1. **B.** 3.14 only *represents* pi. Because it is a terminating decimal, it represents a rational number. **A** means $2 \cdot \sqrt{2}$ (take the square root of 2 first and then multiply by 2). **C** and **D** are both irrational numbers.
2. | | | | |
   |---|---|---|---|
   | $\sqrt{5}$ | irrational | $.\overline{125}$ | rational |
   | $\frac{22}{7}$ | rational | $\sqrt{36}$ | rational |
   | 3.14 | rational | $\frac{3}{7}$ | rational |
   | π | irrational | .135 | rational |
   | −9 | rational | $\frac{1}{9}$ | rational |
3. **C.**
4. −5 is larger than −6 because it is farther to the right on the number line.
5. **D**. Although **B** is actually smaller than −9, it is not an integer and therefore cannot be the correct answer.

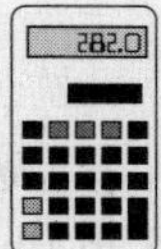

Entering a negative number into the calculator used on the FCAT involves entering the number first and then pressing the +/− key. Try entering a number in your calculator and hitting the +/− key several times. Notice that each time the sign of the number changes in the calculator display.

Whole Numbers

Whole numbers include all natural numbers and zero: {0, 1, 2, 3, 4, . . .}. (The purpose of three dots . . . is to indicate that this set of numbers continues in the same pattern infinitely.)

Natural Numbers

Natural numbers (also called counting numbers) are just like whole numbers except that zero is not included: {1, 2, 3, 4, 5, . . .}.

PLACE VALUE

Place value refers to the position of a digit within a number. For example, in the number 23,456, the number 3 is located in the one-thousands position, and so it has a value of 3 times 1000, or 3000.

Place Value for Whole Numbers

Billions			Millions			Thousands			Ones		
Hundred	Ten	One	Hundred	Ten	One	Hundred	Ten	One	Hundred	Ten	One

Example 1: What is the value of the 7 in the number 10,078,001?

Because the 7 is in the ten-thousands place, the value of the 7 is 7 times 10,000, or 70,000.

Example 2: If the digits for the tens place and the ten thousands place of 34,125 were interchanged, what would be the difference between the two numbers?

"Difference" means subtract.

The difference would be 9,990. By interchanging the 3 and 2 in 34,125, the new number would be 24,135. 34,125 – 24,135 = 9990.

Place Value for Numbers Between Zero and One

You can recognize a number between zero and one because it has a decimal in front of it and its name ends in *-ths*. The *ths* can easily be overlooked when reading a math problem, so read carefully.

Ones (.)	Tenths	Hundredths	Thousandths	Ten thousandths	Hundred thousandths	Millionths

The number 4.563204 has been written in boxes to show each number's place value. Notice that the 4 and its decimal point go in the ones position. This puts the 5 in the tenths position, giving it a value of 5 tenths (5/10). The 6 is in the hundredths position, giving it a value of 6 hundredths (6/100), and so on.

4.5	6	3	2	0	4
Ones (.)	Tenths	Hundredths	Thousandths	Ten thousandths	Hundred thousandths	Millionths

Example: The length of the AIDS virus is 0.00011 millimeter. If 365 AIDS viruses were placed end to end, how long would that be? Round your answer to the ten-thousandths position.

0.00011 • 365 = 0.04015. Rounded to the ten-thousandths position means four places after the decimal: 0.04015. The digit after the 1 is 5, so round up to 0.0402.

Sample Questions

1. What is the value of the 3 in 13,400?
2. What is the value of the 5 in 6.125?
3. Write 67,023,000 in words.
4. Write 4.022 in words.
5. Write ten thousand two in numbers.
6. Write six ten thousandths in numbers.

Answers

1. 3 thousand
2. 5 thousandths
3. Sixty-seven million twenty-three thousand
4. Four and twenty-two thousandths
5. 10,002
6. 0.0006

ROUNDING

Many problems specify exactly how a number is to be rounded or estimated. The problem may say "round to the nearest dollar" or "estimate to the nearest inch." Rounding and estimating both require rounding skills.

To round a number accurately, it is important to understand place value and how it is used in rounding.

Example: Round 2453 to the nearest hundred.

Step 1: Underline the number in the hundreds position: 2453.

Step 2: Examine the digit to the right of the underlined number. In this case, the number is 5.

Step 3: If the digit to the right of the underlined number is less than 5, change all the digits immediately after the underlined number to zero. If the digit to the right of the underlined number is greater than or equal to 5, add 1 to the underlined number and change all the following digits to zero. 2453 rounds to 2500.

Example 1: Round 2413 to the nearest hundred.

Step 1: 2413 Underline the hundreds position.

Step 2: 2400 The digit after the 4 is less than 5, so change the 1 and the 3 to zeros.

Example 2: Round 52,631 to the nearest thousand.

Step 1: 52,631 Underline the thousands position.

Step 2: 53,000 The digit after the 2 is more than 5, so change the 2 to a 3 and change the 6, 3, and 1 to zeros.

Example 3: Round 73.375 to the nearest hundredth.

Step 1: 73.3<u>7</u>5 Underline the hundredths position.
Step 2: 73.380 Add 1 to the 7 and change the 5 to a zero.
Step 3: 73.38 You can drop the zero in this case because it is holding a place that is no longer needed.

Sample Questions

In questions 1–5, round the given number as specified.

1. 48.312 to the tenths position
2. 125.999 to the nearest whole number
3. 4375 to the nearest hundred
4. $17.3421 to the nearest penny
5. 0.1224 to the nearest thousandth
6. Falon used her calculator to find the amount of discount at a sale. The display on her calculator showed $14.7561. What was the savings in dollars?
   A. $14.75
   B. $14.80
   C. $14.00
   D. $14.76

Answers

1. 48.3
2. 126
3. 4400
4. $17.34
5. .122
6. **D**

ESTIMATION

Estimation is a form of rounding. Estimation is extremely useful in checking to see if your answer makes sense.

Test Tip: Always check to see if your answer makes sense.

Generally speaking, when you are asked to *estimate* an answer, you should round the numbers *before* beginning your calculations. This is not only a time-saving device but can also allow you to quickly make a mental prediction about the expected answer. You will know you are supposed to estimate because the question will say "*estimate*" or "about how much" or "approximately how many."

Example 1: If you are paid $7.15 per hour for 20 hours of work, about how much will you make?

Round $7.15 to $7 and multiply by 20. You will make about $140.

Example 2: Andrea spent $2.20 for lunch on Monday, $2.75 on Tuesday, $4.15 on Wednesday, $1.95 on Thursday, and $4.95 on Friday. *Estimate* the total amount she spent.

Round each number to the nearest whole dollar:

Monday	\$2
Tuesday	\$3
Wednesday	\$4
Thursday	\$2
Friday	\$5
Total	\$16

Sample Questions

Estimate each answer as specified.

1. *Estimate*, to the nearest whole number, the following sum: $\frac{11}{12}+\frac{1}{10}+\frac{4}{5}+\frac{2}{5}$.
2. A car traveled 198 miles in 4 hours, 10 minutes. *Estimate* the miles traveled per hour.
3. A quilt pattern requires $1\frac{7}{8}$ yards, $2\frac{1}{2}$ yards, $3\frac{1}{4}$ yards, and $1\frac{1}{8}$ yards of different fabrics. If the fabric costs \$10.99 per yard, *estimate* the cost of the quilt before taxes.
4. The area on either side of a sidewalk is to be sodded with grass. If the shaded area represents 54 square yards of grass, *estimate* the total square yards of grass needed for the entire area outside the sidewalk. Show your work or estimation strategies.

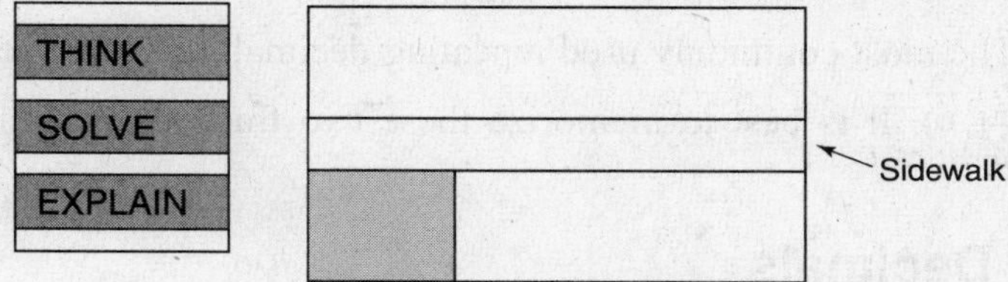

Answers

1. 2. $\frac{11}{12}+\frac{1}{10}$ is close to 1. $\frac{4}{5}+\frac{2}{5}$ is also close to 1.
2. 50 miles in an hour. Round 198 to 200 and divide by 4.
3. \$99. The total yardage is approximately 9 yards $\left(1\frac{7}{8}\right.$ rounds to 2 yards, $2\frac{1}{2}$ rounds to 3 yards, $3\frac{1}{4}$ rounds to 3 yards, and $\left.1\frac{1}{8} \text{ rounds to } 1\right)$. Multiplying 9 by \$11 gives \$99.
4. About 360 square yards. However, your answer may be slightly different. To do this problem, copy the shaded area over several times until you have covered the entire required area. Since each shaded block represents 54 square yards and there are six whole blocks and most of another, you can multiply 54 by 7. This gives 378 square yards. You should deduct yards to show that you are aware that it doesn't cover the entire 378 square yards.

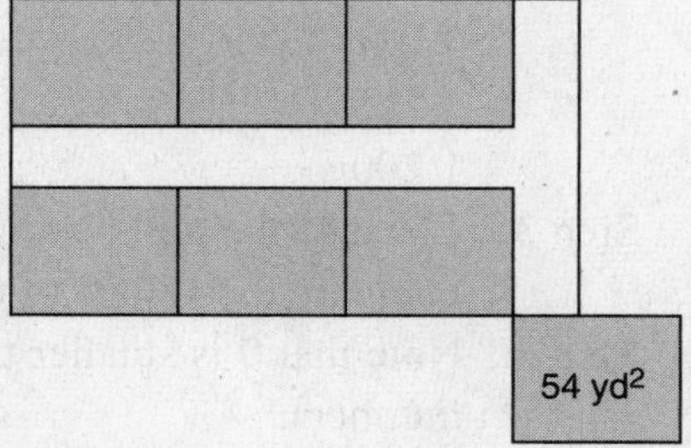

DECIMALS

The term *decimal* usually refers to a number that includes a decimal point. Whole numbers such as 11, 5, and 7 are also called *decimal numbers*. A *decimal system* is based on ten digits (0 through 9). Decimal numbers can be positive or negative, for example, 1.5, 2.09, −7.25, and −12. Decimals may or may not contain leading zeros. The numbers 0.125 and .125 represent the same amount. To give you practice in working with decimals, numbers with and without leading zeros are presented in this book.

Terminating decimals are decimal numbers that terminate or stop. They have a *finite* (you can count them) number of decimal places. Examples of terminating decimals are .5, 7.25, .375, 0.6, −4.3, and .0411.

Repeating decimals are decimal numbers that repeat in a definite pattern to the right of the decimal point. Examples of repeating decimals are $0.\overline{3}$, $1.8\overline{3}$, $.\overline{125}$, and $-1.\overline{6}$. The line over the number(s) indicates that the number (or group of numbers) continues forever.

$$.\overline{3} = .3333333\ldots$$

$$.\overline{125} = .125125125125\ldots$$

$$1.8\overline{3} = 1.833333\ldots$$

The most commonly used repeating decimals are those representing the fractions $\frac{1}{3}(.\overline{3})$ and $\frac{2}{3}(.\overline{6})$. It is best to memorize these two fractions and their decimal equivalents.

Comparing Decimals

It is important to be able to compare the size of one number to that of another. Decimals are one form of numbers that you must be able to compare. Most other types of numbers, such as percents, fractions, and exponents, are relatively easy to compare in size if they are changed to the same form.

Example: Which is smaller, 1.225 or 1.2?

More decimal places does not always mean the number is large.

Step 1: Line the numbers up vertically with the decimal points directly under each other.

1.225
1.2

Step 2: Add zeros to the right of the numbers until each number has the same number of places to the right of the decimal point.

Column 1

↓

1.225
1.200

Step 3: Because the numbers in the first and second columns are the same, compare the values of the numbers in the third column.

Step 4: Note that 0 is smaller than 2; therefore 1.2 is the smaller of the two numbers.

Sample Questions

1. Of the decimals 9.1, 9.25, 9.5, and 9.14 which is the largest?
2. Of the decimals .002, .005, .001 and .0014 which is smallest?
3. Put the decimals 1.41, 1.4, 1.415, and 1.405 in order from largest to smallest.

Answers

1. 9.5. Lined up vertically, the choices look like this:

Column 1

↓

9.10	Check column 1, noting that all are the same.
9.25	Check column 2 and note that 5 is the largest.
9.50	Therefore 9.5 is the largest.
9.14	

2. .001. Compare .0020, .0050, .0010, and .0014.
3. 1.415, 1.41, 1.405, 1.4. Lined up vertically, the choices look like this:

Column 1

↓

1.410	Check column 1, noting that all the numbers
1.400	are the same. For a quick check, compare
1.415	410, 400, 415, and 405 and put in order.
1.405	

FRACTIONS

Fractions are written in the form *a*/*b* (a ratio of two numbers). The denominator *b* of the fraction tells how many numbers are in the total set. The numerator *a* of the fraction tells how many parts there are out of the total set. $\left(\frac{part}{total}\right)$.

Fractions are said to be *equivalent* when they equal the same amount. $\frac{1}{2}$ is equivalent to $\frac{2}{4}$ because 1 part out of 2 represents the same amount (or ratio) as 2 parts out of 4.

Example: 8 out of a total of 10 people ran in a race. $\left(\frac{8}{10}\right.$ of the people ran in the race or, equivalently, $\frac{4}{5}$ of the people ran in the race$\left.\right)$.

A fraction can be changed to its decimal form by dividing the numerator by the denominator.

For example, $\frac{1}{4}$ represents 1 ÷ 4, or 0.25. These forms are equivalent, or have equal value. When comparing fractions and decimals, changing to equivalent forms can make it easier.

Sample Questions

1. One-third of those surveyed at a cafeteria preferred hamburgers to cheeseburgers or hot-dogs. Of those surveyed, .375 preferred cheeseburgers. Which is more popular, hamburgers or cheeseburgers?
2. Put these numbers in order from smallest to largest: 1/4, .2, 0.125, .025.
3. The students at Monroe High School voted on a prom theme: Two-fifths voted for a Hawaiian theme, and the number voting for a Western theme was even lower. Which of the following numbers could represent the number of students voting for a Western theme?

   A. $\frac{6}{10}$
   B. $\frac{1}{2}$
   C. 0.45
   D. 0.25

Answers

1. Cheeseburgers. Convert $\frac{1}{3}$ to a decimal by dividing 1 by 3. Note that .375 is larger than $.\overline{3}$ (or $.33\overline{3}$). When comparing repeating decimals, do not add zeros to the end; simply continue the pattern.
2. .025, 0.125, .2, $\frac{1}{4}$. Before comparing, change the fraction to a decimal.
3. **D.** $\frac{2}{5}$ has a decimal equivalent of .4. Therefore, .25 is a lower number than $\frac{2}{5}$.

An *improper fraction* is a fraction in which the numerator is larger than the denominator. The fraction $\frac{8}{5}$ is an example of an improper fraction. Despite its name, it's perfectly OK to represent a number as an improper fraction. Improper fractions are equivalent forms of mixed numbers. For example, $\frac{8}{5}$ represents the mixed number $1\frac{3}{5}$ and the decimal 1.6.

Test Tip: Do not attempt to put a mixed number on a gridded response form on the FCAT. It is important to change a mixed number to an improper fraction before bubbling, or your machine-scored test might be graded improperly. See Introduction to the FCAT for a more detailed information on gridding answers.

To change a mixed number to an improper fraction, multiply the whole number by the denominator and add the numerator. Keep the same denominator.

Example: Change $5\frac{3}{8}$ to an improper fraction.

Step 1: $5 \cdot 8 = 40$ Multiply the whole number by the denominator.
Step 2: $40 + 3 = 43$ Add the numerator.
Step 3: $\frac{43}{8}$ Keep the same denominator.

Sample Questions

Change the following mixed numbers to improper fractions.

1. $3\frac{4}{5}$

2. $1\frac{2}{3}$

3. $10\frac{3}{8}$

4. $5\frac{1}{3}$

5. $2\frac{1}{2}$

Answers

1. $\frac{3\times 5+4}{5}=\frac{19}{5}$

2. $\frac{1\times 3+2}{3}=\frac{5}{3}$

3. $\frac{10\times 8+3}{8}=\frac{83}{8}$

4. $\frac{5\times 3+1}{3}=\frac{16}{3}$

5. $\frac{2\times 2+1}{2}=\frac{5}{2}$

Lowest Terms

To put a fraction in *lowest terms*, try to find the factors both numbers have in common and divide both numbers by that factor. For example, in $\frac{5}{25}$, both numbers share a factor of 5. Dividing the numerator and the denominator by 5 $\left(\frac{5\div 5}{25\div 5}\right)$, puts the fraction in lowest terms: $\frac{1}{5}$.

You will be more successful with fractions if you review your multiplication tables.

Example: Put $\frac{12}{20}$ in lowest terms.

$$\frac{12}{20}=\frac{12\div 4}{20\div 4}=\frac{3}{5}$$

Addition and Subtraction with Fractions

To add or subtract fractions by hand, you must first have a common denominator, as in the following examples.

Example 1: $\frac{1}{5}+\frac{2}{5}=\frac{3}{5}$

The decimal version is 0.2 + 0.4 = 0.6.

Example 2: $\frac{1}{8}+\frac{5}{8}=\frac{6}{8}=\frac{3}{4}$ (lowest terms)

The decimal version is 0.125 + 0.625 = 0.75.

If the denominators are not the same, find the least common multiple of the denominator and convert one or more of the fractions so that the denominators are the same before adding.

Finding a Common Denominator

If you are not clear about how to find a common denominator, here's a slightly different method that might help.

Follow these steps to find a common denominator for the fractions $\frac{2}{15}$ and $\frac{5}{12}$.

Step 1: Factor the denominators into their prime factors: 15 factors into 3 • 5, and 12 factors into 2 • 2 • 3.

Step 2: Put the denominators in a Venn diagram. Since they both share 3, the 3 is in the center.

More information on Venn diagrams is in Chapter 3.

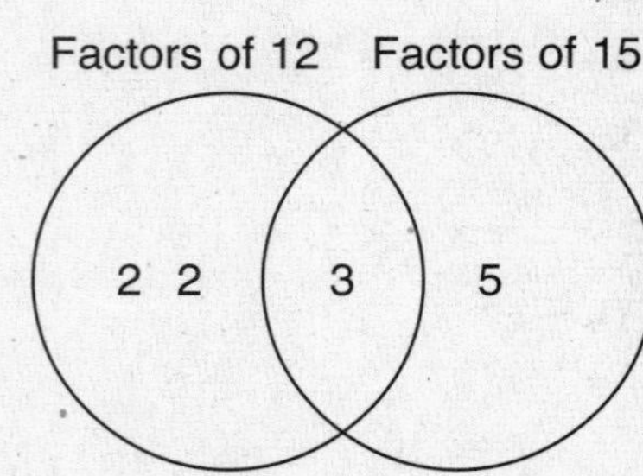

Step 3: Multiply the numbers in the Venn diagram together. The common denominator is 2 • 2 • 3 • 5 = 60.

Converting and Adding or Subtracting

To convert $\frac{2}{15}$ to $\frac{x}{60}$, multiply 2 × 60 and divide by 15: $\frac{2}{15}=\frac{8}{60}$.

To convert $\frac{5}{12}$ to $\frac{x}{60}$, multiply 5 × 60 and divide by 12: $\frac{5}{12}=\frac{25}{60}$.

Add the fractions: $\frac{2}{15}+\frac{5}{12}=\frac{8}{60}+\frac{25}{60}=\frac{33}{60}=\frac{11}{20}$

Sample Question

Add the fractions $\frac{2}{15}$ and $\frac{3}{10}$. First, find a common denominator for these fractions:

Step 1: Factor the denominators into their prime factors. The factors of 15 are _______, and the factors of 10 are _______.

Step 2: Put these factors in a Venn diagram. What (if anything) do they share?

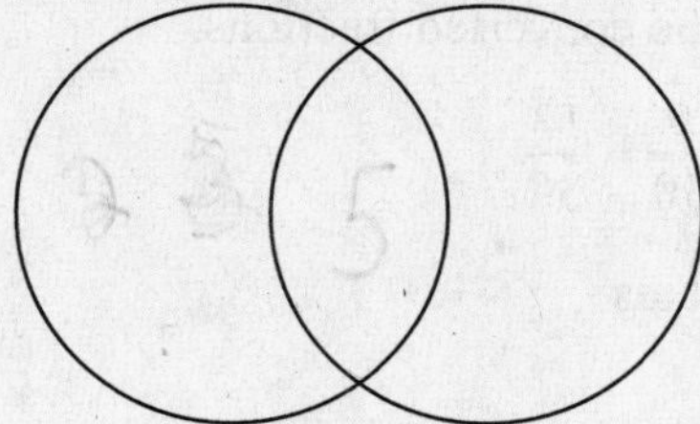

Step 3: Multiply the numbers in the Venn diagram together. The common denominator is _______.

Step 4: To convert the fractions with the new common denominator, set up a proportion using the new denominator in the second ratio. Solve the proportion to find the numerator.

$$\frac{2}{15} = \frac{x}{?} \qquad \frac{3}{10} = \frac{x}{?}$$

Step 5: Add the converted fractions.

Answer

Find a common denominator for the fractions $\frac{2}{15}$ and $\frac{3}{10}$:

Step 1: Factor the denominators into their prime factors. The factors of 15 are 3 and 5, and the factors of 10 are 2 and 5.

Step 2: Put the denominators in a Venn diagram. They share a 5, so the 5 goes in the center.

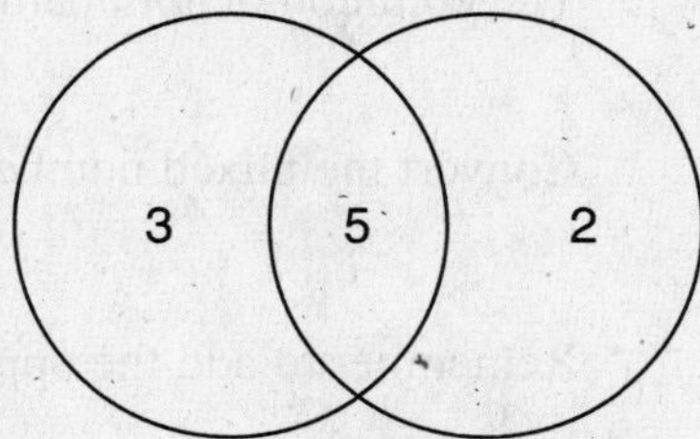

Step 3: Multiply the numbers in the Venn diagram together. The common denominator is $3 \times 5 \times 2 = 30$.

Step 4: To convert the fractions with the new common denominator, set up a proportion using the new denominator in the second ratio. Solve the proportion to find the numerator.

$$\frac{2}{15} = \frac{x}{30} \qquad \frac{2}{15} = \frac{4}{30}$$

$$\frac{3}{10} = \frac{x}{30} \qquad \frac{3}{10} = \frac{9}{30}$$

Step 5: Add the converted fractions.

$$\frac{4}{30} + \frac{9}{30} = \frac{13}{30}$$

Sample Problems

1. $\frac{3}{4} + 1\frac{2}{3}$

2. $\frac{2}{3} - 1\frac{1}{2}$

3. $-2\frac{7}{8} + \left(-\frac{1}{4}\right)$

4. $\frac{3}{5} + \frac{3}{10}$

5. $4\frac{1}{5} - 2\frac{4}{5}$

Sample Answers

1. $\frac{3}{4} + 1\frac{2}{3}$ The common denominator is 12.

$\frac{9}{12} + \frac{20}{12} =$ Convert the mixed number to an improper fraction.

$\frac{9}{12} + \frac{20}{12} = \frac{29}{12}$ Add and grid as an improper fraction.

2. $\frac{2}{3} - 1\frac{1}{2} =$ The common denominator is 6.

$\frac{4}{6} - \frac{9}{6} =$ Convert the mixed number to an improper fraction.

$\frac{4}{6} + \left(-\frac{9}{6}\right) = -\frac{5}{6}$ Remember to add the opposite.

3. $-2\frac{7}{8}+\left(-\frac{1}{4}\right)=$ The common denominator is 8.

$-\frac{23}{8}+\left(-\frac{2}{8}\right)=$ Convert the mixed number to an improper fraction.

$-\frac{25}{8}$ Add the two negatives.

4. $\frac{3}{5}+\frac{3}{10}$ The common denominator is 10.

$\frac{6}{10}+\frac{3}{10}=\frac{9}{10}$

5. $1\frac{1}{5}-2\frac{4}{5}$ Convert the mixed numbers to improper fractions.

$\frac{6}{5}-\frac{14}{5}$ Add the opposite.

$\frac{6}{5}+\left(-\frac{14}{5}\right)=-\frac{8}{5}$ Grid the answer as an improper fraction.

> ***Test Tip***: Before adding or subtracting fractions by hand, ask yourself if you could just as easily use a decimal. If the fraction converts easily to a terminating decimal, you may save yourself a lot of work by doing the conversion and using a decimal, or converting the decimal answer back to a fraction.

Multiplication with Fractions

Multiplying fractions is much simpler than adding or subtracting. To multiply a fraction, simply multiply numerator times numerator and denominator by denominator. Put your answer in lowest terms if necessary.

Example: Find $\frac{1}{2}$ of $\frac{3}{4}$.

$\frac{1}{2}\bullet\frac{3}{4}=\frac{1\bullet 3}{2\bullet 4}=\frac{3}{8}$ The word *of* is generally a signal to multiply.

Change whole numbers or mixed numbers to improper fractions before multiplying.

Example 1: $5\bullet\frac{1}{5}=\frac{5}{1}\bullet\frac{1}{5}=\frac{5}{1}=1$

Example 2: $2\frac{3}{4}\bullet\frac{4}{11}=\frac{11}{4}\bullet\frac{4}{11}=\frac{44}{44}=1$

You can see that $\frac{4}{11}$ is the multiplicative inverse of $2\frac{3}{4}$ because when the numbers are multiplied the product is 1. To find the *reciprocal* of any number, put the number in fraction form and invert (switch the numerator and the denominator).

The fraction $\frac{0}{4}$ has no reciprocal because division by zero $\left(\frac{4}{0}\right)$ is *undefined*. If you attempt to divide by zero on your calculator, you will get an error message.

The reciprocal of 4 is $\frac{1}{4}$.

The reciprocal of $-\frac{2}{3}$ is $-\frac{3}{2}$.

The reciprocal of $1\frac{1}{2}$ is $\frac{2}{3}$.

Division with Fractions

Division with fractions involves one additional step over multiplication. A sample problem might be, "How many students would get candy ifyou had 10 pounds of candy and gave each student $\frac{1}{4}$ pound?"

This problem requires dividing 10 pounds by $\frac{1}{4}$.

Step 1: $10 \div \frac{1}{4} = \frac{10}{1} \div \frac{1}{4}$ Change 10 to a fraction.

Step 2: $10 \div \frac{1}{4} = \frac{10}{1} \bullet \frac{4}{1}$ Multiply 10 by the reciprocal of $\frac{1}{4}$ (4).

$= \frac{40}{1} = 40$ The answer is 40 students.

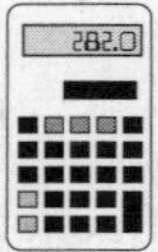

Your FCAT calculator does not have a fraction key. If you are required to grid in a fraction on a gridded-response form, you must be able to perform operations by hand first. If the problem does not require fraction answers, you may be able to use their decimal equivalents, or calculate with decimals and convert back to a fraction. *Do not* attempt to grid a mixed number. You should first change the mixed number to an improper fraction and then grid it. Unless the directions indicate that you may round or estimate, do not attempt to grid nonterminating or repeating decimals.

Examples of fractions or decimals that can be gridded are

$$\frac{1}{3}, \frac{5}{4}, \frac{1}{2}, \frac{-4}{1}\left(-4 \text{ is best}\right), .25, .33, 0.15, .111$$

Examples of fractions or decimals that cannot be gridded are

$$1\frac{2}{3}\left(\text{grid as } \frac{5}{3}\right) \qquad .\overline{6}\left(\text{grid as } \frac{2}{3}\right)$$

$$\pi\left(\text{use } 3.14 \text{ or } \frac{22}{7}\right) \qquad .252525\ldots\left(\text{grid as } \frac{25}{99}\right)$$

PERCENTS

Percents are close relatives of decimals and fractions. You should be able to write equivalent forms of decimals, whole numbers, fractions, and percents. As an example, 50% is equivalent to 0.5 or to the fraction $\frac{1}{2}$. Think of a percent as being part of 100. For instance, 25% means 25 out of 100 or, equivalently, $\frac{25}{100}$ or $\frac{1}{4}$. It also represents 0.25, which is the decimal equivalent of $\frac{1}{4}$.

When changing a percent to a decimal, remove the percent sign and move the decimal point exactly two places to the left.

Example: Changing 42.5% to decimal form gives 0.425.

To change a decimal to a percent, reverse the operation. Move the decimal point exactly two places to the right and add a percent sign.

Example: .125 = 12.5%

When changing a percent to a fraction, put the percentage amount over 100.

Example 1: Change 35% to a fraction.

$35\% = \frac{35}{100}$ or $\frac{7}{20}$

$\frac{7}{20}$ is $\frac{35}{100}$ in lowest terms.

Example 2: Change 15.9% to a fraction.

$15.9\% = \frac{15.9}{100}$, which is correctly expressed as $\frac{159}{1000}$.

(Multiply the numerator and the denominator by 10 so that you will not have a decimal point in the fraction).

Example 3: Change $\frac{2}{3}$ to a decimal and to a percent.

$\frac{2}{3} = .\overline{6} = .66\overline{6} = 66.\overline{6}\% = 66\frac{2}{3}\%$ (Since $\frac{2}{3} = .\overline{6}$, the percent can be expressed as a decimal with an overbar or with a fraction. You may also have to round, expressing $66.\overline{6}\%$ as 66.7%.)

Sample Questions

1. Which of the following are equivalent to 6.5%?

A. 0.65

B. 0.065

C. $\frac{65}{100}$

D. $\frac{6.5}{10}$

2. Convert .2% to decimal form.

3. Complete the following table.

Fraction	Decimal	Percent
$\frac{2}{5}$		40
	$.\overline{3}$	
		75
$\frac{1}{200}$		
$\frac{1}{40}$		2.5
	3	

4. Mr. Gonzalez took his family of five to a soccer game. Each ticket cost $12.00. Mr. Gonzalez bought each person a hotdog for $3.00 and a drink for $2.50. He was charged 7% tax on everything he bought. What was the entire cost for the trip? Round your answer to the nearest penny.

5. Don and Cheryl bought their house 14 years ago. When they sold it they received 210% of the original purchase price. If the original price was $70,000, how much money did they receive for the house?

6. Carla had her car repaired at a cost of $1950. After a deductible is applied (Carla pays the first $500), her insurance company will pay 80% of the remaining balance. How much will the insurance company pay?

Answers

1. **B.** Move the decimal point two places to the left.
2. .2% = .002 Move the decimal point two places to the left and remove the % sign.
3.

Fraction	Decimal	Percent
$\frac{2}{5}$	.4	40
$\frac{1}{3}$	$.\overline{3}$	$33\frac{1}{3}$
$\frac{3}{4}$	.75	75
$\frac{1}{200}$	.005	.5
$\frac{1}{40}$	.025	2.5
$\frac{3}{1}$	3	300

4. **$93.63.** The costs are $12.00 for the ticket, $3.00 for the hotdog, and $2.50 for the drink for a total of $17.50 per person. Multiply by 5 to get $87.50. Add 7% by entering 87.5 into the calculator and hitting + 7%. Round 93.615 to 93.53.
5. **$147,000.** Multiply: 70,000 × 210% = 147,000.
6. **$1160.** $1950 – $500 – $1450. 1450 × 80% = $1160.

ABSOLUTE VALUE

All real numbers can be placed on a number line.

If you were standing on a number line at –2, you would be exactly two spaces away from zero. So we say that the absolute value of –2 is 2. In math notation, that's written as $|-2| = 2$ and read as "the absolute value of negative two is two."

$|2|$, or "the absolute value of 2," also equals 2 because 2 is exactly two spaces away from zero on the number line. Since absolute value represents a distance, it can never be negative.

Note: Do not confuse the absolute value bars | | with parentheses.

Sample Questions

1. What is the absolute value of $\frac{3}{4}$?

2. $|-9.5| = ?$

3. $-|4| = ?$

Answers

1. $\frac{3}{4}$
2. 9.5 (-9.5 is 9.5 spaces away from zero)
3. –4 (read as "the *opposite* of the absolute value of 4"). Because the negative sign is outside the bars, first find the absolute value of 4, and then write down its opposite.

POWERS

There are some *shorthand* methods of representing numbers that you will need to be familiar with for the FCAT. One of these involves powers.

$$5^4$$

Base → 5 ← Exponent (4)

This should be read as "five to the fourth power." What it means is four 5s multiplied together:

$$5 \cdot 5 \cdot 5 \cdot 5 = 625$$

The number 81 written as a *power* is 9^2, and 81 written in *expanded* form is $9 \cdot 9$. When written as 81, we say it is in *standard* form.

Sample Questions

In questions 1–4, write the numbers in expanded form.

1. 10^3

2. 5^2 (also called 5 squared)

3. 2^5

4. 3^3 (also called 3 cubed)

In questions 5–7, write the numbers in standard form (evaluate).

5. 4^3
6. 3^2
7. 5^3
8. Write 49 as a power of 7.
9. Write 16 as a power of 2.
10. Write 81 as a power of 3.

Answers

1. 10 • 10 • 10
2. 5 • 5
3. 2 • 2 • 2 • 2 • 2
4. 3 • 3 • 3
5. 64
6. 9
7. 125
8. 7^2
9. 2^4
10. 3^4

The base number can also be negative. Be careful of numbers that look like $(-2)^5$ or -2^4. For $(-2)^5$, read "negative two to the fifth power." Translated (or written in expanded form), this means (–2) • (–2) • (–2) • (–2) • (–2), which equals –32.

For -2^4 you should immediately notice that there are no parentheses. This number is read as, "the *opposite* of two to the fourth power." Written in its expanded form it means –(2 • 2 • 2 • 2), which equals –16.

Sample Questions

Evaluate the following numbers.

1. -6^2
2. $(-4)^3$
3. -10^4

Answers

1. $-6^2 = -(6 \cdot 6) = -36$ (read "the opposite of 6 squared")
2. $(-4)^3 = (-4) \cdot (-4) \cdot (-4) = -64$ (read "negative 4 cubed")
3. $-10^4 = -(10 \times 10 \times 10 \times 10) = -10{,}000$ (read "the opposite of 10 to the fourth power")

Powers will be discussed in more detail later in this chapter.

Negative Exponents

Note: Any number raised to the zero power equals 1. You can see this demonstrated by carefully examining the pattern shown.

Observe the pattern shown here.

10^3	10^2	10^1	10^0	10^{-1}	10^{-2}	10^{-3}
1000	100	10	1	$\frac{1}{10}$	$\frac{1}{100}$	$\frac{1}{1000}$

Compare 10^3 and 10^{-3}. Notice that 10^3 written in expanded form is equal to 10 • 10 • 10 or 1000, while 10^{-3} written in expanded form equals $\frac{1}{10} \cdot \frac{1}{10} \cdot \frac{1}{10}$, or $\frac{1}{1000}$.

Compare 10^2 and 10^{-2}. Notice that 10^2 written in expanded form is equal to 10 • 10 or 100, while 10^{-2} written in expanded form equals $\frac{1}{10} \times \frac{1}{10}$, or $\frac{1}{100}$.

Example 1: $5^2 = 5 \cdot 5 = 25$

$$5^{-2} = \frac{1}{5} \cdot \frac{1}{5} = \frac{1}{25}$$

Example 2: $2^3 = 2 \cdot 2 \cdot 2 = 8$

$$2^{-3} = \frac{1}{2} \cdot \frac{1}{2} \cdot \frac{1}{2} = \frac{1}{8}$$

Sample Questions

1. Write 4^{-2} in expanded form.
2. Evaluate 3^{-3}.
3. Put these numbers in descending (largest to smallest) order: 5^0, 2, 3^{-1}, 2^{-3}.

Answers

1. $\frac{1}{4} \cdot \frac{1}{4}$
2. $\frac{1}{3} \cdot \frac{1}{3} \cdot \frac{1}{3} = \frac{1}{27}$
3. 2, 5^0 (which equals 1), 3^{-1} $\left(\text{which equals } \frac{1}{3}\right)$, and 2^{-3} $\left(\text{which equals } \frac{1}{8}\right)$.

SCIENTIFIC NOTATION

Scientists use scientific notation as a shorthand way to write numbers that are extremely large or extremely small. Numbers written in scientific notation are always written in the following form.

$$a \times 10^b$$

a is more than 1 but less than 10

Multiplied by a power *b* of 10

Example: 3.5×10^5

Scientific Notation for Numbers Greater Than 1

To change $3.5 \cdot 10^5$ into standard notation, think of 3.5 • 10 • 10 • 10 • 10 • 10. Each time you multiply by 10, the decimal point moves one place to the right. Because you are multiplying by 10 five times (or 10^5), the decimal point moves to the right five times. The number written in standard notation is 350,000.

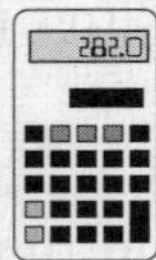

Try this on your calculator. Enter 3.5 and multiply by 10 five times, paying close attention to the result after each multiplication. Your answers should be 35, 350, 3500, 35,000, and 350,000.

To go from the standard form back to scientific notation, reverse the process, starting at the right and moving left. Continue to move the decimal point until it has only one digit in front of it. Count the number of places you moved the decimal and this represents the exponent.

Example: 2,710,000,000,000 (2.71 trillion) in scientific notation is $2.71 \cdot 10^{12}$.

12 decimal places

Scientific Notation for Numbers Less Than 1

Dividing by 10 is the same as multiplying by $\frac{1}{10}$.

Many of the numbers scientists use represent amounts that are extremely small. 4.1×10^{-6} might represent just such a number. To change this to standard notation, follow a procedure similar to that described above. This time think of $4.1 \cdot \frac{1}{10} \cdot \frac{1}{10} \cdot \frac{1}{10} \cdot \frac{1}{10} \cdot \frac{1}{10} \cdot \frac{1}{10}$. Each time you multiply by $\frac{1}{10}$ it is like dividing by 10 and the decimal point moves one place to the left. In this instance, the decimal point moves 6 places to the left. 4.1×10^{-6} written in standard notation is .0000041.

Example: .005 written in scientific notation is $5.0 \cdot 10^{-3}$.

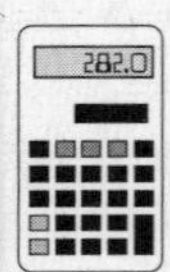

Try this on your calculator. Enter 5 and divide by 10 three times, paying close attention to the result after each division. Your answers should be 0.5, .05, and finally, 0.005. Each division by 10 moves the decimal point one place to the left.

Sample Questions

1. Which of the following numbers is in scientific notation?
   A. 9.1×10^3
   B. 91×10^3
   C. $.91 \times 10^3$
   D. $9.1 \times 10 \cdot 10$

2. Which of the following represents .00203 written in scientific notation?
   A. 2.03×10^3
   B. 20.3×10^{-4}
   C. 2.03×10^{-3}
   D. 0.203×10^3

3. Write 3700 in scientific notation.
4. Write 8,000,000 in scientific notation.
5. Write 1.56922 in scientific notation.
6. Write 10 in scientific notation.
7. Write .0823 in scientific notation.
8. Write 0.103 in scientific notation.
9. Write 0.1234 in scientific notation.
10. Write 0.009 in scientific notation.

Answers

1. **A.** 9.1 is more than 1 but less than 10. It is multiplied by a power of 10.
2. **C.** 2.03 is more than 1 but less than 10. It is multiplied by a power of 10. When the decimal point is moved three places to the left (signified by the –3 exponent), the resulting number is .00203.
3. 3.7×10^3
4. 8×10^6
5. 1.56922×10^0 The decimal point does not move, and so the exponent is 0.
6. 1×10^1
7. 8.23×10^{-2}
8. 1.03×10^{-1}
9. 1.234×10^{-1}
10. 9×10^{-3}

SQUARE ROOTS

The symbol $\sqrt{\ }$ is called a *radical* or square root symbol. We say $\sqrt{25} = 5$ (the square root of 25 is 5) because 5×5 (or 5^2) equals 25. Another instance of this is $\sqrt{4} = 2$. Some numbers, like 1, 4, 9, 16, 25, and 100, are called *perfect squares* because their square root is a whole number.

Always enter the number before the $\sqrt{\ }$ symbol.

To take a square root on an FCAT calculator, first enter the number and then hit the key with the $\sqrt{\ }$ symbol. Try a few times for practice: Take the square root of the perfect squares listed above.

Taking a square root of a number means to work backward from multiplying two identical numbers together. For example, $\sqrt{36}$ can be either 6 or –6 because $6 \times 6 = 36$ and so does -6×-6. Both 6 and –6 are square roots, but positive 6 is called the *primary* square root. Your calculator always gives you the primary square root.

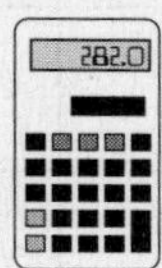

Examples: $\sqrt{1} = 1$ or -1

$\sqrt{4} = 2$ or -2

$\sqrt{9} = 3$ or -3

$\sqrt{16} = 4$ or -4

$\sqrt{81} = 9$ or -9

$\sqrt{121} = 11$ or -11

You can also think of finding a square root as "undoing" a square. This means that it is the inverse operation of squaring. $\sqrt{3^2}$ means to take the square root of 3, which has been raised to the second power. The answer to this problem is 3 because the square root and the power of 2 cancel each other out. Squaring can also undo a square root: $(\sqrt{3})^2 = 3$.

Sample Question

What is $\sqrt{9^2}$?

Answer

If you answered 9, you're right.

Not all square roots are perfect squares. Examples are $\sqrt{0.5625}$, which is 0.75, and $\sqrt{3}$, which is an irrational number because it is nonterminating and nonrepeating. Irrational numbers were covered earlier in this chapter.

Sample Questions

1. What is the primary square root of 81?
2. What is $\sqrt{2} \bullet \sqrt{2}$?
3. Name two perfect squares between 25 and 100.
4. Can you fill in the blanks in the table?

Standard	Scientific Notation	Fraction	Absolute Value	Exponent	Square Root
49	4.9×10^1	$\frac{49}{1}$	|49| or |–49|	7^2	$\sqrt{2401}$
64					
81					

Answers

1. 9 (because $9 \bullet 9 = 81$).
2. 2 (because two identical square roots multiplied together cancel the square root symbol: $\sqrt{2} \bullet \sqrt{2} = (\sqrt{2})^2 = 2$).
3. The perfect squares between 25 and 100 are 36 ($6 \bullet 6$), 49 ($7 \bullet 7$), 64 ($8 \bullet 8$), and 81 ($9 \bullet 9$).
4.

Standard	Scientific Notation	Fraction	Absolute Value	Exponent	Square Root
49	4.9×10^1	$\frac{49}{1}$	|49| or |–49|	7^2	$\sqrt{2401}$
64	6.4×10^1	$\frac{64}{1}$	|64| or |–64|	8^2	$\sqrt{4096}$
81	8.1×10^1	$\frac{81}{1}$	|81| or |–81|	9^2	$\sqrt{6561}$

WORKING WITH POWERS

Addition and Subtraction of Powers

Recall that powers have a base and an exponent.

2^4 ← Exponent } Power

Base ↗

Example 1: $4^2 + 5^2 = 16 + 25 = 41$

When adding or subtracting powers, first put the powers into standard notation and then add or subtract.

Example 2: $7 - 2^2 = 7 - 4 = 3$

Example 3: $x^2 + 3x^2 = 4x^2$

Note that x^2 really means $1x^2$. It is customary in algebraic expressions to "hide" the 1. For example, $6a - 5a = a$, and $4b - 5b = -b$.

Multiplication of Powers

To multiply powers you can use one of two methods.

Method 1: If the bases are the same, keep the same base and add only the exponents.

Example 1: $10^3 \cdot 10^5 = 10^{3+5} = 10^8$

Note that $10^3 = 10 \cdot 10 \cdot 10$
$10^5 = 10 \cdot 10 \cdot 10 \cdot 10 \cdot 10$

In reality, you have a total of eight 10s multiplied together (10^8).

Example 2: $6^5 \cdot 6^{-3} = 6^{5+(-3)} = 6^2$ Add the exponents.

Example 3: $a^2 \cdot a^7 = a^9$ The bases (a) are the same.

Method 2: If the bases are not the same, simplify each power and then multiply.

Example 1: $2^2 \cdot 3^2 = 4 \cdot 9 = 36$

Example 2: $10^3 \cdot 5^2 = 1000 \cdot 25 = 25{,}000$

Example 3: $a^2 \cdot b^3 = a^2b^3$ (cannot be simplified further)

Division of Powers

When dividing powers, you can again use one of two methods.

Method 1: If the bases are the same, keep the base and subtract the exponents.

Example 1: $10^4 \div 10^2 = 10^{4-2} = 10^2$

Shown as a fraction, this problem might look like

$$\frac{10 \cdot 10 \cdot \cancel{10} \cdot \cancel{10}}{\cancel{10} \cdot \cancel{10}}$$

Because each 10 in the denominator *cancels* a 10 in the numerator, you are left with $(10 \cdot 10)/1 = 10^2$.

Example 2: $7^6 \div 7^4 = 7^{6-4} = 7^2$

Example 3: $4^3 \div 4^8 = 4^{3-8} = 4^{3+(-8)} = 4^{-5}$

Note: Recall that $4^{-5} = 1/4^5 = 1/1024$.

Example 4: $\frac{15x^3}{5x} = \frac{15 \cdot x \cdot x \cdot \cancel{x}}{5 \cdot \cancel{x}} = \frac{3 \cdot x \cdot x}{1} = 3x^2$

Method 2: When the bases are not the same, put the numbers in standard notation and divide.

Example 1: $6^2 \div 2^2 = 36 \div 4 = 9$

Example 2: $5^3 \div 10^2 = 125 \div 100 = 1.25$

Example 3: $a^2 \div b^3 = \frac{a^2}{b^3}$ (cannot be simplified further)

Sample Questions

Perform the indicated operations.

1. $10^{10} \div 10^5$
2. $7^5 \div 7^2$
3. $5^2 \div 5^4$
4. $2^3 \div 2^{-7}$
5. $4^3 \div 2^5$
6. $10^9/10^3$
7. $4^5/4^4$
8. $12x^5/4x^2$
9. $14a^3b^2/7ab^2$
10. $3^3/3^{-2}$

Answers

1. $10^{10-5} = 10^5$
2. $7^{5-2} = 7^3$
3. $5^{2-4} = 5^{-2}$
4. $2^{3-(-7)} = 2^{3+7} = 2^{10}$
5. $64 \div 32 = 2$. (The bases were not the same.)
6. $10^9/10^3 = 10^{9-3} = 10^6$
7. $4^5/4^4 = 45^{-4} = 4^1 = 4$
8. $12x^5/4x^2 = 3x^5/1x^2 = 3x^{5-2} = 3x^3 = 3x^3$
9. $14a^3b^2/7ab^2 = 2a^3b^2/ab^2 = 2a^{3-1}b^{2-2} = 2a^2b^0 = 2a^2$
10. $3^3/3^{-2} = 3^{3-(-2)} = 3^{3+2} = 3^5$

WORKING WITH SCIENTIFIC NOTATION

To add or subtract numbers in scientific notation, you must change the numbers to standard form to perform the operations. When performing multiplication or addition operations with numbers in scientific notation, you should work with the power separately from the number in front of it.

Example 1: $3.4 \times 10^3 + 4.91 \times 10^5$ When *adding*, change the numbers to standard form first.
$3400 + 491{,}000 = 494{,}400$

Example 2: $(3.4 \times 10^3) \times (4.91 \times 10^5)$ When multiplying you can rearrange or commute.
$3.4 \times 4.91 \times 10^3 \times 10^5$ Multiply 3.4 by 4.91, and 10^3 by 10^5.
16.694×10^8 Change to scientific notation.
1.6694×10^9

Example 3: $\dfrac{4.9 \times 10^5}{3.5 \times 10^3}$ Divide 4.9 by 3.5 and 10^5 by 10^3.

$\dfrac{4.9}{3.5} \times \dfrac{10^5}{10^3}$

1.4×10^2

> ***Test Tip:*** On the FCAT you may encounter numbers that are too big for your calculator to handle. Using the techniques in the examples shown here will not only save you time but will also allow you much greater accuracy in working problems with very large or very small numbers.

Sample Question

Jupiter's volume is about 3.43×10^{14} cubic miles. Earth's volume is about 2.6×10^{11}. If Jupiter were hollow, about how many Earths would fit into it? Choose the answer closest to the correct answer.

A. 1.32
B. 132
C. 13.2
D. 1320

Answer

D. Divide Jupiter's volume by Earth's volume.

$$\frac{3.43 \times 10^{14}}{2.6 \times 10^{11}} = \frac{3.43}{2.6} \times \frac{10^{14}}{10^{11}} \approx 1.319 \times 10^3 = 1319$$

RADICALS

Radicals include square roots $\sqrt{\ }$, but a radical can be a *cube* root $\sqrt[3]{\ }$ or some other root such as $\sqrt[5]{\ }$. For the FCAT, we will review only operations of square roots. You should be aware that the same rules apply to other roots. Finding a square root is the inverse operation of squaring. Because of this, $(\sqrt{2})^2 = 2$, and $\sqrt{2^2} = 2$. You could say the squaring can undo a square root and vice versa. In the radical $5\sqrt{2}$, the 5 is the *coefficient* and the 2 is the *radicand*, 5 is multiplied by $\sqrt{2}$.

Multiplication with Radicals

Because of the special nature of radicals, it is important to learn how to multiply and simplify them before moving on to addition and subtraction.

When two radicals are multiplied, the procedure is virtually identical to multiplying two whole numbers (with the exception of the radical symbol). Notice that the two numbers under the radical signs are multiplied together under a single radical.

Examples: $\sqrt{2} \cdot \sqrt{3} = \sqrt{2 \cdot 3} = \sqrt{6}$

$\sqrt{5} \cdot \sqrt{2} = \sqrt{5 \cdot 2} = \sqrt{10}$

Sample Questions

1. Find $\sqrt{7} \cdot \sqrt{5}$.

2. Find $\sqrt{8} \cdot \sqrt{3}$.

Answers

1. $\sqrt{7} \cdot \sqrt{5} = \sqrt{35}$
2. $\sqrt{8} \cdot \sqrt{3} = \sqrt{24}$

Now that you are able to multiply radicals, we will work at *simplifying* them. To simplify a radical you must be able to (1) identify perfect squares and (2) factor.

Perfect squares include: 1, 4, 9, 16, 25, 36, 49, 64, 81, and 100.

Perfect square numbers are the result of multiplying two identical numbers together. The perfect squares that you will generally be looking for when simplifying radicals are 1, 4, 9, 16, 25, 36, 49, 64, 81, and 100. There are infinitely more, but these are the most useful.

To simplify the radical $\sqrt{12}$, you need to reverse the multiplication process and factor the 12. There are several ways to factor 12. Always try to select a factor that is a perfect square. $\sqrt{12}$ factors into $\sqrt{4 \cdot 3}$. We continue to reverse the multiplication process by changing $\sqrt{4 \cdot 3}$ to $\sqrt{4} \cdot \sqrt{3}$. Because 4 is a perfect square, $\sqrt{4} = 2$. Therefore, $\sqrt{12} = 2 \cdot \sqrt{3}$ or $2\sqrt{3}$.

Look at the same problem written step by step and review each step.

Step 1: Factor by locating the perfect squares. $\sqrt{12} = \sqrt{4 \cdot 3}$

Step 2: Reverse the multiplication process. $\sqrt{4 \cdot 3} = \sqrt{4} \cdot \sqrt{3}$

Step 3: Take the square root of the perfect square. $\sqrt{4} \cdot \sqrt{3} = 2 \cdot \sqrt{3} = 2\sqrt{3}$

Sample Questions

1. Multiply and simplify $\sqrt{12} \cdot \sqrt{8}$.

2. Multiply and simplify $\sqrt{15} \cdot \sqrt{20}$.

Answers

1. $\sqrt{12} \cdot \sqrt{8}$ Factor.

$\sqrt{4 \cdot 3} \cdot \sqrt{4 \cdot 2}$

$\sqrt{4} \cdot \sqrt{3} \cdot \sqrt{4} \cdot \sqrt{2}$ Take the square root of 4.

$2\sqrt{3} \cdot 2\sqrt{2} = 4\sqrt{6}$

2. $\sqrt{15} \cdot \sqrt{20}$ Factor.

$\sqrt{5} \cdot \sqrt{3} \cdot \sqrt{5} \cdot \sqrt{4}$ Commute.

$\sqrt{5} \cdot \sqrt{5} \cdot \sqrt{4} \cdot \sqrt{3}$ $\sqrt{5} \cdot \sqrt{5} = 5$

$5 \cdot \sqrt{4} \cdot \sqrt{3}$ Take the square root of 4.

$5 \cdot 2 \cdot \sqrt{3} = 10\sqrt{3}$ Multiply 5 by 2.

Addition and Subtraction with Radicals

When adding or subtracting radicals with the same radicand (the number under the radical), simply add the coefficients and keep the same radicand for your answer.

Example: $5\sqrt{2} + 3\sqrt{2} = 8\sqrt{2}$

Sample Questions

1. Add $4\sqrt{2} + (-3\sqrt{2})$.

2. Subtract $2\sqrt{5} - 7\sqrt{5}$.

Answers

1. $\sqrt{2}$ (When no coefficient is shown, it is assumed to be 1.)
2. $-5\sqrt{5}$ (because $2 - 7 = -5$)

SECTION 2: RELATIONS AND FUNCTIONS

Algebra is a way of describing an observed pattern using a general rule. In this section, we will use rules to establish the correct order to perform computations and look at different types of patterns. We will also examine some relationships in which we compare two or more sets of information and inspect a specialized relationship called a function.

ORDER OF OPERATIONS

Now that we've reviewed *how* to perform operations, it's time to take a look at the correct *order* in which to do them. This is called the *order of operations*.

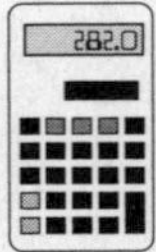

Remember that the FCAT calculator has *not* been programmed with the correct order of operations. Therefore, it is extremely important that you know and are able to use the order of operations.

Example: Evaluate $15 - 9 \div 2 \times b$ when $b = 3$.

A memory clue: PEMDAS (Parentheses, Exponents, Multiplication, Division, Addition, Subtraction) may help.

Working left to right, this is what happens on an FCAT calculator.		If you are following the order of operations rules, you should do the following.	
$15 - 9 \div 2 \cdot 4$	It subtracts.	$15 - 9 \div 2 \cdot 4$	Divide first.
$6 \div 2 \cdot 4$	Then it divides.	$15 - 4.5 \cdot 4$	Multiply.
$3 \cdot 4 = 12$	And last, it multiplies.	$15 - 18 = -3$	Subtract last.
Wrong answer		*Correct answer*	

Operations on numbers or algebraic expressions should be performed in a specific order.

1. Do all operations with grouping symbols. The grouping symbols are parentheses (), brackets [], and the fraction bar $\frac{a}{b}$.
2. Find powers and numbers or expressions under a radical symbol.
3. Do multiplication and division, working from left to right.
4. Do addition and subtraction, working from left to right.

Example: Simplify $29 - 2(3 + 2) + 3^2$.

Step 1: Clear grouping symbols (add 3 + 2)
$29 - 2(5) + 3^2$

Step 2: Find the power.
$29 - 2(5) + 9$

Step 3: Do the multiplication.
$29 - 10 + 9 = 28$

Step 4: Do the addition and subtraction (work from left to right).

The correct solution to simplifying $29 - 2(3 + 2) + 3^2$ is 28.

Sample Questions

Simplify the following expressions.

1. $38 - 9 + 2$
2. $38 - 9 \cdot 2$
3. $(2^3 + 1)5 - 2$
4. $15 \div 3 \cdot 2 \div 10$
5. $(5 + 1)(3 - 4)$

Answers

1. 31. Work from left to right. Do $38 - 9$ first.
2. 20. Multiply first and then subtract 18 from 38.
3. 43. Do the inside parentheses first $(2^3 + 1)$, then multiply by 5. Last, subtract 2.
4. 1. Work from left to right. Do $15 \div 3$ first, next multiply by 2, and last, divide by 10.
5. −6. Multiply 6 by −1.

Grouping Symbols

Fraction bars, parentheses, and brackets are grouping symbols.

Let's look at *grouping symbols* a little more closely. Because they are first in the list of things you must do to simplify an expression or equation, they can have a dramatic effect on the solution.

Example:	$2 \cdot 3 + 1 = 7$	No grouping symbols.
	$2 \cdot (3 + 1) = 8$	Adding parentheses changes the order in which operations are performed.

A *bracket* is simply a differently shaped type of parentheses. It is used when there are groups of parentheses and helps to distinguish between sets of grouping symbols. When there are parentheses and brackets in the same expression, they are said to be *nested.*

Example:	$(4 \cdot 3) + (4 \cdot 5) = 32$	Just parentheses.
	$[4(3 + 4)] = 32$	Nested parentheses.

Sample Question

Write another expression for "twice the expression $(2 \cdot 3) + (4 \cdot 5)$."

Answer

$2[(2 \cdot 3) + (4 \cdot 5)]$

To perform operations with nested parentheses, start at the innermost parentheses, perform that operation first, and work your way outward. Here are the steps for simplifying $[2 \cdot (3 + 4)] \cdot 5$, which was one of the examples shown above.

Step 1: Add $3 + 4$	$[2 \cdot (3 + 4)] \cdot 5$
Step 2: Multiply $2 \cdot 7$	$[2 \cdot 7] \cdot 5$
Step 3: Multiply $14 \cdot 5$	$14 \cdot 5 = 70$

Fraction bars also represent a grouping symbol. Remember that a fraction means that the numerator is divided by the denominator. To evaluate a problem with a fraction, simplify the numerator and denominator separately. Once you have a simple fraction, you can decide whether to put it in lowest terms or change it to a whole number, a mixed number, or a decimal.

Example: $2+\frac{10-1}{-2-1}=2+\frac{9}{-3}=2+(-3)=-1$

Sample Questions

1. Simplify $\frac{5\bullet 2^2}{2^3-3}$.

2. Simplify $\frac{28-16\bullet 2}{2^4-2^2}$.

3. Simplify $\frac{35-4\bullet 5}{-14+3^3}$.

4. Using a fraction bar as a grouping symbol, rewrite the expression 4[12 – 2(1 + 2)] ÷ [9 + (–7)] and simplify.

Answers

1. $\frac{5\bullet 2^2}{2^3-3}$ Find the powers first.

$\frac{5\bullet 4}{8-3}$ Do the operations on the top and bottom.

$\frac{20}{5}=4$ Complete the division.

2. $\frac{28-16\bullet 2}{2^4-2^2}$ Do the multiplication on the top and find the powers on the bottom.

$\frac{28-32}{16-4}$ Do the operations on the top and bottom.

$\frac{-4}{12}=-\frac{1}{3}$ Find the lowest terms.

3. $\frac{35-4\bullet 5}{-14+3^3}$ Do the multiplication on the top and find the power on bottom.

$\frac{35-20}{-22+27}$ Do the operations on the top and bottom.

$\frac{15}{5}=3$ Find the lowest terms.

4. $\frac{4[12-2(1+2)]}{[9+(-7)]}$ Simplify inside the parentheses on the top and do the addition on the bottom.

$\frac{4[12-2(3)]}{2}$ Do the multiplication inside the brackets.

$\frac{4[12-6]}{2}$ Do the subtraction.

$\frac{4[6]}{2}=\frac{24}{2}=12$ Multiply on the top and complete by dividing.

RELATIONSHIPS

If you think back to your elementary school days, you may remember making up a "secret code" with which you could communicate with your friends confidentially. Codes like this are *ciphers* and have been used for centuries to transfer classified information. The simplest forms of ciphers involve a process called *mapping*, in which one letter is substituted for another letter(s) or number(s). We say that the letter *corresponds* to its substitute. In mathematics, this correspondence forms a relation.

Relations are sets of ordered pairs. They represent two sets of information. The members of the set can be ciphers and their substitutes, or they can be something simpler, like cats and dogs, boyfriends and girlfriends, the height of a person and their shoe size, or states and their senators. The *domain* of the relationship is the first item of the pair, and the *range* is the second item. The value in the range can depend on the value in the domain. For example, a person's shoe size can depend on how tall he or she is. This means that the domain of this relationship is the person's height and the range is his or her shoe size.

If you wished to develop a very simple cipher, you might assign a number to each letter of the alphabet. The number assigned would depend on which letter was required. For example, you might assign the number 1 to the letter A, the number 2 to the letter B, and so on. The domain of the relation would be all the letters of the alphabet, and the range would be the numbers 1 through 26.

Relations can take many forms. They can be expressed in words, symbols, variables, tables, and graphs.

Words

Words can be used to describe relationships.

Example: The lowest recorded temperature for Alaska was −80 degrees, 20 degrees below Idaho's lowest temperature. Minnesota's lowest temperature was the same as Idaho's, and Idaho's was 1 degree above Colorado's but the same as North Dakota's.

Symbols

The relations above can be described using symbols or *variables* and written as equations. Let a stand for Alaska, m for Minnesota, c for Colorado, i for Idaho, and n for North Dakota. Then,

$$a = -80$$
$$i = a + 20$$
$$m = i$$
$$i = c + 1$$
$$i = n$$

By substituting −80 for a, you can work your way through these equations and find the lowest temperature for each state listed.

Tables

The information given in the example above can also be represented as a table.

Domain = first coordinate.

Range = second coordinate.

State	Lowest Temperature (°F)
Alaska	−80
Colorado	−61
Idaho	−60
North Dakota	−60
Minnesota	−60

Graphs

Relations come in ordered pairs, and so the order in which they are introduced is important. The domain is the set of the first items or coordinates in an ordered pair. The second coordinates make up the set called the range. In mathematics, the domain contains the *independent variables*, and the range contains the *dependent variables*. When graphed in a Cartesian coordinate system, the domain is graphed on the x-axis and the range is graphed on the y-axis.

For the relation in the table above, the domain and range are as follows.

Domain = {Alaska, Colorado, Idaho, North Dakota, Minnesota}
Range = {−80, −61, −60,} (Note that −60 needs to be listed only once.)

In order to graph these sets, you need to represent them as ordered pairs. The sets can also be modeled as a mapping.

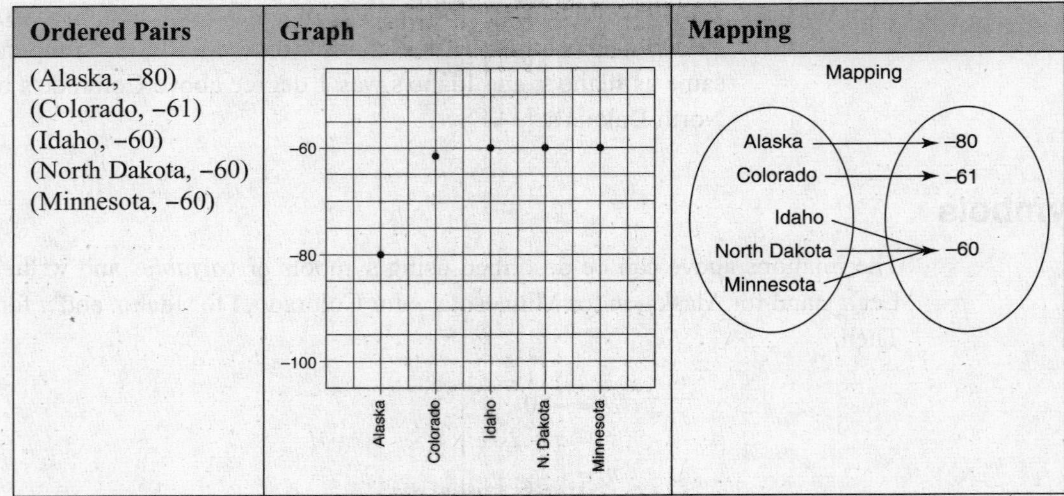

Ordered Pairs	Graph	Mapping
(Alaska, −80) (Colorado, −61) (Idaho, −60) (North Dakota, −60) (Minnesota, −60)		

Example: Represent the relation shown in the graph as

(*a*) A set of ordered pairs
(*b*) A table, and
(*c*) A mapping
(*d*) The domain and range.

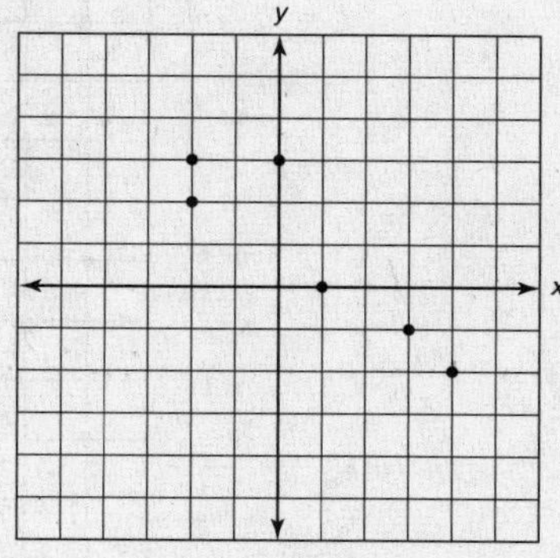

(*a*) The set of ordered pairs is {(−2, 2), (−2, 3), (0, 3), (1, 0), (3, −1) (4,−2)}.

(*b*)

x	*y*
−2	2
−2	3
0	3
1	0
3	−1
4	−2

(*c*)

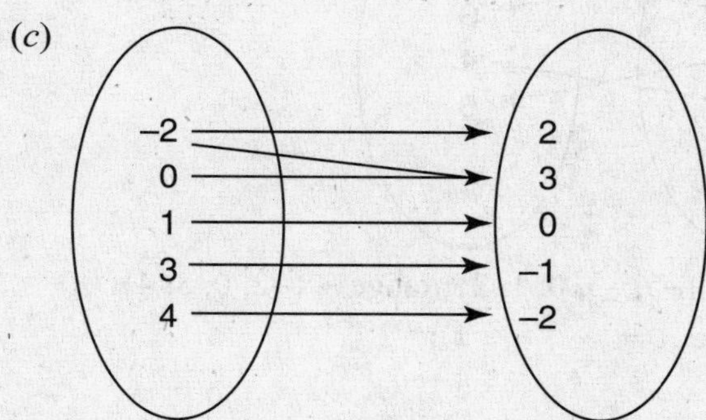

(*d*) The domain (the *x*-coordinates) is {−2, 0, 1, 3, 4}; the range (the *y*-coordinates) is {2, 3, 0, −1, −2}.

Sample Questions

1. State the domain and range of the relation {(0, 2), (−2, 4), (0, −4), (1, 3)}.
2. State the domain and range of the relation in the table representing distance traveled at a speed of 65 miles per hour.

Time (hour)	1.5	2.75	4	4.25	6.5
Distance (mile)	97.5	178.75	260	276.25	422.5

3. Express the relation shown in question 2 as a set of ordered pairs.
4. Create a mapping for the relation {(1, 3), (2, 2), (4, 6), (5, 4)}.
5. State the domain and range of the graph shown, below.

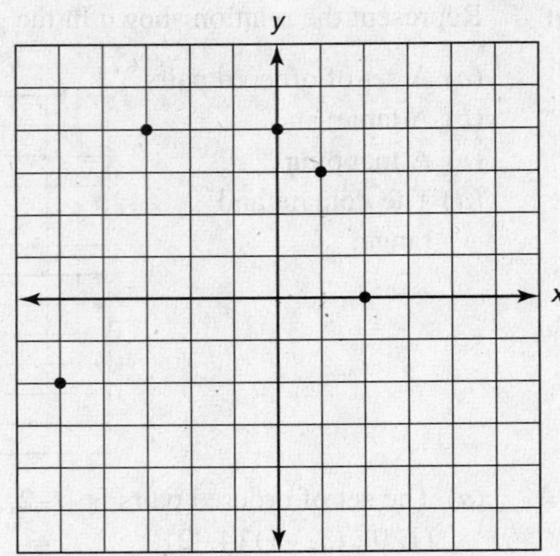

Answers

1. Domain = {−2, 0, 1}; range = {−4, 2, 3, 4}.
2. Domain = {1.5, 2.75, 4, 4.25, 6.5}; range = {97.5, 178.75, 260, 276.25, 422.5}. Remember, range depends on domain and distance depends on time traveled.
3. {(1.5, 97.5), (2.75, 178.75), (4, 260), (4.25, 276.25), (6.5, 422.5)}.
4.

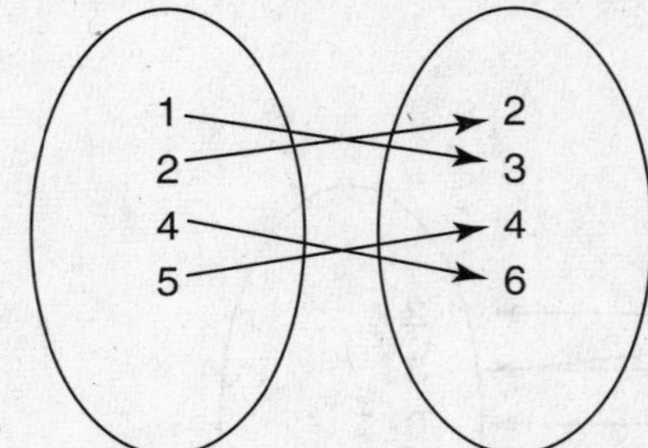

5. Domain = {−5, −3, 0, 1, 2}; range = {−2, 0, 3, 4}.

FUNCTIONS

A function is a special type of relation. In this relation, each element in the domain is paired with exactly one element in the range.

Example 1:

Function

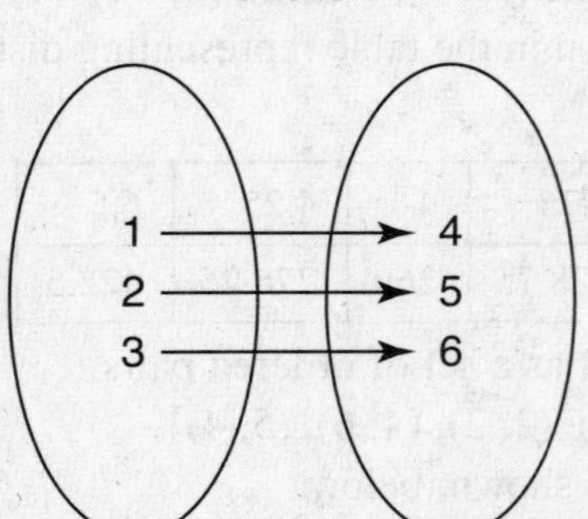

Not a function

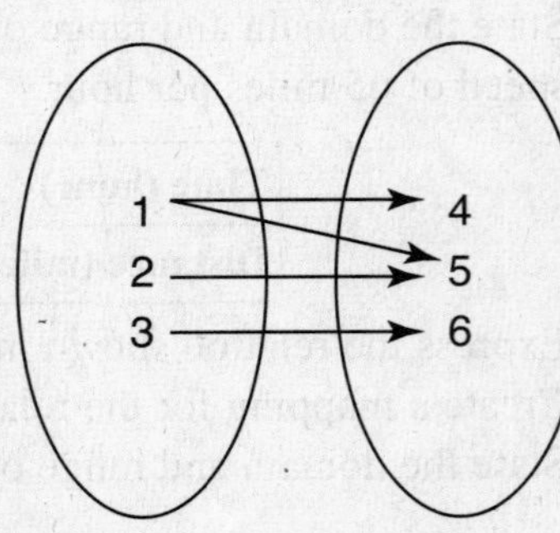

In this case, the 1 is paired with both the 4 and the 5. Therefore it is not a function.

Example 2: A cat and her kittens can represent a function or a nonfunction. Let C stand for cat and k_1, k_2, k_3, and k_4 stand for her four kittens. When the cat is the domain and the kittens are the range, the relationship is not a function. When the roles are reversed, the relationship is a function.

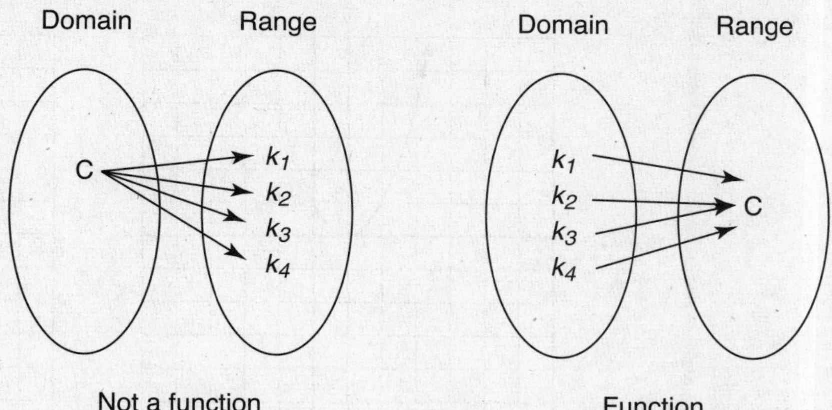

Not a function   Function

In the relation that *is* a function, each kitten (an element in the domain) can have only one mother (an element in the range). In the relation that is not a function, the cat (the element in the domain) has four kittens (elements in the range). If you were to write the nonfunctional mapping as a set of ordered pairs: $\{(c, k_1), (c, k_2), (c, k_3), (c, k_4)\}$, you would see that the ordered pairs all contain the same first element. When this happens, the relation is not a function.

Example 3: Determine if the following set of ordered pairs is a function: $\{(1, 0), (-2, 4), (5, -1), (1, 5), (3, 3)\}$.

This set of ordered pairs is not a function. Notice that of the elements in the domain $\{-2, 1, 3, 5\}$, the 1 is paired with both a 0 and a 5. You can quickly tell if this is a function by looking closely at the ordered pairs.

> If two ordered pairs contain the same first element but different second elements, the relationship is not a function.

Example 4: Determine if the relationship in this table is a function.

x	y
1	2
3	2
5	2

This relationship is a function. Each element in the domain pairs with exactly one element in the range. Again, notice that no two numbers in the domain (x) are the same.

Example 5: The graph of the equations $y = x^2$ and $x = y^2$ are shown here. To determine if each graph is a function, we will use the *vertical line test*. According to this test, if you draw a vertical line through the graph and the vertical line touches the graph in only one place, the graph is a function.

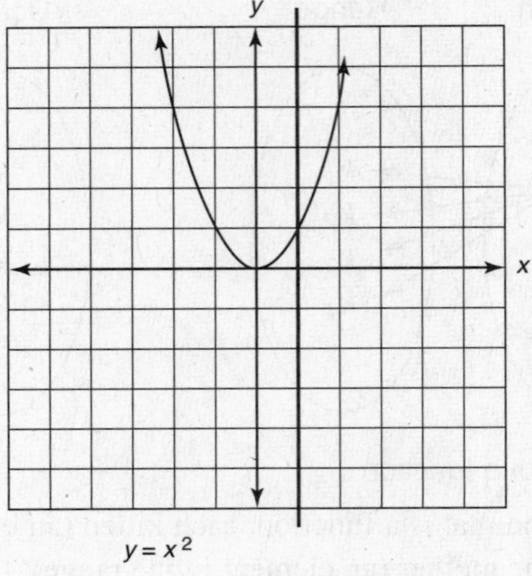

$y = x^2$

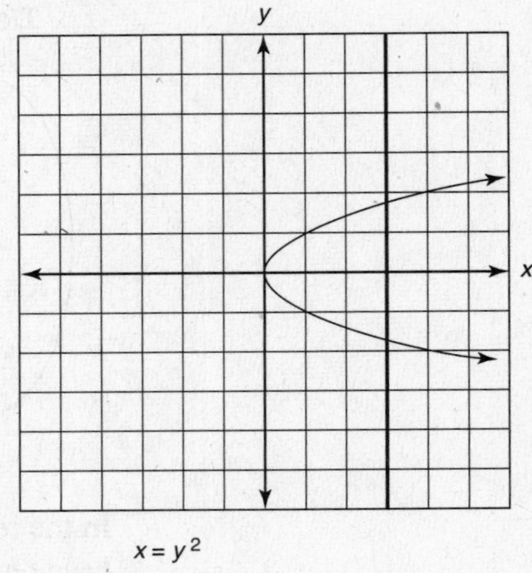

$x = y^2$

Quadratic equations or parabolas are functions.

In the case of $y = x^2$, we can say that y is a function of x. A straight line dropped vertically through the graph touches it in only one place. For each value of x, there is only one value of y. For the graph $x = y^2$, the vertical line touches the graph in more than one place. This means that for one x-value, there is more than one y-value. For example, if $x = 4$, then y can be either 2 or −2, leading to the ordered pairs (4, 2) and (4, −2). Note once more that we have two ordered pairs with the same first element.

As a general rule, equations are functions when they are *linear* (the x is raised to the first power) or *quadratic* (the x is raised to the second power). Equations are *not* functions if the y is raised to the second power. The equations $x = |y|$ and $x = y^2$ are not functions.

Functional Notation

Another way to express the equation $y = 2x + 4$ is to write it as $f(x) = 2x + 4$. This notation can be helpful when preparing tables. $f(x)$ (read "f of x") is substituted for y in the equation. For example $f(3)$, which is read "f of 3," means you should substitute 3 for x in the equation. Thus, for the equation $f(x) = 2x + 4$, $f(3) = 2(3) + 4$ or $f(3) = 10$.

Example: If $f(x) = 5x - 7$, find $f(3)$ and $f(-2)$.

$f(3) = 5(3) - 7 = 8$ Substitute 3 for x.

$f(-2) = 5(-2) - 7 = -17$ Substitute −2 for x.

Sample Questions

For questions 1–8 determine whether or not each set of ordered pairs, table, equation, mapping, or graph is a function.

1. {(2, 1), (1, 2), (4, 3), (3, 4), (2, 2)}
2. {(Tampa, 82°), (Orlando, 85°), (Miami, 95°), (Pensacola, 82°)}
3. $f(x) = 4x + 2$
4.

x	y
2	1
3	2
4	3
5	5
6	5

5.

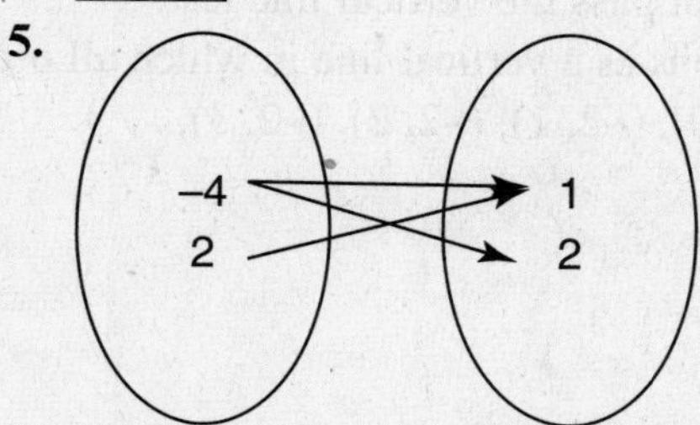

6.

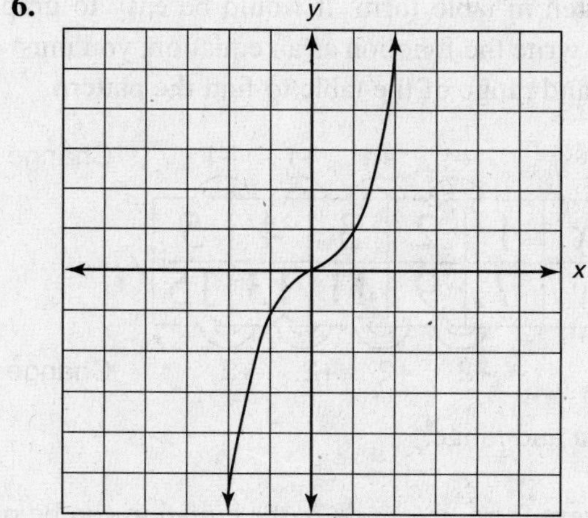

7.

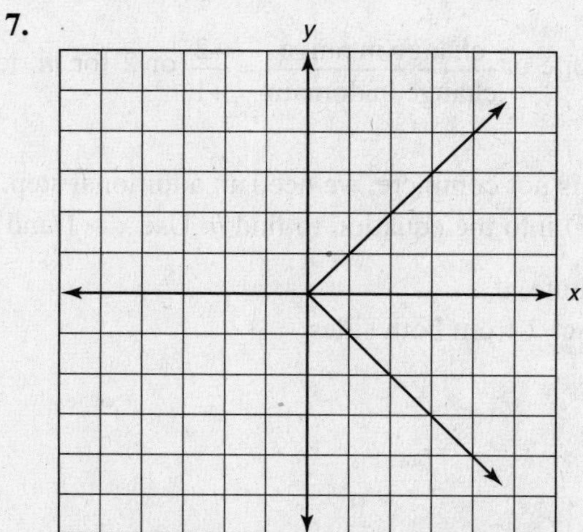

8. $x = -2$
9. If $f(x) = 2x - 7$, find the value of $f(6)$.
10. If $f(x) = -3(x + 1)$, find the value of $f(-2)$.

Answers

1. Not a function. Two different ordered pairs start with 2.
2. Function.
3. Function. This equation graphs as a straight, nonvertical line which passes the vertical line test.
4. Function. Each element in the domain x graphs to one and only one element in the range y.
5. Not a function. The -4 in the domain is paired with both the 1 and the 2 in the range. The ordered pairs for this mapping are $\{(-4, 1), (-4, 2), (2, 1)\}$.
6. Function. This graph passes the vertical line test.
7. Not a function. This graph does not pass the vertical line test.
8. Not a function. This equation graphs as a vertical line in which all ordered pairs have a domain of -2, for example $\{(-2, 0), (-2, 1), (-2, 2), (-2, 3), \ldots\}$.
9. $f(6) = 2(6) - 7 = 12 - 7 = 5$
10. $f(-2) = -3(-2 + 1) = (-3)(-1) = 3$

EQUATIONS FROM PATTERNS

The following function is written in table form. It would be easy to graph if converted to ordered pair form. However, to write the function as an equation, you must examine the relationships between the domain and range of the table to find the pattern.

Change in y-values = range.
Change in x-values = domain.

Change in x values = +1

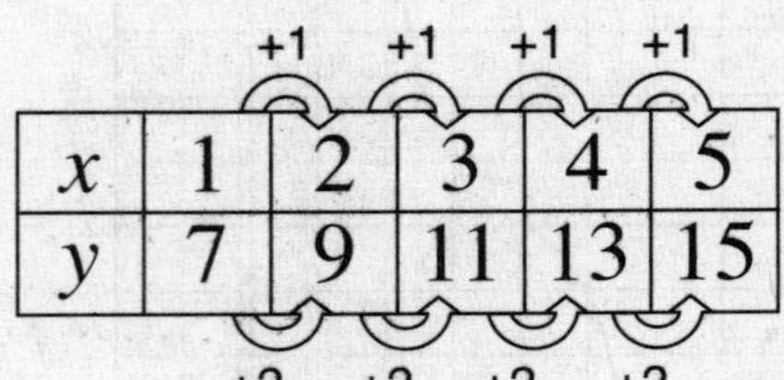

x	1	2	3	4	5
y	7	9	11	13	15

Change in y values = +2

Note the changes in the domain and range.

Step 1: Using the slope-intercept form, $y = mx + b$, the equation can be partially described by substituting the slope or $\frac{\text{change in range}}{\text{change in domain}} = \frac{+2}{+1}$ or 2 for m, leading to the new equation: $y = 2x + b$.

Step 2: Because the equation is not complete, we need an additional step. Substitute one of the ordered pairs (1, 7) into the equation to find b. Use $x = 1$ and $y = 7$.

$7 = 2(1) + b$ Substitute.
$7 = 2 + b$ Subtract 2 from both sides.
$5 = b$

Step 3: Put all the parts together. Your new equation is $y = 2x + 5$.

Try this method on more ordered pairs in the table. Notice that every ordered pair "works" in the equation. In other words, when the ordered pair is substituted for x and y, both sides are equal.

$$9 = 2(2) + 5$$
$$11 = 2(3) + 5$$
$$13 = 2(4) + 5$$

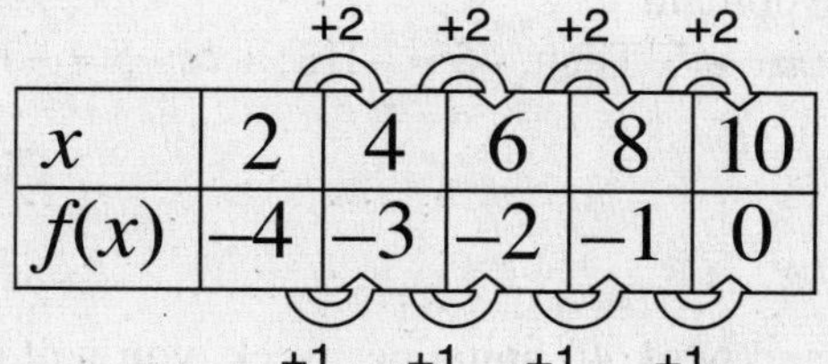

x	2	4	6	8	10
$f(x)$	–4	–3	–2	–1	0

Change in x values = +2

Change in y values = +1

Step 1: Using $y = mx + b$, substitute $\dfrac{\text{change in range}}{\text{change in domain}} = \dfrac{+1}{+2}$ for m: $y = \dfrac{1}{2}x + b$.

Step 2: Substitute one of the ordered pairs from the table into the equation and solve.

$$-4 = \frac{1}{2}(2) + b$$
$$-4 = 1 + b$$
$$-5 = b$$

Step 3: Write the equation: $y = \dfrac{1}{2}x - 5$.

Test this equation with more ordered pairs from the table:

$$-3 = \frac{1}{2}(4) - 5 \qquad -2 = \frac{1}{2}(6)$$
$$-3 = 2 - 5 \qquad -2 = 3 - 5$$

Sample Questions

Find the equation for each table.

1.

x	2	4	6	8	10
y	4	6	8	10	12

2.

x	–2	–1	0	1	2
y	0	3	6	9	12

3.

x	1	2	3	4	5
y	–3	–4	–5	–6	–7

Answers

1. $y = x + 2$. $\dfrac{\text{change in range}}{\text{change in domain}} = \dfrac{2}{2} = 1$. Substitute the slope (1) and one ordered pair into the equation $y = mx + b$. Then, $6 = 1(4) + b$; $6 = 4 + b$; $b = 2$.

2. $y = 3x + 6$. $\dfrac{\text{change in range}}{\text{change in domain}} = \dfrac{3}{1} = 3$. Substitute the slope (3) and one ordered pair into the equation $y = mx + b$. Then, $0 = 3(-2) + b$; $0 = -6 + b$; $b = 6$.

3. $y = -x - 2$. $\dfrac{\text{change in range}}{\text{change in domain}} = \dfrac{-1}{1} = -1$. Substitute the slope (−1) and one ordered pair into the equation $y = mx + b$. Then, $-5 = -1(3) + b$; $-5 = -3 + b$; $b = -2$.

CHANGING PARAMETERS

If you earn $8 per hour and work 40 hours per week, you will earn $320 in a week. A $.75 raise per hour would mean that you would earn $8.75 per hour and $350 per week. A question on the FCAT about *changing parameters* might ask you to find *how much more* you would make per week if you got $.75 more per hour. You could find the new weekly salary and subtract or you could simply multiply $.75 times 40 to arrive at the $30 answer.

This type of problem allows a little more freedom in deciding how to answer the question. No one way is right or wrong but, as in most problems, one way may be a lot quicker and involve fewer steps. The fewer steps you use to solve a problem, the fewer chances for error.

Example 1: Nancy's garden is 15 feet long and 6 feet wide. If she doubles the width of the garden, how will that affect the area?

Method 1: Multiply 15 by 6 to get the area of the old garden, 90 square feet. Multiply 15 by 12 to get the area of the new garden, 180 square feet. You can see that the new garden has twice the area of the old garden.

Method 2: Understand that doubling one dimension (the width only) doubles the area. Doubling two dimensions (the length and the width) quadruples (multiply by four) the area. No calculations are required if you have this understanding.

Example 2: Jeff purchased 4 pairs of jeans at Jiffy Jean for $18.00 each. The next day, Jiffy Jean put the same jeans on sale for 20% less. How much would the jeans have cost if he had waited one day?

Learning to use your FCAT calculator saves time and reduces errors.

Method 1: Total cost of jeans: $18.00 × 4 = $72.00.
Calculate the 20% discount:
$72.00 × .2 = $14.40. Subtract $14.40 from $72.00. $72.00 − $14.40 = $57.60 (total cost).

Method 2: Total cost of jeans: $18.00 × 4 = $72.00.
Cost of jeans after discount: 72 − 20% = $57.60 (total cost).

You can see that method 2 saves time and has fewer steps. Both methods are correct and yield the same answer, but method 2 is simpler.

GRAPHS

> Histograms work well when you have grouped data.

Once you have organized your data into a table, it is a simple matter to prepare a graph. Here we will look at bar graphs, histograms, line graphs, scatterplots, and circle graphs.

When you make a bar graph, histogram, line graph, or scatterplot, you begin by using a grid and labeling the axes. It is important to check ahead of time to see what the lowest numbers and highest numbers of your data are. The most pleasing arrangement of a bar or line graph shows the highest numbers reaching almost to the top of the grid, or in the case of a horizontal bar graph, almost to the right of the grid.

Bar Graphs and Histograms

In a bar graph, each bar represents data. Bar graphs are used to *compare* data. The data is categorized on the x-axis for a vertical bar graph, with the height of the bars shown on the y-axis. For a horizontal bar graph, the data is categorized on the y-axis and the height of the bars is shown on the x-axis.

As you design your bar graph, decide on the labels for the axes before beginning. Always label the graph's axes and give your graph a title.

All bars should begin either directly on the x-axis or, in the case of a horizontal bar graph, directly on the y-axis. Each bar should be exactly the same width as the others with equal spaces between them. You will need to give a lot of thought to any scales you use along the axis. If your data is spread out, you may need to count by 2s, 5s, 10s, or 100s (the most common) so that all the numbers will "fit" into the grid. It is not a good idea to add lines to the grid you are given, so check ahead of time to make certain the scale you choose will fit.

Example: Ms. Sturges gave a science test. The following bar graph shows the results of the test given to 60 students: 16 received an A, 15 received a B, 26 received a C, 2 received a D, and 1 received an F.

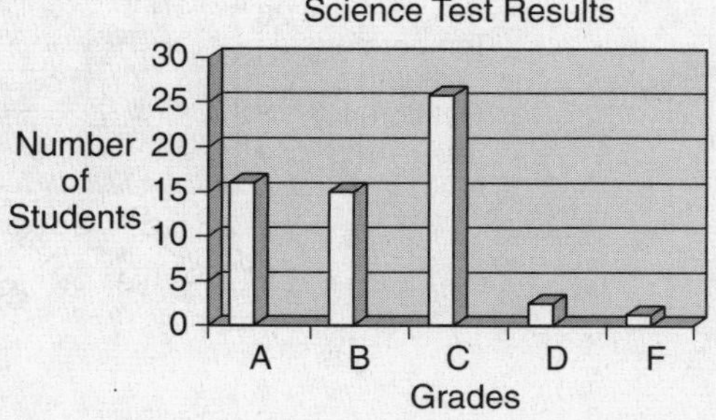

The graph above could also be represented as a horizontal bar graph.

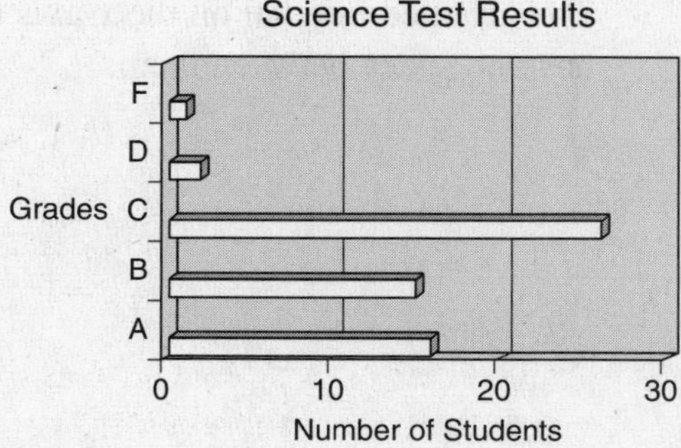

Double bar graphs compare two similar sets of data. In the double bar graph shown here, high and average wind speeds for several US cities are compared.

Example:

A *legend* to the right of the graph explains what each bar represents.

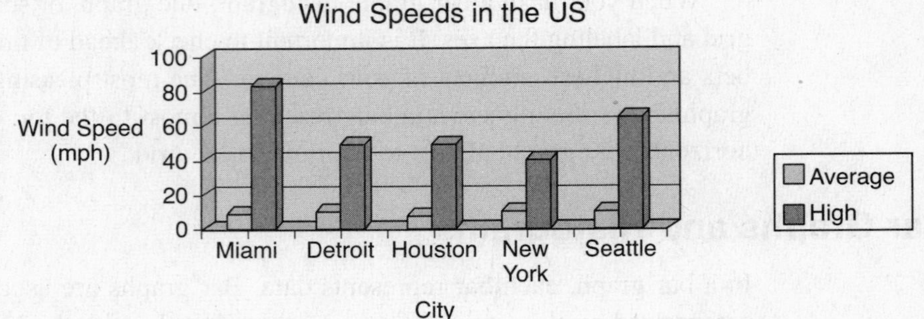

Notice that related pairs of bars are placed together with equal spaces between.

Histograms are similar to bar graphs in that each bar represents data. However, in a histogram the data is grouped in intervals such as 2005–2009, or 0–200 as shown in the graph below. Because the data is grouped, there is no space between the bars. Remember that the grouped data must be in equal intervals. For example, 2005–2009 represents an interval of 5 years; therefore, the next expected intervals should be 2010–2014 and 2015–2019.

Tropical ocean waters have certain physical characteristics, one of which includes temperature. The histogram shown here relates average temperature to depth in warm tropical waters during the month of March.

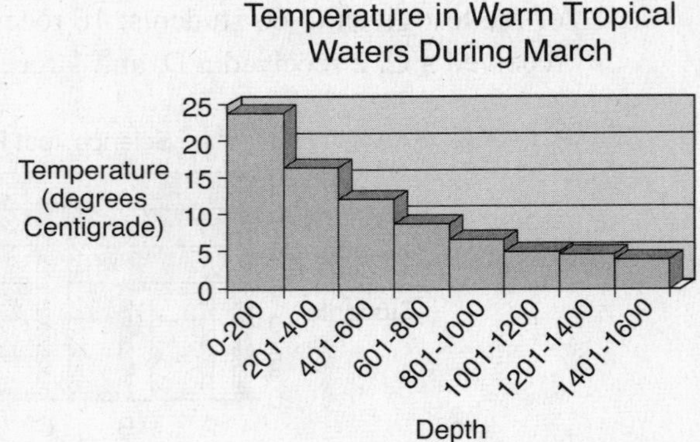

Example: In the histogram notice that all numerical intervals on both axes are equal. The grouped interval on the x-axis representing depth causes the bars to have no space between them.

Sample Questions

1. In the histogram shown in the example above, at approximately what depth does the water temperature begin to stabilize (level off)?
2. The public high school graduation rate for six US states is as follows: Colorado, 71.5%; Florida, 57%; Louisiana, 55.4%; Missouri, 71.7%; New York, 61%; and Oregon, 67.2%. Construct a bar graph using this data.

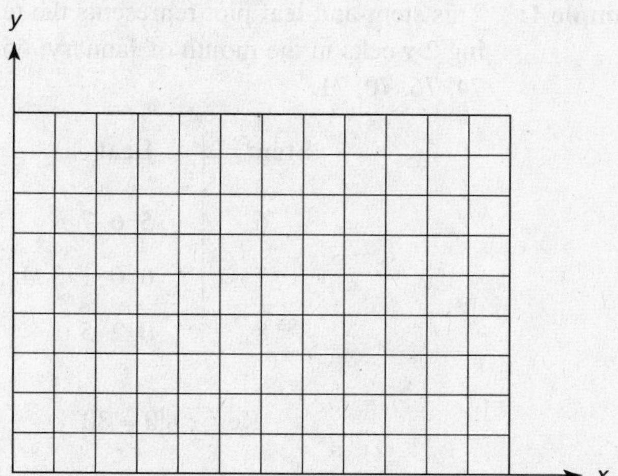

3. Which of the six states listed in question 2 has the highest graduation rate?
4. In question 2 how much higher is the graduation rate in Colorado than in Florida?
5. If you were to compare the preferred music style of teenage boys and teenage girls, what type of graph would be most appropriate?

Answers

1. Water temperatures begin to stay roughly the same after 1200 feet.
2.

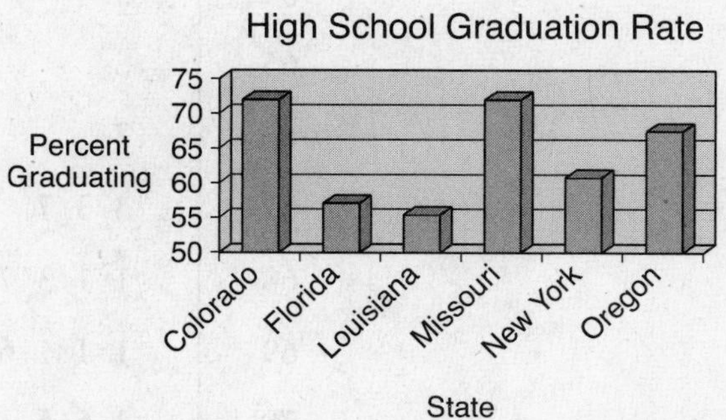

> ***Test Tip***: If your data does not start at zero and you need more space for the scale, you should consider starting the scale at a number other than zero. In the bar graph above, the data on the vertical axis begins at 50%.

3. Missouri
4. 71.5% – 57% = 15.5%
5. A double bar graph with different music categories. Use differently shaded bars for boys and girls.

Stem-and-Leaf Plots

A stem-and-leaf plot is a way of organizing numerical data that allows you to see how the data falls into intervals while still showing every number in the set. The leaf portion of the plot is one digit representing the lowest place value, and the stem is the higher place value. The stem can be more than one digit.

Example 1: This stem-and-leaf plot represents the temperatures in Miami recorded during 2 weeks in the month of January: 65, 66, 73, 75, 70, 67, 78, 82, 80, 85, 74, 76, 70, 71.

Stem	Leaf
6	5 6 7
7	0 0 1 3 4 5 6 8
8	0 2 5

Key : 8|0 = 80

Notice that the stem for this data represents the tens place and that the leaf represents the ones place. The leaves are put in order from least to greatest.

Example 2: Mr. Bailey wanted to see how his remedial class performed on a standardized test. Their scores were as follows: 705, 705, 723, 681, 689, 681, 691, 658, 677, 673, 643, 713, 683, 723, 696, 713, 703, 730, 691, 673, 713, 696, 687, 735, 691, 688, and 667. Arrange these scores as a stem and leaf plot.

Stem	Leaf
64	3
65	8
66	7
67	3 3 7
68	1 1 3 7 8 9
69	1 1 1 6 6
70	3 5 5
71	3 3 3
72	3 3
73	0 5

Key : 73|0 = 730

This stem-and-leaf plot could easily be made into a histogram or bar graph by changing the stems to intervals and using the leaves to count the number of students scoring in that interval.

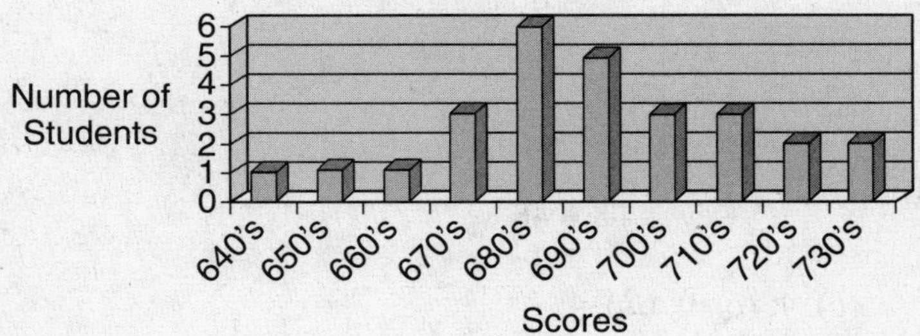

Sample Questions

1. The stem-and-leaf plot shown here the scores for a 100-point exam in history.

Stem	Leaf
5	1 3 3 2
6	4 7 9
7	2 4 5 5 5 8
8	4 5 7
9	0 1 2 6 7

(*a*) What were the high and low scores?
(*b*) What test score occurred most frequently?
(*c*) Most of the students scored in what interval?
(*d*) How many students scored 80 or better?
(*e*) What was the difference between the highest and lowest score?
(*f*) What is the key for this stem-and-leaf plot?

2. Some students were asked, "How many states outside Florida have you visited?" Their answers were 5, 6, 0, 10, 22, 8, 7, 5, 5, 1, 4, 3, 17, 29, 36, 48, 15, 19, 0, 3, 5.
(*a*) Make a stem-and-leaf plot from this data.
(*b*) How many students have visited more than 20 states?
(*c*) What is the fewest number of states these student have visited?
(*d*) How many students have visited 10–19 states?

Answers

1. (*a*) Low score, 51; high score, 97; (*b*) 75; (*c*) 70s; (*d*) 8; (*e*) 46; (*f*) answers may vary; one answer is key: 9|1 = 91.

2. (*a*)

Stem	Leaf
0	0 0 1 3 3 4 5 5 5 5 6 7 8
1	0 5 7 9
2	2 9
3	6
4	8

Key : 4|8 = 48

(*b*) 4; (*c*) 0; (*d*) 4.

Line Graphs

A line graph is a line drawn to connect, in order, various data pairs on a graph. Double line graphs show two related sets of data graphed together on the same grid. They are used to show a comparison of the change over time in two related sets of data.

The following double line graph compares the mean SAT scores of college-bound students over the period 2000–2005.

Line graphs are best used to show change over time, with time represented on the horizontal axis and the vertical axis displaying the amount of change.

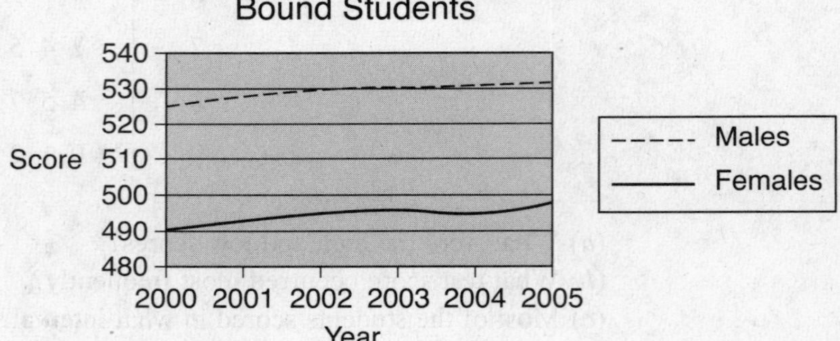

Sample Questions

1. What conclusion can you draw about the comparison between male and female SAT scores for 2000–2001?
2. Over the period shown, what happens to the scores of both males and females?
3. What is the approximate difference between the scores of males and the scores of females?
4. The sales of recorded music (in millions of dollars) are shown in the accompanying table. Draw and label a double line graph using this data.

Type of Recording	2000	2001	2002	2003	2004	2005	2006	2007	2008
Compact discs	333	407	495	662	723	779	783	847	939
Cassettes	360	366	340	345	273	225	173	159	124

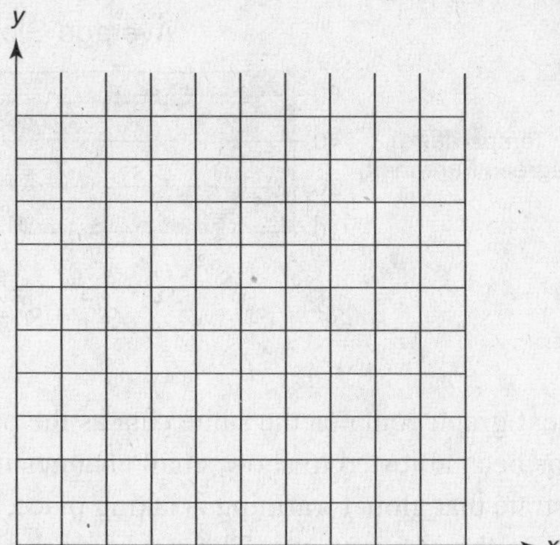

5. Approximately when does the number of compact discs sold overtake the sale of cassettes?

6. According to the table on the previous page, what was the percent of decrease in cassette sales from 2007 to 2008?

Answers

1. Males score higher than females.
2. The scores are gradually increasing.
3. 35 points.
4.

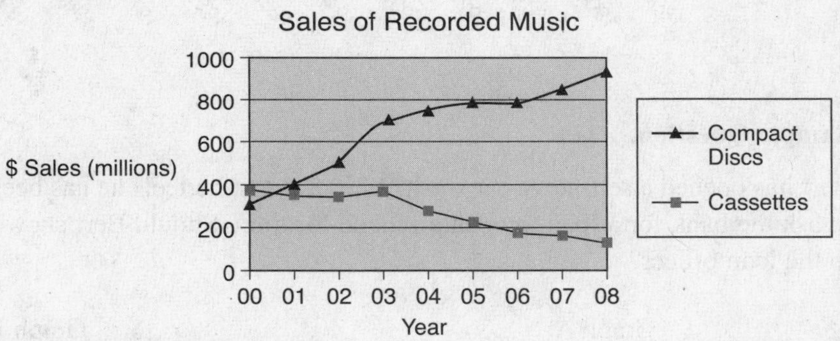

5. Approximately mid-2000.
6. $\frac{\text{Amount of change}}{\text{Original amount}} = \frac{-35}{159} = -.220 = -22\%$

Misleading Graphs

The accompanying graph represents the average global temperatures for 1880–1999. It appears from looking at the graph that global temperatures have not changed in the last 100 years. Keep in mind that scientists believe that seemingly small temperature changes can have dramatic effects. Because of the scale used on the vertical axis, the temperature changes during the last 100 years, although accurately graphed, are not clearly defined.

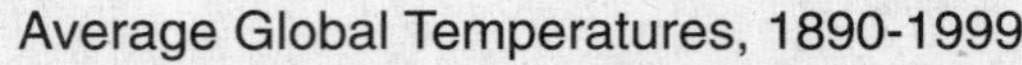

When drawing a graph, be sure to include a title and all labels.

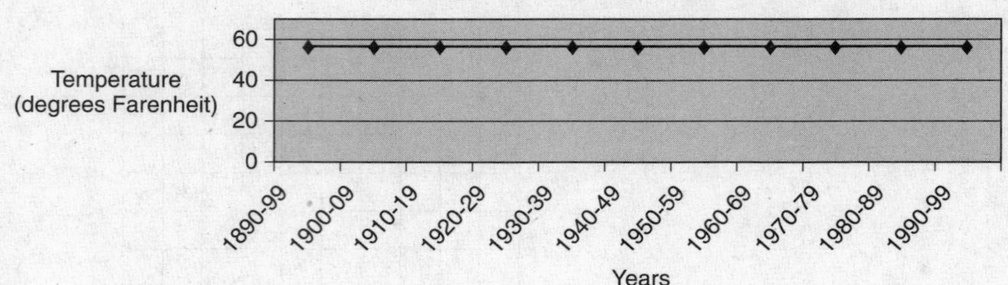

The next graph contains the same data as the one above. However, the scale on the vertical axis has been adjusted to show small changes in the graph. If a scientist were attempting to demonstrate that global warming is taking place, the second graph would certainly be more effective than the previous one. Those who wish to show that global warming is *not* taking place would be more inclined to use the first graph. Simply changing the scale of the vertical axis causes changes in the graph to appear more or less dramatic, as the designer of the graph wishes.

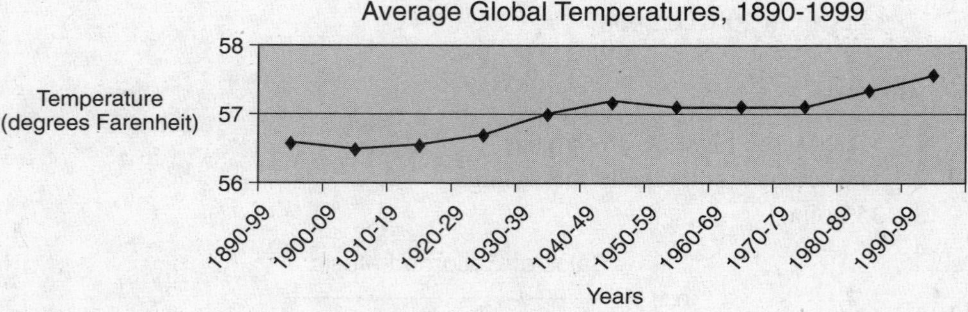

Sample Question

Bert has opened a self-serve car wash. After 5 years, he feels he has been successful enough to ask the bank for a loan to open a second location. Should Bert show graph A or graph B to the loan officer?

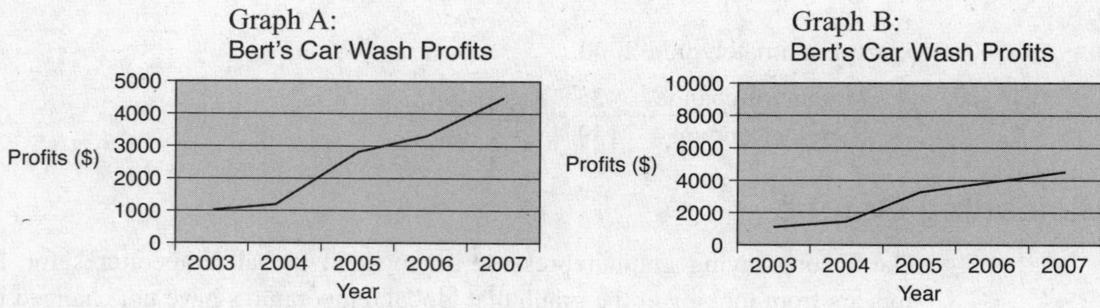

Answer

Bert should use Graph A. Although both graphs show that profits have increased by a little over $4000, the first graph makes the increase appear more prominent.

Scatterplots

Scatterplots are a set of points showing a relationship between two sets of data. When the points form a pattern, it is called a *trend*. Scatterplots can have a positive trend, a negative trend, or no trend.

When the data forming a scatterplot rises from left to right, it is said to have a *positive trend*. By this we mean that one set of data rises as the other rises. For example, you would expect test scores to increase as study time increases.

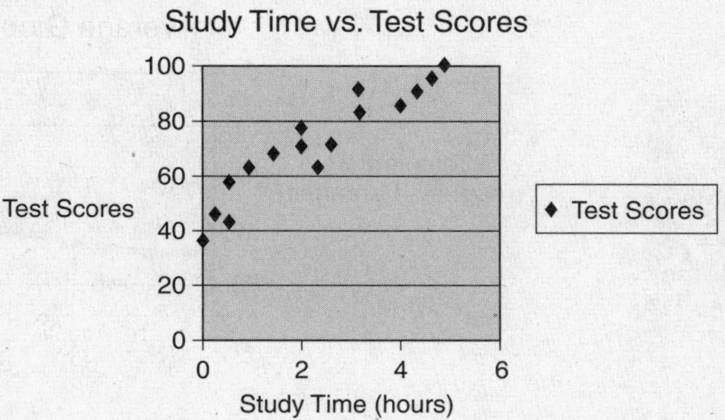

If the data forming a scatterplot goes down from left to right, it has a *negative trend*. One set of data rises while the other decreases. In the graph shown here, the cost per ounce of cereal is compared to the size of the box. As you can see, the larger the box of cereal, the cheaper the cereal is per ounce.

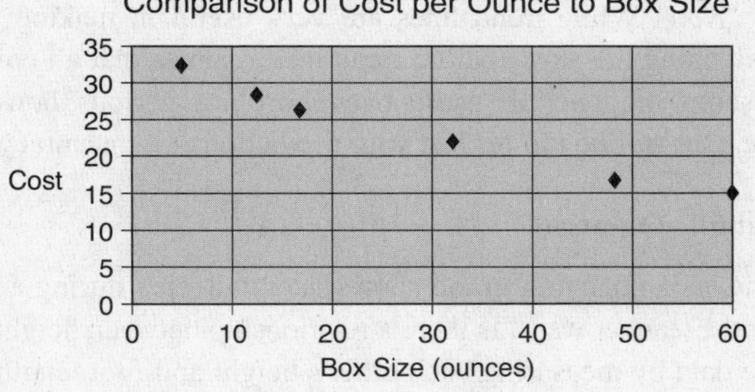

A scatterplot with no trend has a scattering of points that have no specific direction. The two sets of data have *no relationship* to each other. Notice in the scatterplot shown here that there is no relationship between a person's height and their IQ.

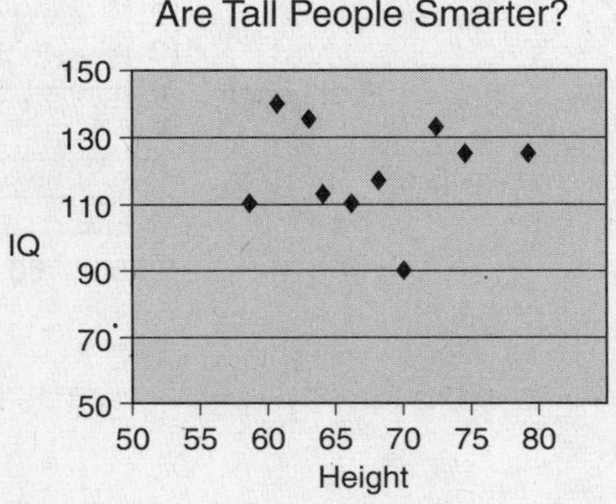

Trend Lines

A trend line is a straight line that, when drawn through a scatterplot, travels in the same general direction as the data points and passes through the center of the data. Trend lines can be used to make predictions about future values. When the data is strongly *correlated*, the trend line fits the data very closely with very few (if any) points wandering away from the vicinity of the trend line. A prediction made from such a trend line tends to be more accurate than one based on a trend line in which the data points stray away from the line.

Test Tip:
If you are given a trend line on a graph, ignore the data and use only the trend line to make predictions.

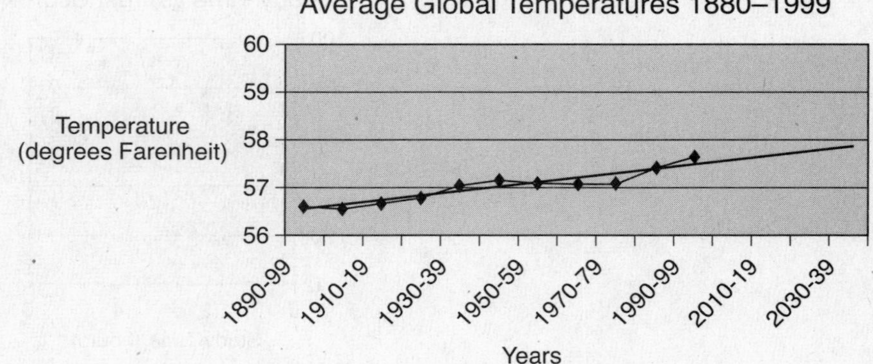

The graph above uses the same data as the line graph containing data about global temperatures. By using a trend line, you could make a prediction that temperatures would continue to rise. If the vertical and horizontal axes were extended, you might be able to make a prediction how long it would take for average global temperatures to rise to 60 degrees.

Note: While trend lines are very useful in making predictions, you need to use care. Extending the axes and the trend line assumes that all conditions will remain the same. This assumption generally works over short time periods, however, it is not a good idea to extend the axes or line too far lest your predictions lose accuracy.

Sample Question

The accompanying graph shows data collected during a class exercise. The question posed by the teacher was, "Is there a relationship between height and shoe size?" Students gathered the data by measuring each other's height and foot length.

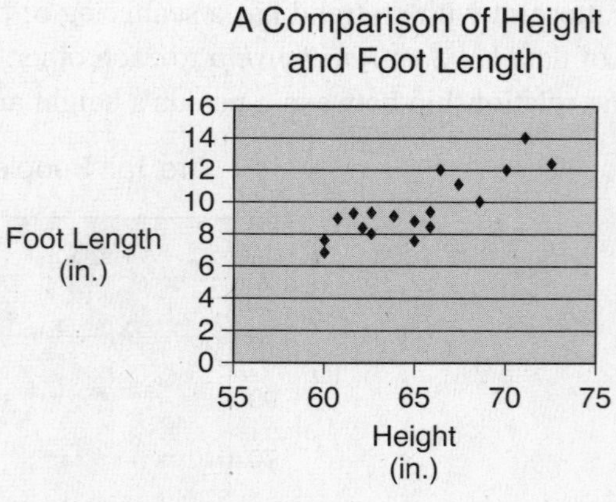

1. Draw a trend line for the data.
2. Predict, using your trend line, what the approximate foot length would be for a person 75 inches tall.
3. Complete the following statement: The taller the person, the ______ the shoe size.
4. Is shoe size positively, negatively, or not related to height?

Answers

1.

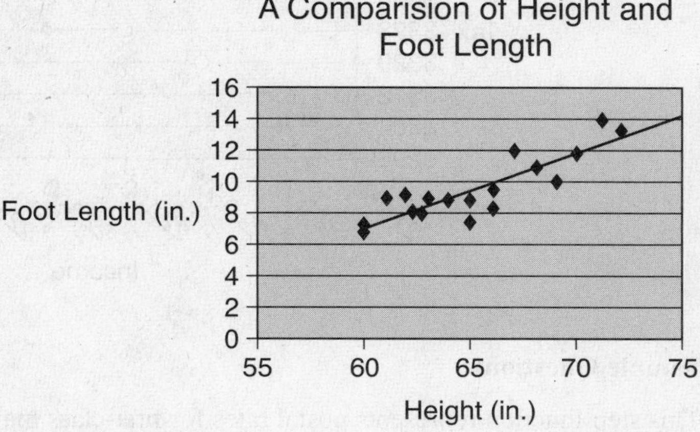

2. Approximately 14 inches.
3. The taller the person, the larger the shoe size.
4. There is a positive relationship between height and shoe size.

Step Functions

A step function (sometimes called a step graph), is a particular type of function that, when graphed, forms a stair step. Each step is formed by one open circle and one closed circle joined by a line segment. The open circle indicates the absence of data.

Example 1: Cellular Phone Company charges $.60 per minute for long distance charges outside its service area. Partial minutes are rounded up to the next whole minute.

Read the graph by looking at the line segment in the graph.

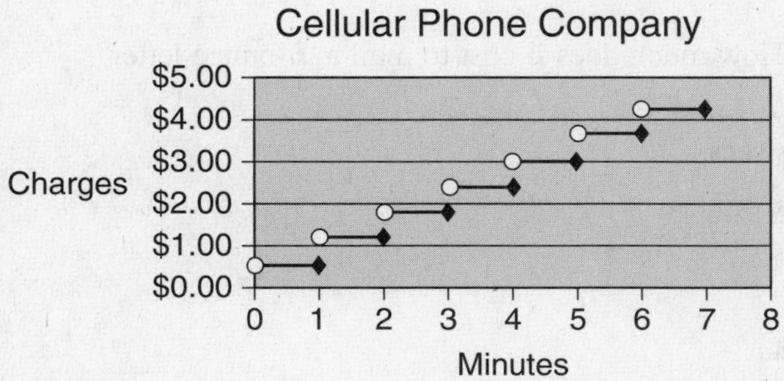

The charge for a 4-minute, 30-second call is $3 because 4 minutes, 30 seconds has been rounded to 5 minutes.

Example 2: The Internal Revenue Service's tax tables form a step function because taxes are figured from the table in intervals of $50. In the step function shown here, a portion of the tax tables is shown graphically.

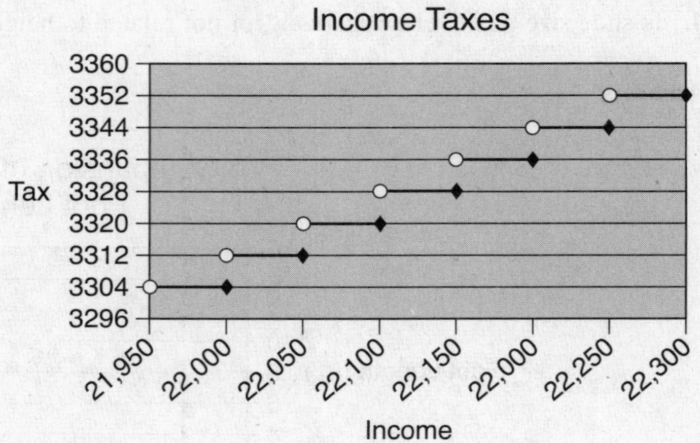

Sample Questions

This step function represents postal rates for first-class mail. A first-class letter costs $.37 to mail for the first ounce and $.23 for each additional ounce.

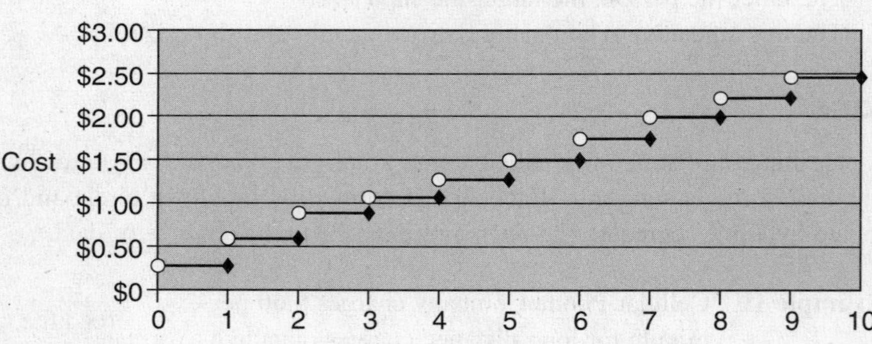

1. How much does it cost to mail a $3\frac{1}{2}$-ounce letter?

2. How much does it cost to mail a $\frac{3}{4}$-ounce letter?

Answers

1. $1.06
2. $.37

Circle Graphs

A circle graph can show how a whole is broken into parts. Frequently, circle graphs are used to show percentages. On the FCAT, you may be asked to display data in a circle graph.

Example: Ann's monthly budget is broken down as follows: rent, $650; food, $210; car payment, $320; entertainment, $100; and other expenses, $245.

Before making the circle graph, set up a table.

Budget Item	Cost ($)	Percent of Total	Angle (degrees)
Rent	650		
Food	210		
Car payment	320		
Entertainment	100		
Other	245		
Total	1525	100	360

Calculate the percent of total for each item by dividing that item by 1525. For example, the percent of rent = 650 ÷ 1525 = .42 = 42%.

You will always need a total in order to find a percent.

Budget Item	Cost ($)	Percent of Total	Angle (degrees)
Rent	650	42	
Food	210	14	
Car payment	320	21	
Entertainment	100	7	
Other	245	16	
Total	1525	100	360

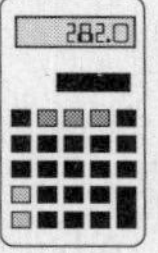

Each piece of the pie chart is formed by an angle. A portion representing 42% of the circle forms a 153-degree angle. To calculate the angles for this circle, simply change the percent to a decimal and multiply by 360 degrees. For example, 42% = .42 × 360 ≈ 153 degrees.

Calculator Hint: 42% of 360 means multiply 360 by 42% on your calculator.

Budget Item	Cost ($)	Percent of Total	Angle (degrees)
Rent	650	42	153
Food	210	14	50
Car payment	320	21	76
Entertainment	100	7	24
Other	245	16	57
Total	1525	100	360

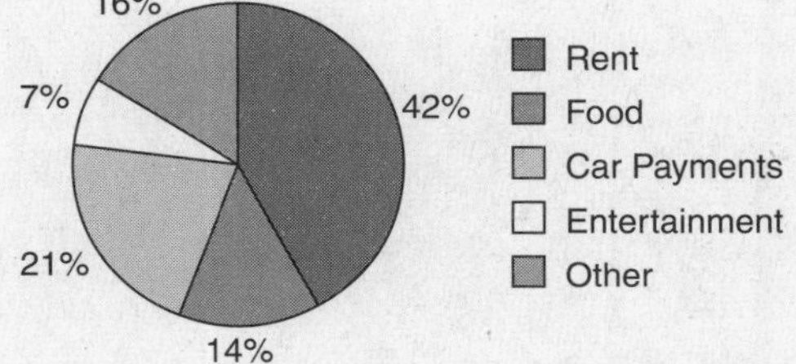

Sample Questions

In 1997 the United States signed an international treaty to set limits on emissions of "greenhouse gasses." These gasses trap thermal radiation in the atmosphere, which increases warming of the earth and its atmosphere. The graph below shows the world carbon dioxide emissions from the use of fossil fuels in 2007.

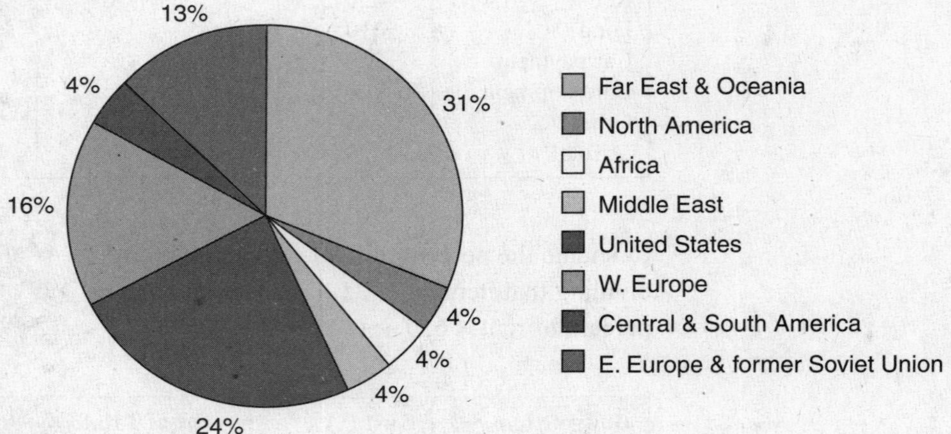

1. US greenhouse gas emissions equal those of which three combined areas of the world?

2. If the greenhouse gas emissions of the United States in 2007 equal 1625 million metric tons of carbon equivalent, what were the approximate total greenhouse gas emissions of the entire world during that year?

A. 6770 metric tons
B. 390 metric tons
C. 39,000 metric tons
D. 6000 metric tons

Answers

1. Western Europe and any two of the following: North America, Africa, Central and South America, and the Middle East.
2. A. You can use your estimation skills or use ratio and proportion to solve this problem.

Method 1: Use estimation. 24% is close to 25%. This means that the US emissions are approximately $\frac{1}{4}$ of those of the entire world, or, put another way, the world's emissions are 4 times those of the United States. Multiply: $1625 \times 4 = 6500$ metric tons. **A** is the closest answer.

Method 2: Use ratio and proportion. $\frac{24}{100} = \frac{1625}{x}$. Solve: $24x = 162{,}500$. Divide by 24; $x \approx 6770$ metric tons.

SECTION 3: EXPRESSIONS AND EQUATIONS

ALGEBRAIC EXPRESSIONS

Algebraic expressions are used to translate English statements into equations. You can think of an algebraic expression as math shorthand. Some examples are listed in the accompanying table.

English Expression	Algebraic Expression
Twice a number	$2n$
5 more than a number	$n + 5$
4 less than a number	$n - 4$
7 less a number	$7 - n$
The sum of twice a number and 6	$2n + 6$
The product of a number and 1.5	$1.5n$
A number squared	n^2
A number cubed	n^3
The sum of the squares of two different numbers	$a^2 + b^2$
The square of the sum of two different numbers	$(a + b)^2$
Twice the sum of 5 and a number	$2(5 + n)$
The volume of a cube with side lengths s	s^3
The area of a rectangle with sides of w and $w + 3$	$w(w + 3)$

***Test Tip*:** Notice that the words *less than* reverse the order of the numbers in an expression.

Example: "3 less than a number" is expressed as $n - 3$, while "3 less a number" is expressed as $3 - n$.

Expressions can be *evaluated* but not solved. When you evaluate an expression, you substitute a number for the variable and perform whatever operations need to be done to find a "value" for the expression.

Example: Evaluate $2x + 5$ when $x = 3$.

$2(3) + 5$	Substitute 3 for x.
$6 + 5$	Multiply first.
11	Then add.

So, the value of $2x + 5$ when $x = 3$ is 11.

Sample Questions

1. Evaluate $-3x + 9$ when $x = 4$.
2. Evaluate $-2x^2$ when $x = -5$.
3. Evaluate $7(x - 8)$ when $x = -2$.
4. Evaluate $V = \frac{1}{3}\pi r^2$ when $r = 6$.
5. Write an expression for 2 more than the product of 3 and a number.

Answers

1. $-3(4) + 9 = -12 + 9 = -3$
2. $-2(-5 \bullet -5) = -2(25) = -50$
3. $7(-2 - 8) = 7(-10) = -70$
4. $\frac{1}{3} \bullet 3.14 \bullet 6^2 = \frac{1}{3} \bullet 36 \bullet 3.14 = 12 \bullet 3.14 = 37.68$
5. $2 + 3x$ (or $3x + 2$)

ALGEBRAIC EQUATIONS

When two algebraic or mathematical expressions are equal to each other, they are called an *equation*.

Example: $5x + 1 = 26$

Equations can be *solved*. When solving an equation, you *undo* the operations that are done in the equation by using inverses and reversing the order of operations. For example, to undo the subtraction of 3, you add 3 because addition is the *inverse* operation of subtraction.

Solving One-Step Equations

Math operations usually include adding, subtracting, multiplying, dividing, and finding square roots.

A one-step equation can be solved using only one operation. Select the operation that is the inverse of the operation on the same side as the variable in the equation.

Example 1:

$x - 6.5 = 4.2$

$x - 6.5 + 6.5 = 4.2 + 6.5$

$x = 10.7$ 6.5

Solve by adding 6.5 to both sides. (The inverse of subtracting is adding 6.5.)

Example 2:

$5.6 + a = 12.9$

$5.6 - 5.6 + a = 12.9 - 5.6$

$a = 12.9 - 5.6$

$a = 7.3$

Because $5.6 + a = a + 5.6$, you can solve by subtracting 5.6 from both sides.

Example 3:

$3x = 51$

$\frac{3x}{3} = \frac{51}{3}$

$x = 17$

Solve by dividing both sides by 3. (The inverse of multiplying by 3 is dividing by 3.)

Example 4:

$\frac{2}{3}x = 12$

$\frac{\frac{2}{3}x}{\frac{2}{3}} = \frac{12}{\frac{2}{3}}$

$\frac{3}{2} \bullet \frac{2}{3}x = 12 \bullet \frac{3}{2}$

$x = 18$

Solve by dividing both sides by $\frac{2}{3}$.

Remember that dividing by a fraction is the same as multiplying by its reciprocal.

Example 5: $\frac{x}{4} = 15$ — Solve by multiplying both sides by 4.

$4 \bullet \frac{1}{4}x = 4 \bullet 15$

$x = 60$

Sample Questions

1. Solve $x - 2.5 = 6.2$ for x.

2. Solve $\frac{2}{3}a = 16$ for a.

3. Solve $11 = \frac{1}{3}b$ for b.

4. Solve $c^2 = 121$ for c.

5. Solve $4.5y = 90$ for y.

Answers

1. $x = 8.7$. Add 2.5 to both sides of the equation.
2. $a = 24$. Multiply both sides of the equation by $\frac{3}{2}$.
3. $b = 33$. Multiply both sides of the equation by 3.
4. $c = 11$. Take the square root of both sides of the equation (recall that the square root is the inverse operation of a square).
5. $y = 20$. Divide both sides of the equation by 4.5.

Solving Multistep Equations

Multistep equations require more than one step to solve. In general, follow these steps to solve multistep equations:

> Work left to right on each side of the = sign.

Step 1: If needed, simplify by adding like terms or by distributing.

Step 2: Deal with addition and subtraction first.

Step 3: Work with multiplication and/or division last.

Example 1: Solve $2x + 5.5 = 13.7$ — Not necessary to simplify.

Step 1: $2x + 5.5 - 5.5 = 13.7 - 5.5$ — Subtract 5.5 from both sides.

$2x = 8.2$

Step 2: $\frac{2x}{2} = \frac{8.2}{2}$ — Divide both sides by 2.

$x = 4.1$

Example 2: Solve $4x - 3 = 13$

Step 1: $4x - 3 + 3 = 13 + 3$ — Add 3 to both sides.

Step 2: $4x = 16$ — Divide both sides by 4.

$\frac{4x}{4} = \frac{16}{4}$

$x = 4$

Example 3: Solve $4.9 + \frac{x}{3} = 1.1$ — Not necessary to simplify.

Step 1: $4.9 - 4.9 + \frac{x}{3} = 1.1 - 4.9$ — Subtract 4.9 from both sides.

$\frac{x}{3} = -3.8$

Step 2: $3 \cdot \frac{x}{3} = 3 \cdot -3.8$ — Multiply both sides by 3.

$x = -11.4$

Example 4: Solve $2(x - 8) + 5 = 15$ — Simplify first.

Step 1: $2(x - 8) + 5 = 15$ — Use the distributive property.

$2x - 16 + 5 = 15$ — Combine like terms.

Step 2: $2x - 11 = 15$ — Add 11 to both sides.

$2x - 11 + 11 = 15 + 11$

Step 3: $2x = 26$ — Divide both sides by 2.

$x = 13$

Example 5: $4x - 7 = 7x + 14$ — Simplify first by placing all variables on one side.

Step 1: $4x - 4x - 7 = 7x - 4x + 14$ — Subtract $4x$ from both sides.

$-7 = 3x + 14$

Step 2: $-7 - 14 = 3x + 14 - 14$ — Subtract 14 from both sides.

Step 3: $-21 = 3x$ — Divide both sides by 3.

$\frac{-21}{3} = \frac{3x}{3}$

$-7 = x$

Example 6: $-6x + 14 = 2(-3x + 7)$ — Simplify first by distributing and placing all variables on one side.

Step 1: $-6x + 14 = -6x + 14$ — Distribute the 2. Notice that both sides are the same.

Step 2: $14 = 14$ — Add $6x$ to both sides. The variable disappears. This equation is called an *identity* because both sides are identical for all values of the variable. Any number can be substituted for x.

Sample Questions

Solve each of the following equations for the variable.

1. $2x + 3 = 21$
2. $15 = 9 - \frac{1}{3}b$
3. $4 - 3a = 19$
4. $\frac{4}{3}c + 4 = -12$
5. $-10x - 45 = 5x$
6. $3x - (4 - 2x) = 11$
7. $6(x - 2) = 18$
8. $7a = -16 - 9a$
9. $9x + 2 = 4x$
10. $-4(3x - 5) = -12x + 20$

Answers

1. $2x + 3 = 21$	Subtract 3 from both sides.
$2x = 18$	Divide by 2.
$x = 9$	
2. $15 = 9 - \frac{1}{3}b$	Subtract 9 from both sides.
$6 = -\frac{1}{3}b$	Multiply both sides by –3.
$-18 = b$	
3. $4 - 3a = 19$	Subtract 4 from both sides.
$-3a = 15$	Divide both sides by –3.
$a = -5$	
4. $\frac{4}{3}c + 4 = -12$	Subtract 4 from both sides.
$\frac{4}{3}c = -16$	Multiply both sides by $\frac{3}{4}$.
$c = -12$	
5. $-10x - 45 = 5x$	Add $10x$ to both sides.
$-45 = 15x$	Divide both sides by 15.
$-3 = x$	
6. $3x - (4 - 2x) = 11$	Distribute the minus sign.
$3x - 4 + 2x = 11$	Add like terms.
$5x - 4 = 11$	Add 4 to both sides.
$5x = 15$	Divide by 3.
$x = 3$	
7. $6(x - 2) = 18$	Distribute the 6.
$6x - 12 = 18$	Add 12 to both sides.
$6x = 30$	Divide both sides by 6.
$x = 5$	

8. $7a = -16 - 9a$ — Add $9a$ to both sides.
   $16a = -16$ — Divide both sides by 16.
   $a = -1$
9. $9x + 2 = 4x$ — Subtract $9x$ from both sides.
   $2 = -5x$ — Divide both sides by -5.
   $-\frac{2}{5} = x$
10. $-4(3x - 5) = -12x + 20$ — Distribute the -4.
    $-12x + 20 = -12x + 20$

This equation is identical on both sides. It is an identity, and any number can be substituted for x.

Solving Equations Containing More Than One Variable

You have been solving equations in which there is only one variable. If you solve an equation that contains two or more variables, such as $3x + y = 5$, the value of x is said to come out *in terms of y* (or whatever other letter or letters are in the equation). In this case, following the rules you have learned will allow you to solve the equation.

When solving for y in terms of x, the idea is to move everything except y to the opposite side of the equation.

Example 1: Solve for y in terms of x.

$2y = 12x$ — Divide both sides by 2.
$y = 6x$

Example 2: Solve for a in terms of b.

$a - 9 = 2b$ — Add 9 to both sides.
$a = 2b + 9$

Example 3: Solve for c in terms of d.

$6d + 2c = 14$ — Subtract $6d$ from both sides.
$2c = 14 - 6d$ — Divide both sides by 2.
$c = 7 - 3d$

Example 4: Solve for x in terms of y.

$3x + y = 5$ — Subtract y from both sides.
$3x = 5 - y$ — Divide each term on both sides by 3.
$x = \frac{5}{3} - \frac{y}{3}$ — The equation is solved for x in terms of y.

Sample Questions

In problems 1–10, solve for y.

1. $6y = 18x$

2. $\frac{y}{2} = 10x$

3. $y - 6 = x$
4. $8 - y = x$
5. $2y + 1 = 3x$
6. $5y - 2 = x$
7. $2(y + 4) = 6x$
8. $-3(y - 1) = 12x$
9. $6x - 3y = 12$
10. $2x + 9y = 5$
11. Solve $A = lw$ for l in terms of A and w.
12. Solve $V = lwh$ for h in terms of V, l, and w.
13. Solve $P = 5000T/V$ for V in terms of P and T.
14. Solve $C = \pi r^2$ for r in terms of C and π.
15. Solve $3a = 4b + 6c$ for c in terms of a and b.

Answers

1. $6y = 18x$ — Divide both sides by 6.
   $y = 3x$
2. $\frac{1}{2}y = 10x$ — Multiply both sides by 2.
   $y = 20x$
3. $y - 6 = x$ — Add 6 to both sides.
   $y = x + 6$
4. $8 - y = x$ — Subtract 8 from both sides.
   $-y = x - 8$ — Divide both sides by –1.
   $y = -x + 8$ — (or $y = 8 - x$).
5. $2y + 1 = 3x$ — Subtract 1 from both sides.
   $2y = 3x - 1$ — Divide both sides by 2.
   $y = \frac{3}{2}x - \frac{1}{2}$
6. $5y - 2 = x$ — Add 2 to both sides.
   $5y = x + 2$ — Divide both sides by 5.
   $y = \frac{1}{5}x + \frac{2}{5}$ — Notice that $x/5 = \frac{1}{5}x$.
7. $2(y + 4) = 6x$ — Distribute the 2.
   $2y + 8 = 6x$ — Subtract 8 from both sides.
   $2y = 6x - 8$ — Divide both sides by 2.
   $y = 3x - 4$
8. $-3(y - 1) = 12x$ — Distribute the –3.
   $-3y + 3 = 12x$ — Subtract 3 from both sides.
   $-3y = 12x - 3$ — Divide both sides by –3.
   $y = -4x + 1$
9. $6x - 3y = 12$ — Subtract $6x$ from both sides.
   $-3y = 12 - 6x$ — Divide both sides by –3.
   $y = -4 + 2x$

10. $2x + 9y = 5$ — Subtract $2x$ from both sides.
$9y = 5 - 2x$ — Divide both sides by 9.
$y = \frac{5}{9} - \frac{2}{9}x$

11. $A = lw$ — Divide both sides by w.
$A/w = l$

12. $V = lwh$ — Divide both sides by lw.
$h = V/lw$

13. $P = 5000T/V$ — Multiply both sides by V.
$PV = 5000T$ — Divide both sides by P.
$V = 5000T/P$

14. $C = \pi r^2$ — Divide both sides by π.
$C/\pi = r^2$
$\sqrt{C/\pi} = \sqrt{r^2}$ — Take the square root of both sides.
$\sqrt{C/\pi} = r$

15. $3a = 4b + 6c$ — Subtract $4b$ from both sides.
$3a - 4b = 6c$ — Divide both sides by 6.
$\frac{3}{6}a - \frac{4}{6}b = c$ — Put in lowest terms.
$\frac{1}{2}a - \frac{2}{3}b = c$

TIPS FOR SOLVING WORD PROBLEMS

One of the most difficult things for students to do is to solve word problems. Most of the time students say they are unable to figure out what the problem is asking them to do. You can make this process easier by using the following strategies.

- **Read the problem carefully**. It may be necessary to read the problem more than once. As you read, identify what the problem is asking and in what units your answer should be expressed (feet, inches, miles). *It can be very helpful to underline important information.*

> ***Test Tip:***
> Try to identify what information is not needed to solve the problem, and mark it out.

- **Identify what you need to do.** If a formula is needed, write the formula down before beginning to solve the problem. Try to *estimate* what a reasonable answer would be so that you can tell if your final answer makes sense. If you need to write an equation, select a variable and assign it to an unknown quantity.
- **Solve the problem.** Work through the problem step by step. If you are using a calculator, you will make fewer mistakes if you write the steps down. Check what your calculator is telling you because it is easy to punch the wrong key. On the FCAT, most think-solve-explain problems require you to show your work or explain how you got your answer. Writing steps down can count as showing work or as an explanation.
- **Check your answer.** Many students skip this step, but it is probably the most important. This is where you compare your *estimated* answer to the answer you got when you solved the problem. Does your answer make sense? For example, have you calculated that a dog weighs 900 pounds instead of 90 pounds? Is your answer rounded to the designated number of places? (Incorrect rounding is a common error.)

Sample Questions

1. One-third of the students at Robinson High School are freshmen. If there are 350 freshmen, how many students are there?
2. The principal at Lemon Bay High School wants to order tee-shirts for all the teachers. The tee shirt company tells him that there is a \$50 charge for the silk screen plus \$18 for each shirt. Which of the following is the equation used to calculate the total cost C if t represents the number of tee shirts?
   A. $C = 18(50 + t)$
   B. $C = 18t + 50$
   C. $C = 50t + 18$
   D. $C = 18t$
3. A bacteria colony begins with 10,000 bacteria, and the population doubles every hour. The pattern of growth follows the equation $p = 10{,}000(2^h)$, where p is the population and h is the number of hours.

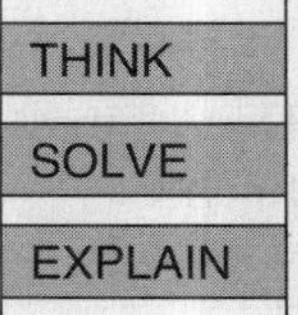

(*a*) What is the population after 4 hours? Show or explain how you got your answer.
(*b*) What is the population after 1 day? Show or explain how you got your answer.

4. To the nearest meter, find the height of the trapezoid in the accompanying diagram if it has an area of 72.

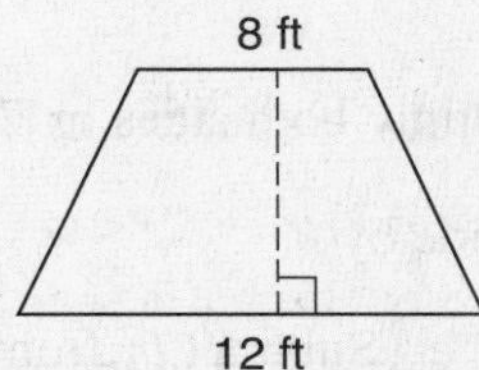

5. Which of the following is the solution of the equation $A = \frac{1}{2}bh$ for h in terms of A and b?
   A. $\frac{1}{2}Ah = b$
   B. $\frac{A}{b} = h$
   C. $\frac{A}{\frac{1}{2}b} = h$
   D. $\frac{2A}{b} = h$
6. Which of the following is the solution of the equation $c^2 = a^2 + b^2$ for a in terms of b and c?
   A. $c^2 - b^2 = a^2$
   B. $c^2 + b^2 = a^2$
   C. $\sqrt{c^2 - b^2} = a$
   D. $\sqrt{c^2 + b^2} = a$
7. Xia weighs 94 pounds. If she goes on a weight-gain diet and gains $\frac{1}{4}$ pound per day, how long will it be before she weighs 115 pounds?

Answers

1. Let s stand for the number of students; then $\frac{1}{3}s = 350$ and $s = 1050$ students.
2. **B**
3. (*a*) $10{,}000 \bullet 2^4 = 10{,}000 \bullet 16 = 160{,}000$.
   (*b*) $10{,}000 \bullet 2^{24} = 10{,}000 \bullet 16{,}777{,}216 = 167{,}772{,}160{,}000$.

   Your calculator allows you to calculate 2^{24}. Multiply 2×2 and hit the = key 23 times. Your calculator does *not* allow you to multiply this number by 10,000 because it will exceed the number of digits the display can hold. You can do this by hand simply by adding the four zeros from 10,000 to the end of the number in the display.

4. $A = \frac{1}{2}h(b_1 + b_2)$. Substitute the given values and solve for h:

Don't forget to check your work!

$$72 = \frac{1}{2}h(12+8)$$

$$72 = \frac{1}{2} \bullet 20h$$

$$72 = 10h$$

$h = 7.2$ Round to the nearest whole number.

$h = 7$

5. **D.** $A = \frac{1}{2}bh$

   $2A = bh$ Multiply both sides by 2.

   $\frac{2A}{b} = h$ Divide by b.

6. **C.** $c^2 = a^2 + b^2$ Subtract b^2 from both sides.

   $c^2 - b^2 = a^2$ Take the square root of both sides.

   $\sqrt{c^2 - b^2} = \sqrt{a^2}$

   $\sqrt{c^2 + b^2} = a$

7. 84 days. Let d stand for the number of days. Then, $\frac{1}{4}d + 94 = 115$. Subtract 94 from both sides and multiply by 4.

SLOPE

The slope of a line is an indication of its steepness. It tells you how the y-values of the points on the line change as the corresponding x-values change. Slope is expressed as:

$$\frac{\text{change in } y}{\text{change in } x}$$

Algebraically, the change in y is expressed as $y_2 - y_1$ (the second y-coordinate minus the first y-coordinate). The change in x is expressed as $x_2 - x_1$, where (x_1, y_1) and (x_2, y_2) are two points on the graph. The slope of a line containing the points (x_1, y_1) and (x_2, y_2) can be expressed as

$$m = \frac{y_2 - y_1}{x_2 - x_1} \text{ or } m = \text{change in } y \text{ divided by change in } x$$

The change in y is commonly referred to as the *rise* or vertical change, and the change in x is referred to as the *run* or horizontal change.

The slope is also referred to as the **rate of change**.

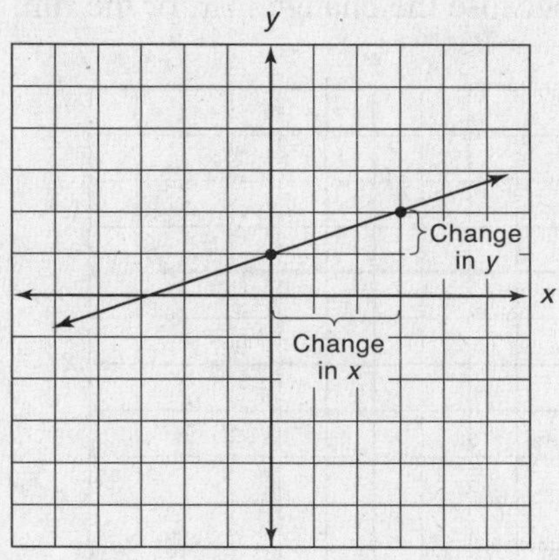

The slope of a line can be positive, negative, zero, or undefined.

Lines with a positive slope slant upward from left to right as shown in the diagram. In lines with a positive slope, the *rise* and *run* have the same sign. Both the rise and run are positive, or both are negative.

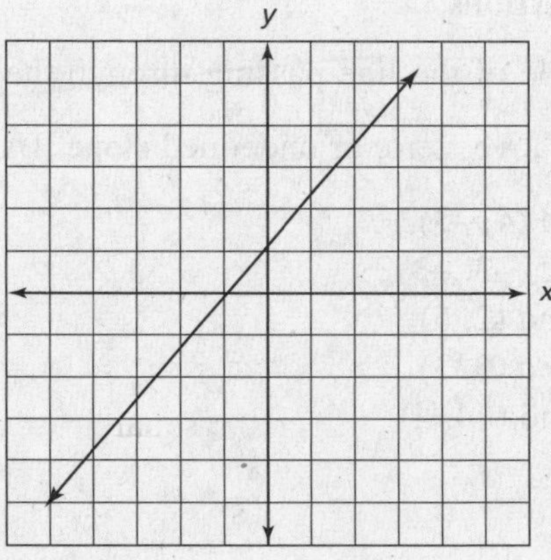

Positive slope

Lines with a *negative* slope slant downward from left to right as in the diagram shown here. The rise and the run have different signs: One is positive and one is negative.

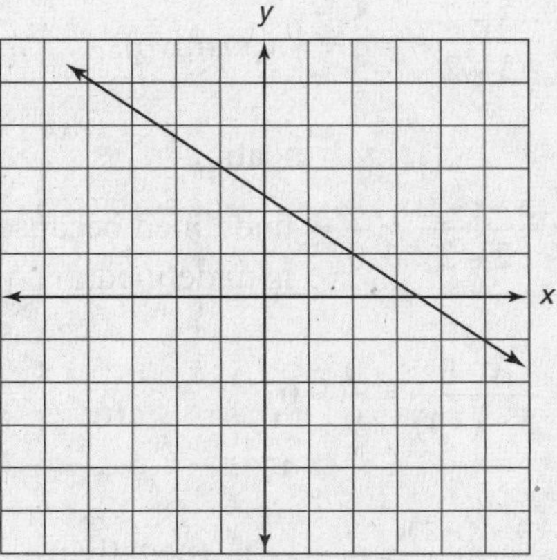

Negative slope

Lines with *zero* slope are flat or horizontal lines. You can recognize these lines algebraically because the change in y, or the rise, is zero: $0/(x_2 - x_1)$. Vertical lines have an *undefined* slope because the change in x, or the run, is zero: $(y_2 - y_1)/0$.

Division by zero is undefined.

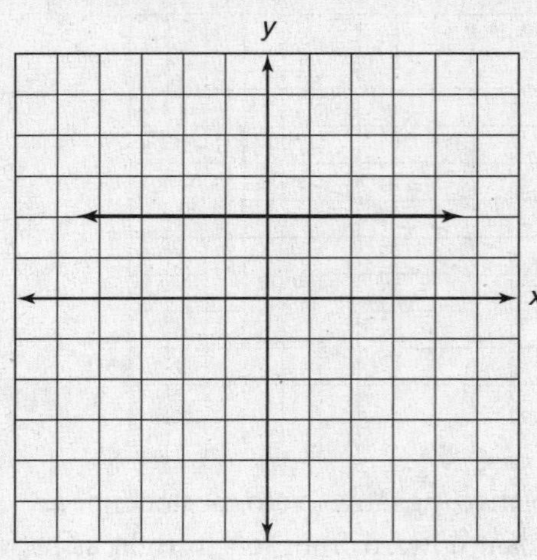

Zero slope

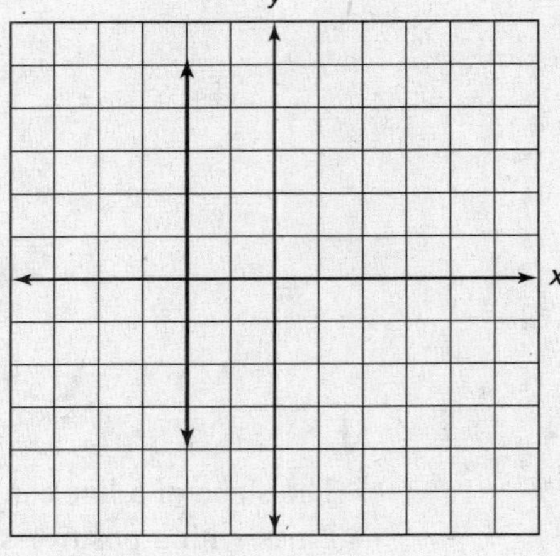

Undefined slope

Sample Questions

Find the slope of the line passing through the given points. Decide whether the line has a positive, negative, zero, or undefined slope. Use the formula $\frac{y_2 - y_1}{x_2 - x_1}$.

1. (1, 5) and (4, –3)
2. (2, 2) and (–3, –3)
3. (2, –3) and (2, 5)
4. (–4, 3) and (5, 3)
5. (2, –2) and (–4, 4)

Answers

1. $\frac{y_2 - y_1}{x_2 - x_1} = \frac{-3-5}{4-1} = \frac{-8}{3}$ (negative)

2. $\frac{y_2 - y_1}{x_2 - x_1} = \frac{-3-2}{-3-2} = \frac{-5}{-5} = 1$ (positive)

3. $\frac{y_2 - y_1}{x_2 - x_1} = \frac{5-(-3)}{2-2} = \frac{8}{0} =$ undefined because division by zero is undefined.

4. $\frac{y_2 - y_1}{x_2 - x_1} = \frac{3-3}{5-(-4)} = \frac{0}{9} = 0$

5. $\frac{y_2 - y_1}{x_2 - x_1} = \frac{4-(-2)}{-4-2} = \frac{6}{-6} = -1$ (negative)

Finding Slope on a Graph

To find the slope of a line on a graph, select two points on the graph. Use the line formed by these two points as the hypotenuse of a right triangle. The height of the triangle is the rise, and the base is the run.

If the line travels upward from left to right, the slope is positive. If the line travels downward from left to right, the slope is negative. Horizontal lines have a zero slope, and vertical lines have an undefined slope.

Example 1: In the accompanying diagram, the slope is positive because the line slants upward from left to right.

$$m = \frac{\text{change in } y}{\text{change in } x} = \frac{4}{1} = 4$$

Another way to remember slope is $\frac{\text{rise}}{\text{run}}$.

The slope of the line is 4.

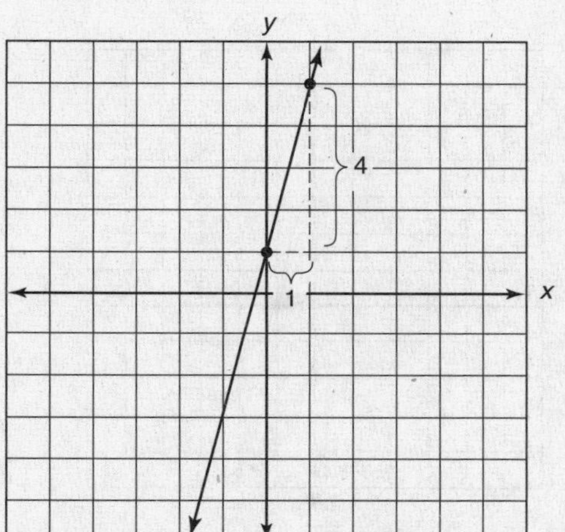

Example 2: In the accompanying diagram, the slope is negative because the line slants downward from left to right.

$$m = \frac{\text{change in } y}{\text{change in } x} = \frac{-3}{1} = -3$$

(Note that the rise is negative because the line is going downward from left to right. The slope of the line is −3.)

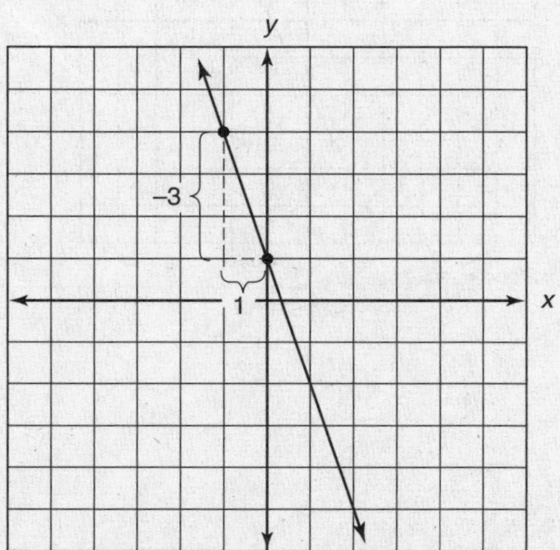

Sample Questions

Use the graphs shown here to find the slopes of the following lines.

1 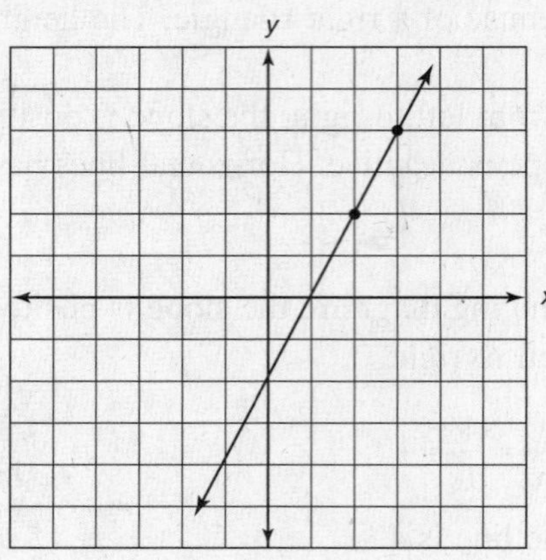

2 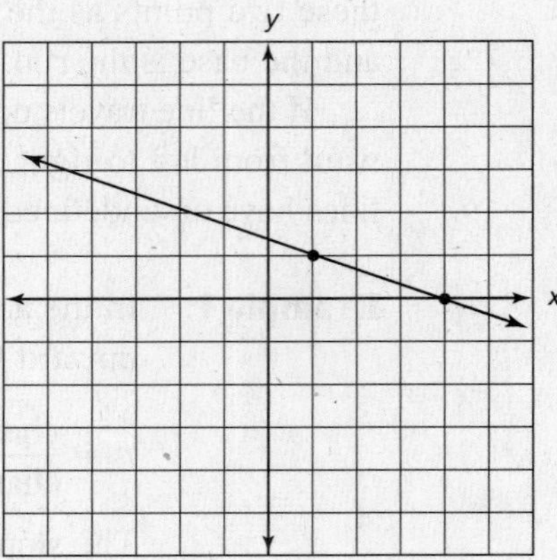

3 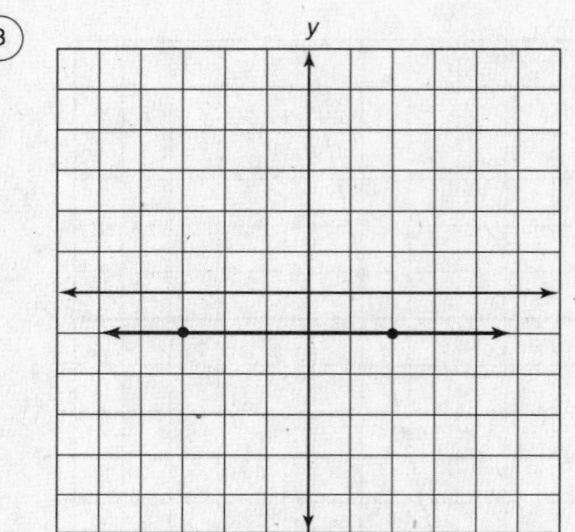

4 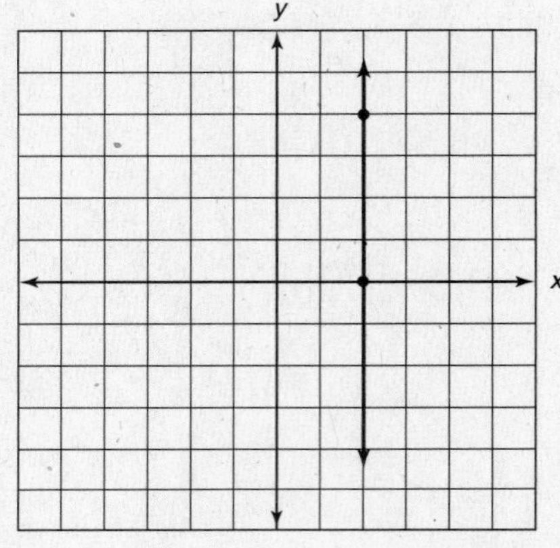

Answers

1 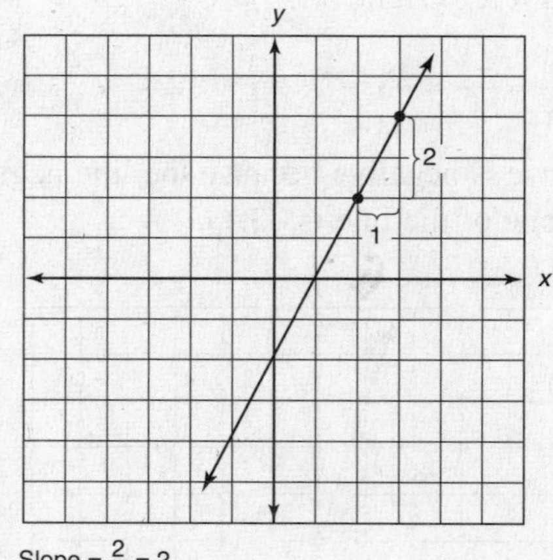

Slope = $\frac{2}{1} = 2$

2 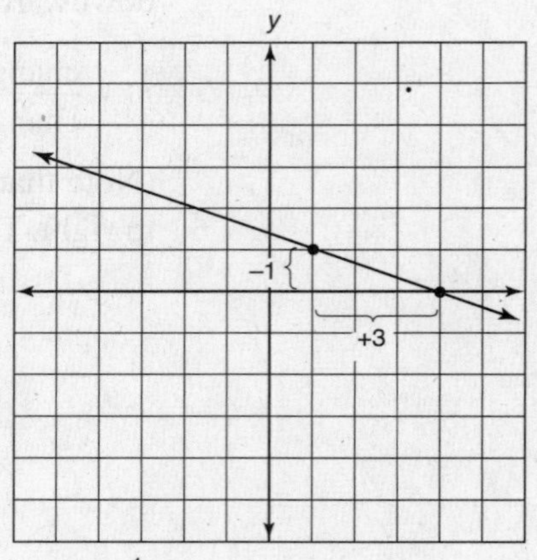

Slope = $-\frac{1}{3}$

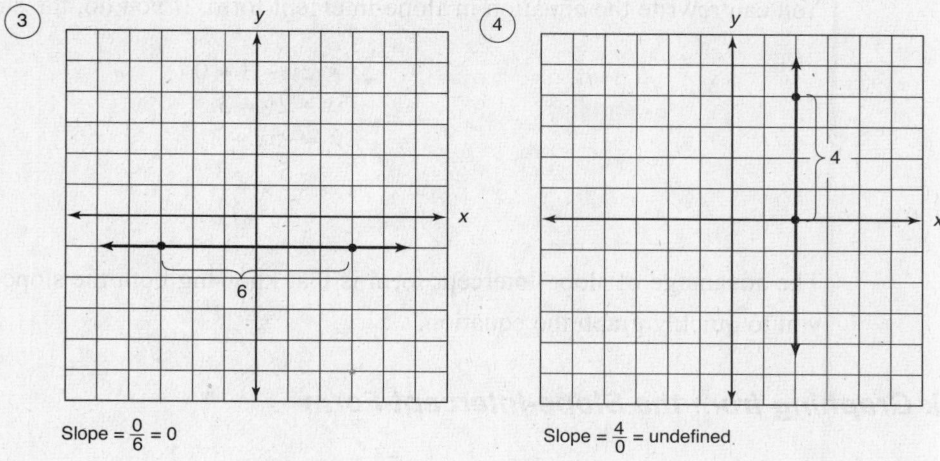

Finding the Slope from an Equation

Slope-Intercept Form

If you are given the equation of a line, it is fairly simple to find its slope. An equation written in the form $y = mx + b$ is in *slope-intercept* form. Examine the number in the equation in place of m and you have the slope. The number for b represents the y-intercept, or the place where the line crosses the y-axis.

Example 1: Find the slope of the line $y = -\frac{1}{2}x - 3$.

Since the line is in slope-intercept form

($y = mx + b$), $m = -\frac{1}{2}$ and $b = -3$.

The slope of the line is $-\frac{1}{2}$, and the line crosses the y-axis at -3.

> Remember that the number that is multiplied by x is the coefficient.

Example 2: A local cable company charges $35.00 for basic service plus $14.95 for each additional premium channel. This can be represented by the equation $y = 14.95x + 35$. If y represents a customer's total cost for cable service and x represents the number of premium channels, what is the rate per premium channel?

> The slope also represents the rate of change.

Since the slope represents the rate of change, the rate per premium channel is $14.95. Every time the customer adds or removes a premium channel, the total cost changes by $14.95.

Standard Form

The slope is not always the coefficient of x. Sometimes equations are given in the form $ax + by + c = 0$. When an equation is in this form, the slope is represented by $-a/b$, where a is the coefficient of x and b is the coefficient of y. For example, in the equation $5x + 2y - 3 = 0$, $a = 5$ and $b = 2$. The slope is $-5/2$ or -2.5.

> Slope = $\frac{-a}{b}$ in $ax + by + c = 0$

You can rewrite the equation in slope-intercept form. If you do, the slope will still be –5/2:

$$5x + 2y - 3 = 0$$
$$5x + 2y = 3$$
$$2y = -5x + 3$$
$$y = \frac{-5}{2}x + \frac{3}{2}$$

The advantage of slope-intercept form is that knowing both the slope, $\frac{-5}{2}$, and b, $\frac{3}{2}$, allows you to quickly graph the equation.

Quick Graphing from the Slope-Intercept Form

To quickly graph the equation $y = \frac{-3}{2}x + 2$, begin at the y-intercept, 2. This represents the first point (0, 2). From this point the slope ($\frac{-3}{2}$) will direct your movements. Since the

$$\frac{\text{change in } y}{\text{change in } x} = \frac{-3}{2}$$

go *down* three spaces (change in y) and *right* two spaces (change in x). This new point (2, –1) represents the second point for the graph. A third point can be found by going down three additional spaces and right two more spaces to the point (4, –4). You now have three points to make your line.

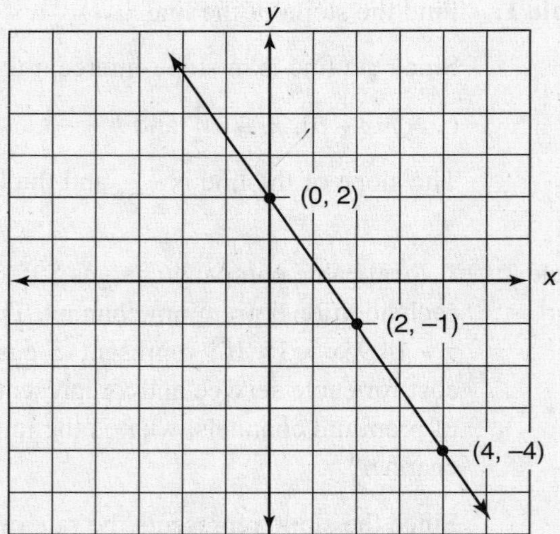

Sample Questions

1. Find the slope of the line represented by the equation $y = 4x + 5$.
2. Find the slope of the line represented by the equation $y = 5 - 2x$.
3. Find the slope of the line represented by the equation $3x + 4y - 5 = 0$.
4. Find the slope of the line represented by the equation $2y - 3x + 1 = 0$.
5. Find the slope of the line represented by the equation $x + 4 = y$.

Answers

1. The equation is in slope-intercept form, and so the slope is represented by the coefficient of x (4).
2. In slope-intercept form the equation is $y = -2x + 5$. The coefficient of x (-2) is the slope.
3. The equation is in standard form, $ax + by + c = 0$, where $a = 3$ and $b = 4$. The slope is represented by $-a/b$ or $-3/4$.
4. The coefficient of $x = -3$, and the coefficient of $y = 2$. Therefore, $a = -3$ and $b = 2$. $-a/b = -(-3)/2 = 3/2$.
5. If there is no written coefficient in front of x, the coefficient is understood to be 1.

Slope and Parallel Lines

Lines that are parallel have the same slope. Both equations have a slope of 2. Thus, $y = 2x - 5$ and $y = 2x + 3$ graph as parallel lines. The only difference is the y-intercept.

> The y-intercept is the point at which a line crosses the y-axis.

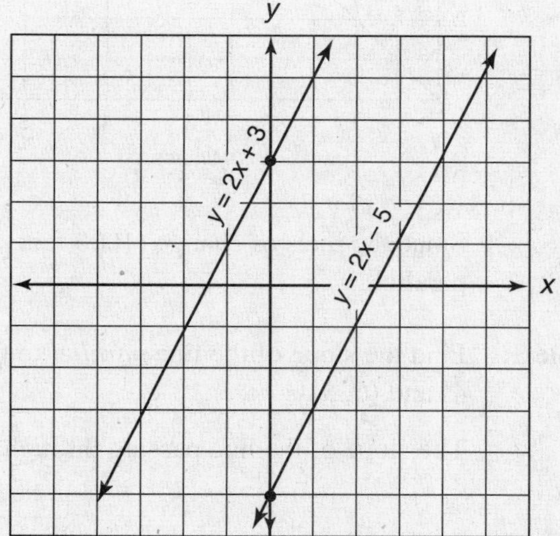

Example 1: Demonstrate by graphing that $y = \frac{1}{3}x - 1$ and $y = \frac{1}{3}x$ are parallel.

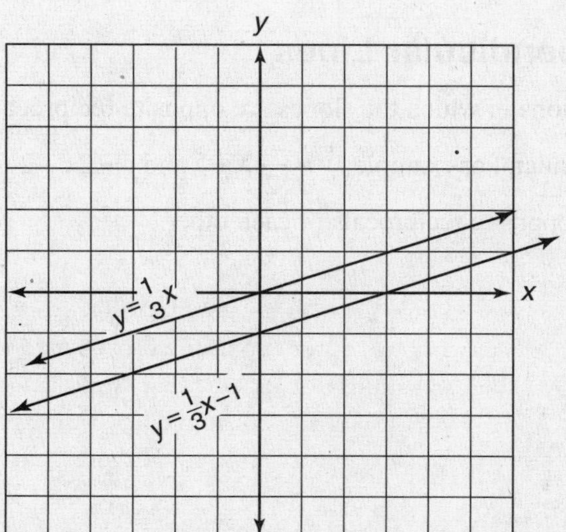

Example 2: Denver, Colorado, is at an altitude of 4000 feet. An airplane leaves Denver and climbs at a rate of 1000 feet per minute. At the same time, an airplane leaves Miami, Florida, which is at sea level, climbing at a rate of 1000 feet per minute. Write the equations of the lines that model each airplane's path. Are the paths parallel?

Denver: $y = 1000x + 4000$
Miami: $y = 1000x$

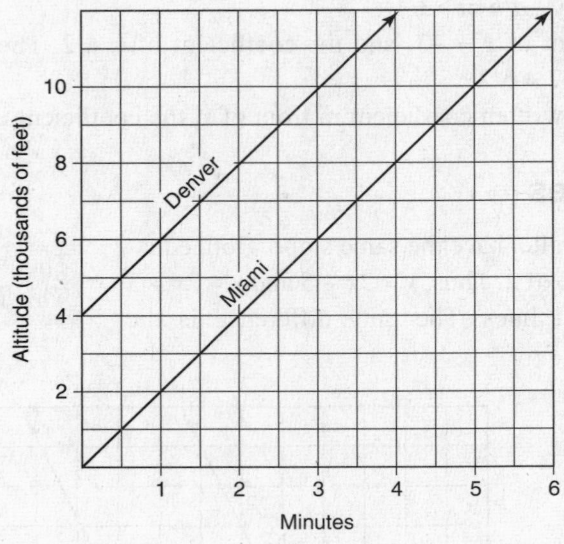

Since the rates of change (1000 feet per minute) are the same, the lines are parallel.

Example 3: Find the slope of the line *parallel* to the line passing through the points (–2, 4) and (0, 5).

The slope of the line passing through the points (–2, 4) and (0, 5) is

$$\frac{5-4}{0-(-2)} = \frac{1}{2}$$

Slope $= \dfrac{\text{change in } y}{\text{change in } x}$

Slopes of parallel lines are equal; therefore the slope of a line parallel to this line is also $\frac{1}{2}$.

Slope and Perpendicular Lines

Reciprocal: exchange numerator and denominator.

Equations in which the slopes are opposite reciprocals of each other form lines that are perpendicular. For example, $y = \frac{-1}{2}x + 5$ and $y = 2x + 2$ form perpendicular lines because $\frac{-1}{2}$ and 2 are opposite reciprocals of each other.

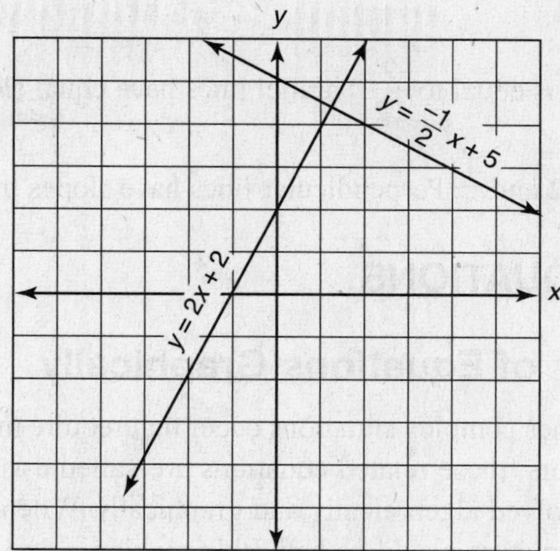

Example 1: Find the slope of the line perpendicular to the line passing through the points (3, −3), and (4, −5).

The slope of the line passing through the points (3, −3), and (4, −5) is

$$\frac{-5-(-3)}{4-3}=\frac{-2}{1}=-2$$

The slope of the line perpendicular to this line is the opposite reciprocal of −2, or $\frac{1}{2}$.

Sample Questions

The slopes of six lines are $\overleftrightarrow{AB}=\frac{3}{4}$, $\overleftrightarrow{BC}=-2$, $\overleftrightarrow{CD}=\frac{8}{-4}$, $\overleftrightarrow{DE}=\frac{1}{2}$, $\overleftrightarrow{EF}=\frac{9}{12}$, and $\overleftrightarrow{FG}=\frac{4}{3}$.

1. Which line is parallel to $\overleftrightarrow{AB}$?

A. $\overleftrightarrow{BC}$

B. $\overleftrightarrow{EF}$

C. $\overleftrightarrow{FG}$

D. $\overleftrightarrow{CD}$

2. Which line is perpendicular to BC?

A. $\overleftrightarrow{CD}$

B. $\overleftrightarrow{BC}$

C. $\overleftrightarrow{DE}$

D. $\overleftrightarrow{EF}$

Answers

1. **B.** $\frac{3}{4}$ is equal to $\frac{9}{12}$. Parallel lines have equal slopes.

2. **C.** –2 and $\frac{1}{2}$. Perpendicular lines have slopes that are opposite reciprocals of each other.

SYSTEMS OF EQUATIONS

Solving Systems of Equations Graphically

Sometimes complex situations occur that require more than one equation to be solved. When this occurs, these related equations are called a *system of equations*. Systems of equations can be solved algebraically and graphically. When an exact answer is not required, or when a visual picture would be helpful in making a decision, a graphical model is best.

Graphing Linear Equations

Systems of equations can have one solution, no solution, or an infinite number of solutions.

One Solution

When two or more graphed equations intersect at one point, the point is the *solution* for that system of equations. This system is called *consistent* and *independent.*

Example 1: $2x = y - 5$ and $y = -1$

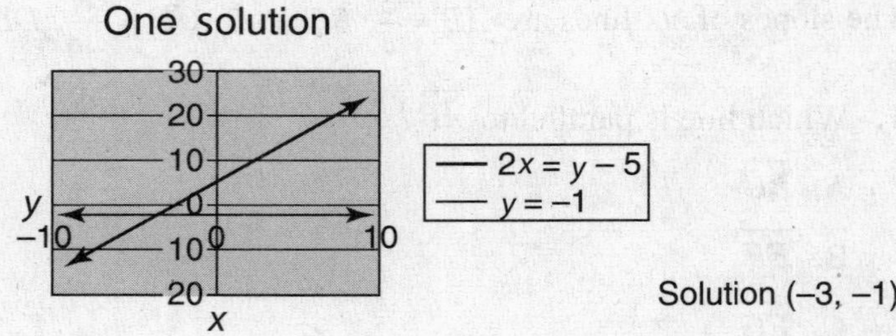

Example 2: BestRate Telephone Company offers two different rates. Rate A costs \$.05 per minute for long distance plus a \$10 monthly access charge. Rate B costs \$.07 per minute with a \$3 monthly access charge. Let x represent the number of minutes and y represent the total cost.

Rate A's plan gives the equation $y = .05x + 10$.
Rate B's plan gives the equation $y = .07x + 3$.

The solution to these equations is the point where the graphs of the two equations intersect. On examination of the graph you find that at the intersection of the point (350, 27.50) the cost of both plans is the same. Under both plan A and plan B it costs \$27.50 per month for 350 minutes of usage. If you use less than 350 minutes of long distance a month, plan B would be the least expensive. If you use more than 350 minutes per month, plan A would be the best plan.

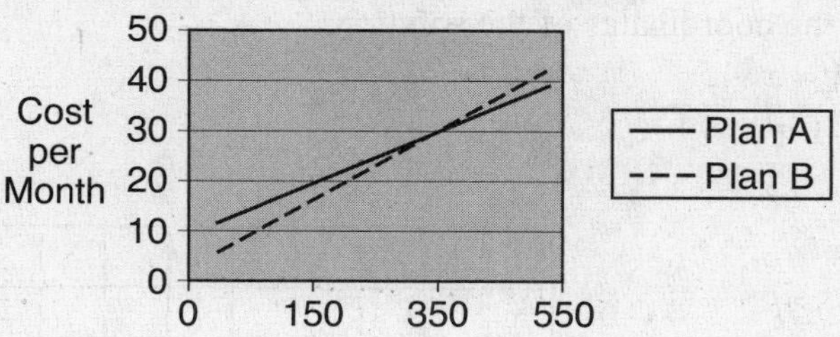

No Solution

When two or more graphed linear equations do not intersect at any point, they are parallel and have no common solution. They are called *inconsistent*.

Remember that equations with the same slope graph parallel lines.

Example: $y = -2x - 4$ and $y = 2 - 2x$

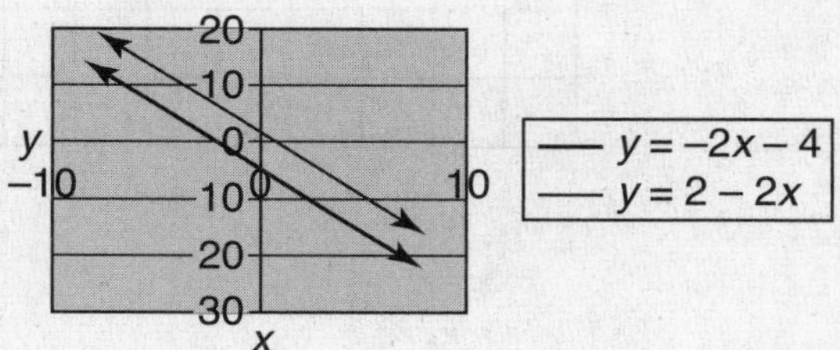

Infinite Solutions

When two or more graphed linear equations are collinear (represent the same line), there is an infinite number of solutions to these equations. All values of x along this shared line work in both equations. The equations are said to be *consistent* and *dependent*.

Example: $y = 3x - 1$
$2y - 6x = -2$

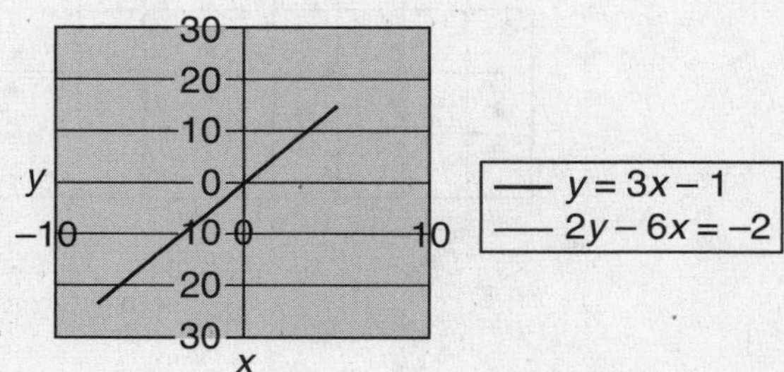

Sample Questions

Graph each system of equations on the same set of axes and determine whether the system has no solution, one solution, or infinitely many solutions. If there is only one solution, identify the coordinates of the solution.

1. $y = x - 5$
$2y = x - 4$

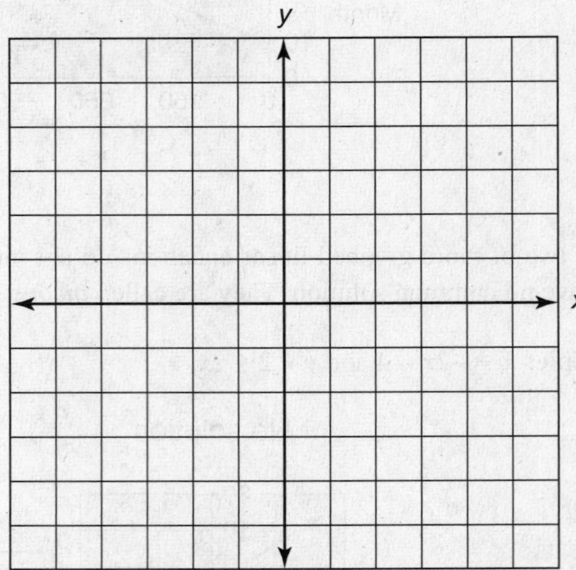

2. $y = 2x + 3$
$y = 2x$

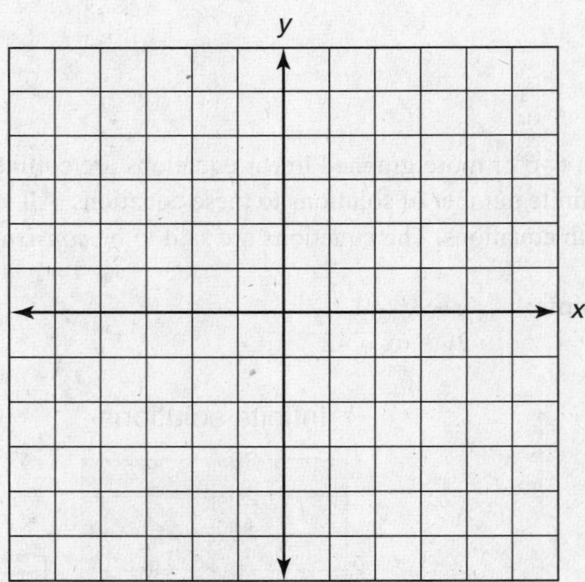

3. $y = x - 2$
$2x - 2y = 4$

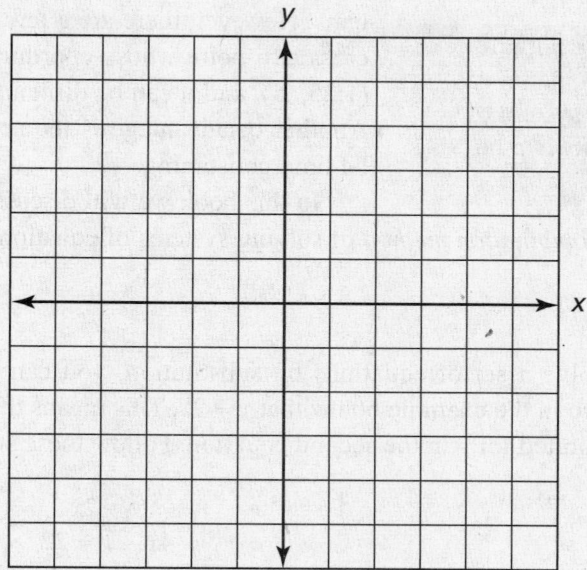

Answers

1. The graphs intersect at (6, 1).

For $y = x - 5$:

x	0	5	10
y	–5	0	5

For $2y = x - 4$:

x	0	2	4
y	–2	–1	0

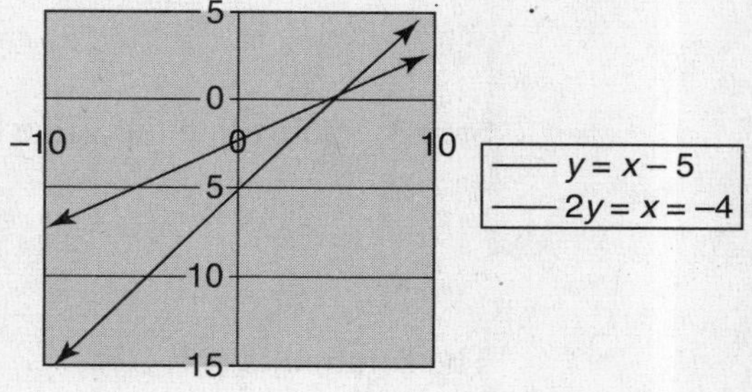

2. For $y = 2x + 3$:

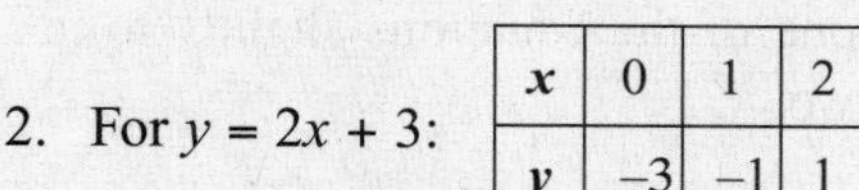

x	0	1	2
y	–3	–1	1

For $y = 2x$:

x	0	1	2
y	0	2	4

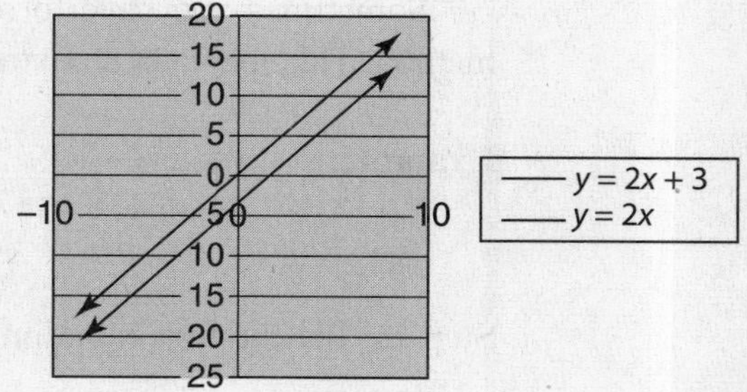

3. For $y = x - 2$:

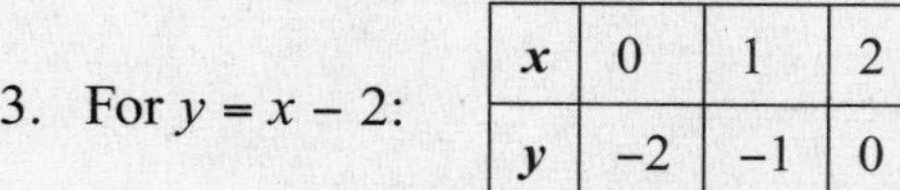

x	0	1	2
y	–2	–1	0

For $2x - 2y = 4$:

x	0	1	2
y	–2	–1	0

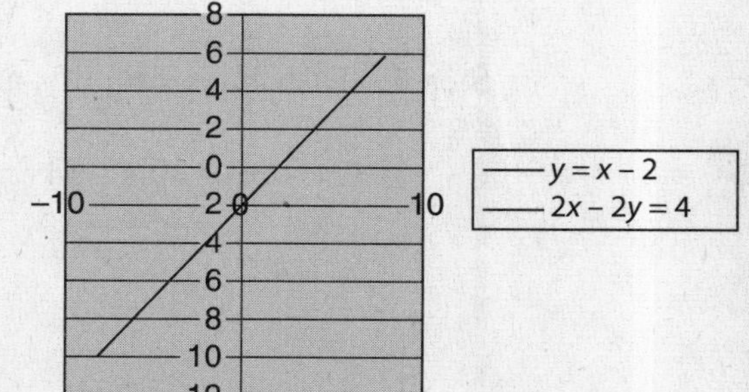

Solving Systems of Equations Algebraically

> On the FCAT, you are not allowed to use a graphing calculator, so being able to solve a system of equations algebraically can be very advantageous.

Finding the solution to a system of equations by graphing is excellent for some types of problems. However, there are a few disadvantages. First, the graphs may cross at a point whose coordinates are not whole numbers, such as (1.25, .5), and it can be difficult to pinpoint the location graphically. Another disadvantage is the actual process of graphing, which can be time-consuming.

In this book we will discuss the *substitution method* and the *linear combination method* of solving systems of equations.

Substitution

To solve a set of equations by substitution, you can substitute one equation for another. Notice in the example below that $y = 2x$. This means that because y is equal to $2x$, $2x$ can be substituted for y in the second equation. Follow these steps to solve the system of equations.

$$y = 2x$$
$$4x - y = 22$$

Step 1: Substitute $2x$ for y in the second equation, yielding $4x - 2x = 22$.

Step 2: Simplify and solve:

$2x = 22$

$x = 11$ The x-coordinate is 11.

Step 3: Substitute the value of x back into one of the *original* equations:

$y = 2(11)$

$y = 22$ The y-coordinate is 22.

The solution is (11, 22).

Sometimes you have to rearrange one of the equations slightly to use the substitution method. This may add an extra step or two.

Example: $y - x = 10$

$2x - 3y = -34$

Step 1: Put the first equation into the slope-intercept form: $y = x + 10$.

Step 2: Substitute $x + 10$ into the second equation in place of y: $2x - 3(x + 10) = -34$.

Step 3: Distribute, simplify, and solve:

$2x - 3x - 30 = -34$

$-x - 30 = -34$

$x = 4$ The x-coordinate is 4.

Step 4: Replace the value of x in one of the original equations:

$y - 4 = 10$
$y = 10 + 4$
$y = 14$ The y-coordinate is 14.

The solution is (4, 14).

Sample Questions

In questions 1–3, solve each system by substitution.

1. $y = 3x$
$2x - y = 5$

2. $y = x - 3$
$5x + y = 27$

3. $x + y = -1$
$2x - 3y = -7$

Answers

1. $2x - 3x = 5$ Substitute $y = 3x$ into $2x - y = 5$.
$-x = 5$, so $x = -5$ Substitute -5 for x in $y = 3x$.
$y = 3(-5) = -15$
The solution is $(-5, -15)$.

2. $5x + x - 3 = 27$ Substitute $y = x - 3$ into $5x + y = 27$.
$6x = 30$
$x = 5$
$y = 5 - 3 = 2$ Substitute 5 for x in $y = x - 3$.
The solution is (5, 2).

3. $y = -x - 1$ Put one equation in slope-intercept form.
$2x - 3(-x - 1) = -7$ Substitute into the second equation.
$2x + 3x + 3 = -7$ Simplify.
$5x + 3 = -7$ Solve by subtracting 3 and dividing by 5.
$x = -2$ The x-coordinate $= -2$.
$-2 + y = -1$ Substitute into one of the original equations.
$y = 1$ The y-coordinate $= 1$.
The solution is $(-2, 1)$.

Linear Combination

The *linear combination* method of solving systems of equations has also been called the *addition* or *subtraction* method. The best time to use a linear combination is when the system you need to solve does not contain an equation with a coefficient of 1. In this case, it would be tedious to solve for one variable in terms of the other.

Example 1: Examine the following system of equations.

$3a - 2b = 8$
$3a + 2b = 28$

You cannot easily substitute one equation for another. However, notice that one of the terms in the first equation, $-2b$, is the opposite of the term in the second equation, $2b$. If you add the two equations together, the terms will cancel each other out, leaving only one variable to solve for.

$3a - 2b = 8$	
$+\ 3a + 2b = 28$	Add the equations.
$6a = 36$	Divide by both sides by 6.
$a = 6$	

Substitute 6 for a in one of the original equations:

$3(6) - 2b = 8$	
$18 - 2b = 8$	Subtract 18 from both sides.
$-2b = -10$	Divide both sides by -2.
$b = 5$	

The solution is (6, 5) for (a, b).

Example 2: Sometimes when you compare two equations, you find you cannot readily eliminate one of the variables because there are no opposites to cancel. In this case, you can multiply both sides of the equation by a constant in order to match up terms. Examine these two equations:

$7x + 8y = 23$ $3x - 2y = -1$	No term in the first equation is an opposite of one in the second equation.
$7x + 8y = 23$ $4(3x - 2y = -1)$	If you multiply all the terms in the second equation by 4, then $(-2y)(4) = -8y$. $-8y$ is the opposite of $8y$ in the first equation.

$7x + 8y = 23$	
$12x - 8y = -4$	Add.
$19x = 19$	Divide both sides by 19.
$x = 1$	

$3(1) - 2y = -1$	Substitute in an original equation.
$3 - 2y = -1$	Subtract 3 from both sides.
$-2y = -4$	Divide both sides by -2.
$y = 2$	

The solution is (1, 2).

Example 3: In this case, both equations are multiplied by constants in order to "force" them to cancel a term. Examine the equations:

$5x + 2y = 4$
$4x + 3y = -1$

If we multiply the top equation by 3 and the bottom equation by −2, we will be able to add the equations and "cancel" the terms.

$3(5x + 2y = 4)$	Multiply all terms by 3.
$-2(4x + 3y = -1)$	Multiply all terms by −2.
$15x + 6y = 12$	
$-8x - 6y = 2$	Add.
$7x = 14$	Divide by 2.
$x = 2$	
$5(2) + 2y = 4$	Substitute 2 for x in an original equation.
$10 + 2y = 4$	Subtract 10 from both sides.
$2y = -6$	Divide by 2.
$y = -3$	

The solution is (2, −3).

Sample Questions

In questions 1–3, solve the systems using either the substitution or the linear combination method.

1. $y - x = 2$
$4x - 3y = 18$

2. $2x - 3y = -8$
$3x + 3y = 33$

3. $2x + 4y = 0$
$3x + 2y = -8$

Answers

1. The system can be solved by the substitution method by putting the first equation into the slope-intercept form and substituting it into the second equation.

$y - x = 2$	
$y = x + 2$	Put in the slope-intercept form.
$4x - 3(x + 2) = 18$	Substitute into the second equation.
$4x - 3x - 6 = 18$	Distribute.
$x - 6 = 18$	Add like terms.
$x = 24$	
$y - 24 = 2$	Substitute into an original equation.
$y = 2 + 24 = 26$	Solve.

The solution is (24, 26).

2. The system can be solved by the linear combination method because none of the coefficients is 1. Also, when the two equations are added, $-3y$ in the equation cancels $3y$ in the 2nd equation.

$2x - 3y = -8$
$\underline{3x + 3y = 33}$ Add.
$5x = 25$ Divide by 5 and solve for x.
$x = 5$

$2(5) - 3y = -8$ Substitute 5 for x in one equation and solve for y.
$10 - 3y = -8$
$-3y = -18$
$y = 6$

The solution is (5, 6).

3. The system can be solved by the linear combination method because none of the coefficients is 1. Also, after the second equation is multiplied by -2, $4y$ in the first equation cancels $-4y$ in the second equation.

$3x + 2y = -8$ Multiply the second equation by -2.
$-6x - 4x = -8$

$2x + 4y = 0$ Add the first and the new second equations and solve for x.
$\underline{-6x - 4y = 16}$
$-4x = 16$
$x = -4$

$2(-4) + 4y = 0$ Substitute -4 for x in one equation.
$-8 + 4y = 0$ Solve by adding 8 to both sides.
$4y = 8$ Divide by 4.
$y = 2$

The solution is (4, 2).

INEQUALITIES

An inequality is generally used when making statements involving terms such as *at most*, *at least*, *between*, *greater than*, or *less than*. These statements are called *inequality* statements. An inequality is a statement that two expressions may or may not be equal. Based on the sign of inequality used, one expression may be larger than another.

The tables below show inequality signs and their meanings.

Sign	Meaning
$>$	Greater than
$<$	Less than
$\geq$	Greater than or equal to
$\leq$	Less than or equal to
$\neq$	Unequal

Statement	Sample Numbers	Algebraic Expression
A number greater than 4	{4.01 5, 20}	$n > 4$
At most 6	{0, 1, 5.8, 6}	$n \leq 6$
Between 4 and 8, non-inclusive	{4.5, 6, 7.9}	$4 < n < 8$
At least 3	{3, 4, 92}	$n \geq 3$
A number less than –5	{–6, –6.9, –12}	$n < -5$
A number between 3 and 9	{3, 5, 8.9}	$3 \leq n \leq 9$

Solving Inequalities

Solving an inequality is almost like solving an equation.

Example 1: Solve the inequality $2y > 10$ for y.

$2y > 10$ Divide both sides by 2.
$y > 5$ y can be any number larger than 5.

Example 2: Solve the inequality $\frac{3}{4}x \leq 12$.

$\frac{3}{4}x \leq 12$ Multiply both sides by $\frac{4}{3}$.
$x \leq 16$ x can be any number less than or equal to 16.

Example 3: Solve the inequality $3x - 6 > -18$.

$3x - 6 > -18$ Add 6 to both sides.
$3x > -12$ Divide both sides by 3.
$x > -4$ x can be any number greater than –4.

Important: When an inequality statement is multiplied or divided by a negative number, the inequality is reversed.

When multiplying or dividing by a negative, the inequality will reverse.

Example 4: Solve the inequality $-5x + 7 < 22$.

$-5x + 7 < 22$ Subtract 7 from both sides.
$-5x < 15$ Divide both sides by –5.
$x > -3$ Notice that the inequality is reversed.

Example 5: Solve the inequality $-\frac{2}{3}x \geq 6$.

$-\frac{2}{3}x \geq 6$ Multiply both sides by $-\frac{3}{2}$.
$x \leq -9$ The inequality is reversed.

Number Line Graphs

Inequality statements with one variable can be graphed on a number line. The symbols < and > are represented by an open circle. The solution to an inequality is the set of all numbers that satisfy (work in) the inequality statement. The solution to the statement $n < -3$ is all numbers that are less than -3.

The statement $n > -3$ or $-3 < n$ is graphed as

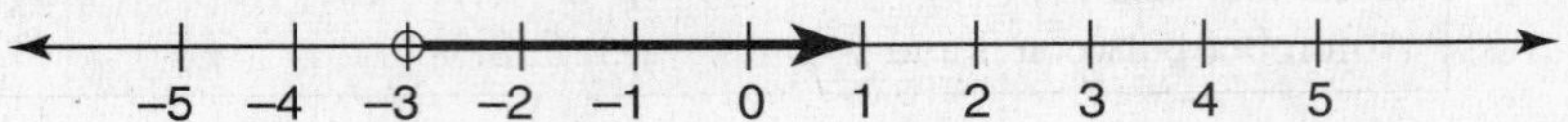

The arrow in the diagram points to the right because the numbers possible for n are greater than -3. *Note*: These numbers do not have to be whole numbers; they could be -2.999 or $.000001$. The only requirement is that they be greater than -3.

The symbols ≤ and ≥ are represented by a closed circle. The statement $n \leq -1$ or $-1 \geq n$ is graphed as

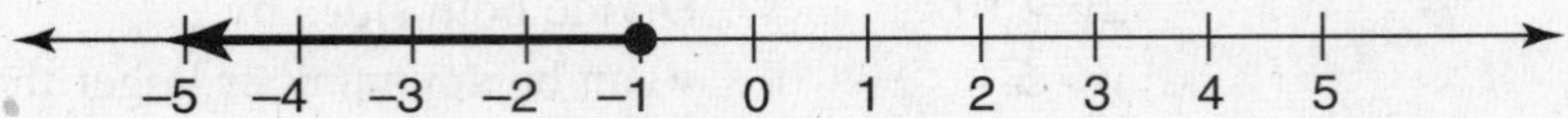

Compound inequalities combine two inequality statements. Two related inequality symbols can be used to describe this relationship such as $-5 \leq c < 1$. This statement says that c can be between -5 and 1, including -5. The graph looks like

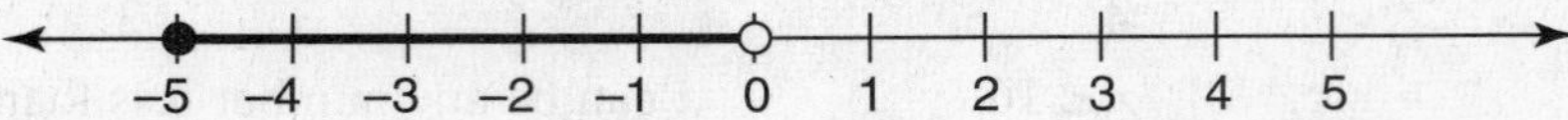

Consider the statement "Senior citizens and children under six pay half price for breakfast." If senior citizens must be 60 or older, the statement could read:

$n \geq 60$ or $n < 6$, where n is the age of the individual

Shown as a graph, it looks like

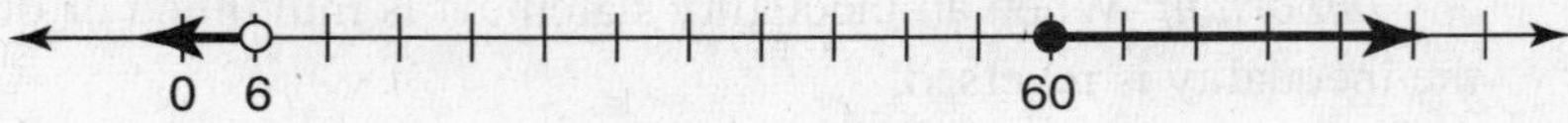

Sample Questions

1. Graph the inequality statement $n \leq 0$.
2. Solve for the variable: $2x - 5 \geq 10$.
3. Solve for the variable: $6 - n < 12$.
4. If $\frac{n}{3} + 2 < 13$ find all possible values for n.
5. To get on a ride at a theme park you must be at least 3 ft, 4 in. tall. Write an inequality statement representing the restriction in inches.
6. Marissa will be paid \$7.12 per hour at most. Write an inequality statement to represent the restriction.

Answers

1.

–4 –3 –2 –1 0 1 2 3 4

2. $2x - 5 \geq 10$ Add 5 to both sides.
   $2x \geq 15$ Divide by 2.
   $x \geq 7.5$
3. $6 - n < 12$ Subtract 6 from both sides.
   $-n < 12 - 6$ Divide both sides by -1.
   $n > -6$ Or $6 < n$

   Remember, multiplying or dividing by a negative flips the inequality. Also, it is generally best to express an inequality statement with the variable on the left.

4. $\frac{n}{3} + 2 < 13$ Subtract 2 from both sides.

   $\frac{n}{3} < 11$ Multiply both sides by 3.

   $n < 33$
5. Height ≥ 40.
6. Salary ≤ 7.12.

Graphing Inequalities with Two Variables

When an inequality contains two variables, it is still possible to graph it. In this case, the inequalities are graphed much like linear equations. In fact, it is usually best to pretend at first that you are graphing a linear equation and make adjustments to the graph to reflect the inequality. The solution to the inequality is all numbers that satisfy the inequality, represented as the shaded portion of the graph.

Example: $y = \frac{2}{3}x + 2$

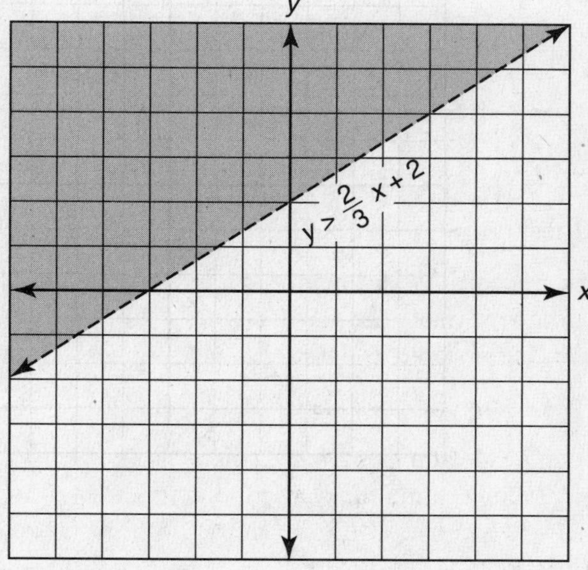

Test Tip: If the equation is in the slope-intercept form, and the inequality statement says $y >$ or $y \geq$, shade the area above the line. When the inequality statement says $y <$ or $y \leq$, shade the area below the line.

In this graph, the equation is graphed as though it were $y = \frac{2}{3}x + 2$. When the inequality symbol is < or >, the boundary line is dotted, not solid. If the inequality symbol had been ≤ or ≥, the line would have been a solid line.

For the inequality $y \le 2x - 3$, use a solid boundary line and shade the area below the line. The graph for all values of x is the entire area below the line.

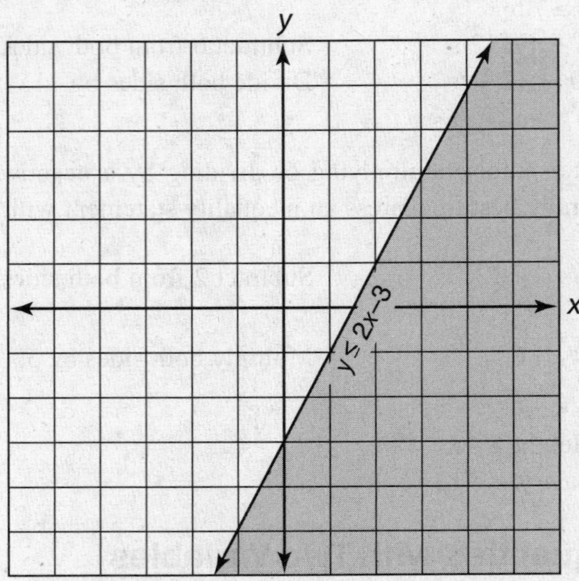

Sample Questions

1. Graph the inequality $y > \frac{3}{4}x - 2$

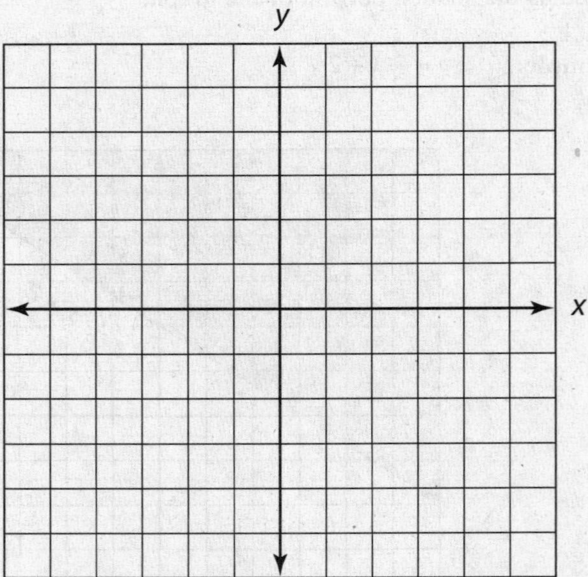

2. Graph the inequality $x + 4y \le -8$.

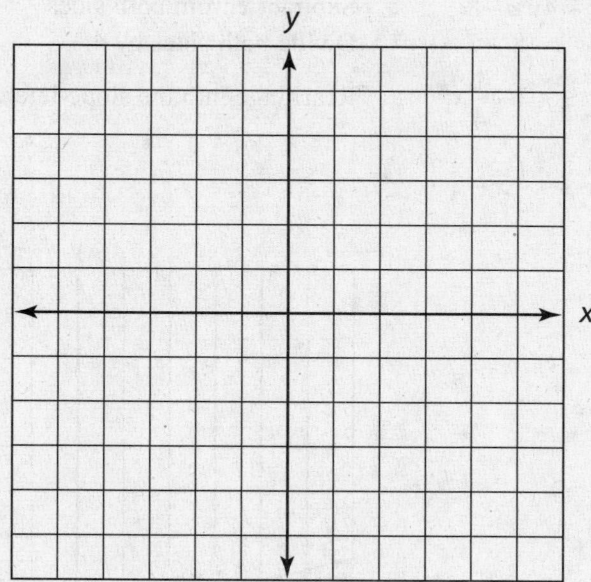

Answers

1.

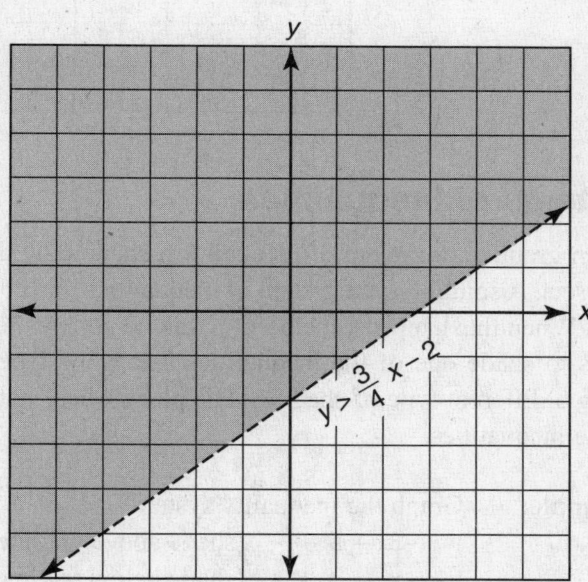

2. To more easily graph the inequality, put it into the slope-intercept form:

$x + 4y \leq -8$ Subtract x from both sides.

$4y \leq -8 - x$ Divide both sides by 4.

$y \leq -2 - \frac{1}{4}x$ Rearrange into the slope-intercept form.

$y \leq -\frac{1}{4}x - 2$

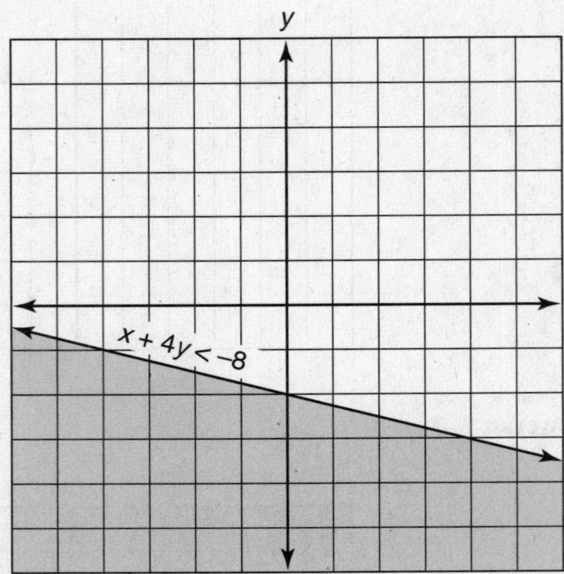

Graphing Systems of Inequalities

When graphing a system of inequalities, shade only the portion where the two inequalities intersect. A solution for a system of inequalities must be a point that works in all the inequalities. When this happens, the point is said to *satisfy* the system of inequalities. Sometimes it helps to shade one of the inequalities using one type of shading and the other inequality using a different type of shading. The place where both shadings overlap is the solution set to the inequalities.

Example: Graph the inequality system

$y > -3x + 5$ (shade above the line)

$y \leq -x$ (shade below the line)

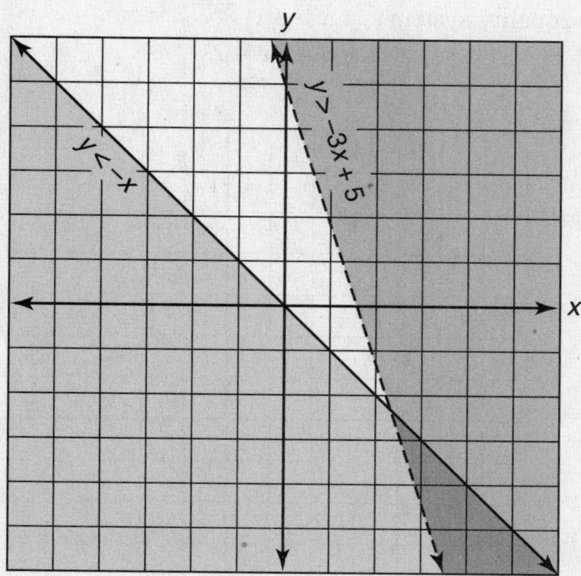

The point (4, −5) is located in the overlapping area. If you substitute the point into both equations, it will work.

$y > -3x + 5$	and	$y \leq -x$
$-5 > -3(4) + 5$		$-5 \leq -(4)$
$-5 > -12 + 5$		$-5 \leq -4$
$-5 > -7$		

Sample Questions

1. Graph the inequality system $y + 2x < 6$
$y \geq -2x + 1$

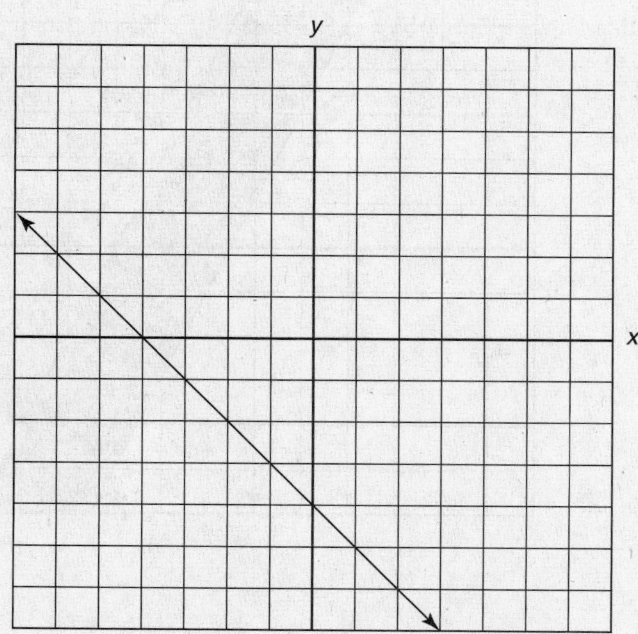

2. Graph the inequality system $x \geq 3$
$y < 2x - 2$

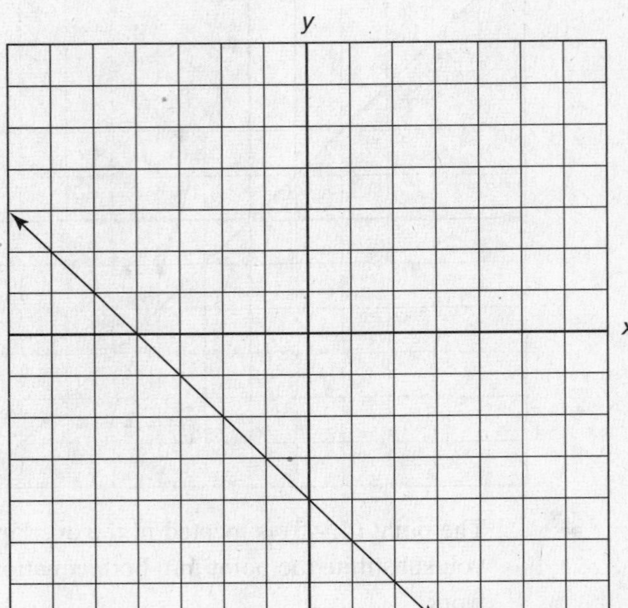

Answers

1. If you put the two equations into the slope-intercept form, you would see that the lines are parallel because they have the same slope:
$y < -2x + 6$ (shade below)
$y \geq -2x + 1$ (shade above)

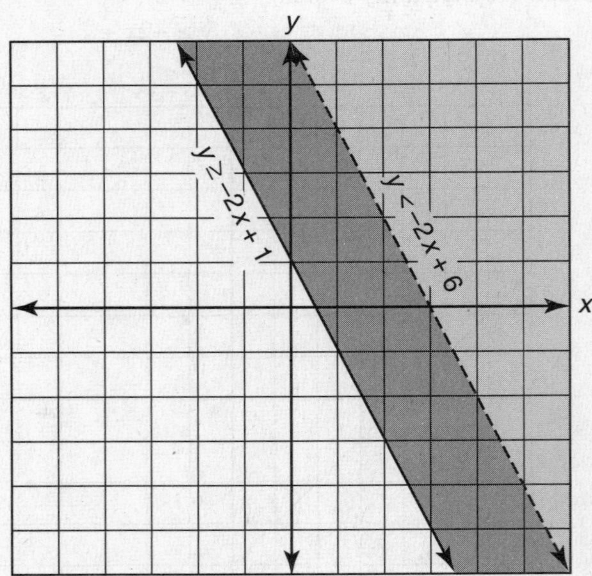

2.

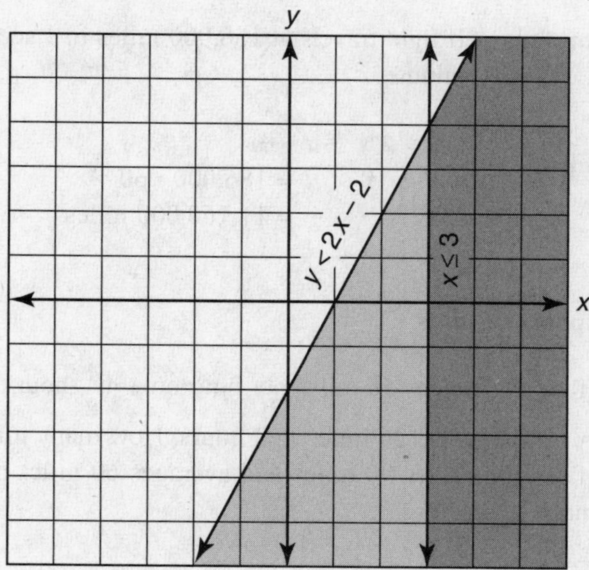

RATE AND RATIO

A *ratio* is a comparison of two measures or numbers that represent essentially the same type of quantities. For example, if there are 25 girls at a party and 20 boys, we say the ratio of girls to boys is 25 to 20, or 5 to 4. Another way to express this ratio is 5 : 4 or $\frac{5}{4}$.

Examples: Bright Light College has 1500 students, 800 of whom are men.

The ratio of men to women is 800 : 700 or $\frac{8}{7}$.

The ratio of men to all students is 800 : 1500 or $\frac{8}{15}$.

The ratio of women to men is 700 : 800 or $\frac{7}{8}$.

A *rate* is a comparison of two numbers or measures that represent different types of quantities. For example, 8 oranges for $1.00 represents a rate. It can be expressed as 8 : 1, 8/1, or 8 to 1. Because the denominator is a 1, it is specifically called a *unit rate*. Other example are working 4 hours for $100, expressed as 4 : 100 or 1 : 25.

Another common example of a rate is traveling 110 miles in 2 hours. The rate is $\frac{110 \text{ miles}}{2 \text{ hours}}$. When you use rates that involve distance, time, and speed, you can use the formula $d = rt$, where d = distance, r = rate (speed), and t = time.

$d = rt$
$110 = 2r$ Divide both sides by 2.
$50 = r$ The rate is 50 miles per hour.

You can use the formula $d = rt$ to find the distance traveled.

Example: If light travels at 186,000 miles in 1 second, how far does it travel in 1 minute?

1 minute = 60 seconds

$$d = rt$$
$$d = 186{,}000 \bullet 60$$
$$= 11{,}160{,}000 \text{ miles}$$

Sample Questions

1. If a car averages 56 miles per hour on a $4\frac{1}{2}$-hour trip, how far does it travel?
2. A train travels 126 miles in 3 hours. How many miles per hour is that?
3. If Tiffany drives 90 miles and averages 60 miles per hour, how long will it take her to make the trip?

Answers

1. $d = rt$
   $d = 56 \bullet 4.5$
   $= 252$ miles
2. $d = rt$
   $126 = 3r$ Divide both sides by 3.
   $42 = r$
3. $d = rt$
   $90 = 60t$ Divide both sides by 60.
   $1.5 = t$ It took $1\frac{1}{2}$ hours.

PROPORTIONS

When two ratios are equal, they are called *proportions*. A fairly common example of this might be when you double a recipe to feed more people. Each ingredient must be doubled or the recipe will be out of proportion and the food will not taste right. In a recipe for a cake, you might double the flour but forget to double the sugar. The cake would not be sweet enough because the ratio of flour to sugar would not be the same in the new recipe as it was in the old.

Proportions are usually expressed in equation form with one ratio or rate equaling another: $\frac{a}{b} = \frac{c}{d}$. We say, "$a$ is to b as c is to d." Proportions can also be written in the form $a : b = c : d$. When solving these equations the *cross-products* method is generally used.

Example: $\frac{5}{12} = \frac{c}{60}$

To solve, multiply diagonally (cross-multiply). Thus, $5 \bullet 60 = 12 \bullet c$ or $300 = 12c$. Dividing both sides of the equation by 12 yields $c = 25$.

Proportions are extremely useful in solving many types of word problems.

Example 1: Sandi can type 223 words in 5 minutes. What is her typing speed in words per minute?

$$\frac{223 \text{ words}}{5 \text{ min}} = \frac{x \text{ words}}{1 \text{ min}}$$

(*Note*: Each rate should be set up exactly the same.) Multiply diagonally, $223 = 5x$, or $x = 44$. The rate is 44 words per minute.

Example 2: Mr. Dumesnil drove his car 375 miles in 7.5 hours. He used 10.5 gallons of gas. How far could he drive in 10 hours? (*Note*: The amount of gas used is not needed to solve the problem.)

$$\frac{375 \text{ miles}}{7.5 \text{ hours}} = \frac{m \text{ miles}}{10 \text{ hours}}$$

$375 \cdot 10 = 7.5 \cdot m$ Cross-multiply.
$3750 = 7.5m$ Divide both sides by 7.5.
$500 = m$

He can travel 500 miles in 10 hours.

> ***Test Tip:*** Many word problems give you more information than you need. Always read carefully.

Sample Questions

1. The ratio of length to width in the American flag is 19 to 10. How wide is a flag that is 95 inches long?

2. The Vehicle Assembly Building at Cape Canaveral was completed in 1965. It is 40 stories or 552 feet high. The Bank of America Tower in Jacksonville, Florida, was completed in 1990 and is 42 stories or 617 feet high. If a scale model of the Vehicle Assembly Building is 20 inches tall, how tall would a scale model of the Bank of America Tower be?

A. 22 inches
B. 23 inches
C. 17,029 inches
D. 40 inches

3. David is making Bananas Foster for his guests. The recipe calls for $\frac{1}{4}$ cup of butter to 1 cup of brown sugar. If he accidentally puts in $\frac{1}{3}$ cup of butter, how much sugar will he need to keep the same proportion?

Answers

1. $$\frac{19 \text{ long}}{10 \text{ wide}} = \frac{95 \text{ long}}{w \text{ wide}}$$

$95 \cdot 10 = 19 \cdot w$ Cross-multiply.
$950 = 19w$
$50 = w$ The flag is 50 inches wide.

2. **A.** Set up a ratio for each building and make a proportion.

$$\frac{552 \text{ actual}}{20 \text{ scale}} = \frac{617 \text{ actual}}{s \text{ scale}}$$

$552s = 20 \cdot 617$

$552s = 12340$

$s \approx 22.355$

3. You can set up a proportion with the old quantities and the new quantities:

Old New

$$\frac{\frac{1}{4}}{1} = \frac{\frac{1}{3}}{x}$$

$\frac{1}{4}x = \frac{1}{3}$ Cross-multiply.

$x = \frac{1}{3} \div \frac{1}{4}$ Divide both sides by $\frac{1}{4}$.

$= \frac{1}{3} \cdot \frac{4}{1} = \frac{4}{3} = 1\frac{1}{3}$ cups

INTEREST

Another commonly used formula is the interest formula. This formula allows you to solve for interest amounts (I), principal amounts (p) (the amount of money deposited or borrowed), the annual interest percentage or *rate* (r), or time (t) (the amount of time the principal was deposited or borrowed).

The formula for interest is $I = prt$, where I = interest, p = principal, r = rate, and t = time.

Example: Find the amount of interest charged if Max borrows \$1500 at an annual interest rate of 9.5% for 6 months.

> ***Test Tip***: If you have difficulty changing percents to decimals, this might be a good time to quickly review Chapter 1.

Note: 6 months represents 6 months out of a year or $\frac{6}{12}$ of a year. This can also be expressed as $\frac{1}{2}$ or .5.

Substitute for t in the $I = prt$ formula.

$I = prt$ Substitute the given values.

$= 1500 \cdot 9.5\% \cdot \frac{6}{12}$ Multiply by 9.5%.

(6 months = $\frac{6}{12}$ of a year).

$= \$71.25$ Max pays \$71.25 interest.

Sample Questions

1. Find the total deposit at the end of 2 years if Andy deposits \$2000 for 2 years into a savings account that pays 3% interest per year.
2. Find the original amount deposited if a 5.5% certificate of deposit pays \$143 interest over a period of 6 months.

3. What interest rate was charged if Jayne paid \$45 interest over 9 months on a \$900 loan?

4. How long will you have to leave a \$10,000 deposit in the bank before it doubles if the bank is paying 4% interest?

Answers

1. $I = prt$, where $p = 2000$, $r = 3\%$, $t = 2$
   $I = 2000 \cdot 3\% \cdot 2$
   $= \$120$ — Add the interest to the principal
   $\$2000 + \$120 = \$2120$ — The total deposit is \$2120.
2. $I = prt$, where $I = 143$, $r = 5.5\%$, $t = \frac{6}{12}$ or .5
   $143 = 5.5\% \cdot .5 \cdot p$ — Change 5.5% to a decimal and
   $= 2.75\%p$ — express 6 months as .5
   $\$5200 = p$ — Divide both sides by 2.75%.
3. $I = prt$, where $I = 45$, $p = 900$, $t = \frac{9}{12}$ or .75
   $45 = 900 \cdot r \cdot .75$ — Change 9 months to .75.
   $45 = 675r$ — Divide both sides by 675.
   $.0\bar{6} = r$
   $.067 = r$ — The rate is about 6.7%.
4. In order for the deposit to double, it must receive a total of \$10,000 interest.
   $I = prt$, where $I = 10{,}000$, $p = 10{,}000$, $r = 0.04$
   $10{,}000 = 10{,}000 \cdot t \cdot .04$
   $10{,}000 = 400t$ — Divide by 400.
   $25 = t$ — It will double in 25 years.

SECTION 4: POLYNOMIALS

A monomial is a single term with no addition or subtraction sign. For example, 12, $5x$, x^2, -10, $4ab$, or $2a^2b^5$ are all monomials. Polynomials (*poly* meaning many) can be one or more monomials joined together with addition or subtraction signs.

Name	Number of Terms	Example
Monomial	1 (*mono* means one term)	$-3x^2$
Binomial	2 (*bi* means two terms)	$2x^2 + 4x$
Trinomial	3 (*tri* means three terms)	$x^2 - 2x + 1$

SIMPLIFY MONOMIALS AND MONOMIAL EXPRESSIONS

When a monomial has the same variable as another monomial, they are called *like terms*. Like terms can be added or subtracted. For example, $5x^2$ and $2x^2$ are like terms because each term has x^2 as a common factor. All you have to do is add or subtract the numbers in front of the x^2 (called *coefficients*). $5x^2 + 2x^2 = 7x^2$ and $5x^2 - 2x^2 = 3x^2$. You cannot add $5x^2$ and $2x$ because they do not have the same variable (one is x and other other is x^2).

ADD, SUBTRACT, AND MULTIPLY POLYNOMIALS

Adding Like Terms

Add like terms by adding the numbers in front of the terms. Be sure to use the rules for adding integers.

Example: Add $(4x^2 - 2) + (x^2 + 2x - 1)$.

Method 1. Use a horizontal method to add like terms: Remove parentheses. Identify like terms. Group the like terms together.

When adding, you can simply drop the parentheses.

Add the like terms.

$(4x^2 - 2) + (x^2 + 2x - 1)$

$= 4x^2 - 2 + x^2 + 2x - 1$ Commute terms.

$= 4x^2 + x^2 + 2x - 2 - 1$ Add like terms.

$= 5x^2 + 2x - 4$

Method 2: Using a vertical method to add like terms: Arrange the like terms so that they are lined up under each other in vertical columns. Add the like terms in each column following the rules for adding signed numbers.

Use 0 to hold the place of a missing variable.

$$\begin{array}{r} 4x^2 + 0x - 2 \\ + x^2 + 2x - 1 \\ \hline 5x^2 + 2x - 3 \end{array}$$

Subtracting Like Terms

Subtract like terms by first changing the subtraction sign to an addition sign and then changing the signs of the terms being subtracted. Follow the rules for adding polynomials.

Example: Simplify: $(4x^2 - 2) - (x^2 + 2x - 1)$.

Use a horizontal method to subtract like terms: Change the subtraction sign to an addition sign, and then change the signs of all of the terms being subtracted (terms that come after the subtraction sign).

Because this is not addition you cannot drop the parentheses until you have changed signs.

$(4x^2 - 2) - (x^2 + 2x - 1)$

$(4x^2 - 2) + (-x^2 - 2x + 1)$

$= 4x^2 - 2 + -x^2 - 2x + 1$

$= 4x^2 - x^2 - 2x - 2 + 1$

$= 3x^2 - 2x - 1$

Sample Questions

Simplify by adding or subtracting like terms.

1. $(5x - 3) + (4x^2 + 3x)$
2. $(6x^2 + 1) - (2x^2 - 8)$

3. $(3a - 2b - 6) + (12b + 3) - (-4)$
4. $(x + 2y) + (-2x - 3y - 1)$
5. $(x + 1) + (x - 3) - (-x + 4)$
6. Subtract $(-4x^2 - 2y^2 - 7)$ from $(2x^2 + 5y^2 + 7)$.

Answers

1. $(5x - 3) + (4x^2 + 3x)$
   $0x^2 + 5x - 3$ Line up vertically with highest powers first.
   $+ 4x^2 + 3x$
   $4x^2 + 8x - 3$ Combine like terms.
2. $(6x^2 + 1) - (2x^2 - 8)$ Subtraction Problem
   $(6x^2 + 1) + (-2x^2 + 8)$ Change subtraction sign to addition and change signs on remaining terms.
   $6x^2 + 1 + -2x^2 + 8$ Remove parentheses and add like terms.
   $6x^2 - 2x^2 + 1 + 8$
   $4x^2 + 9$
3. $(3a - 2b - 6) + (12b + 3) - (-4)$
   $3a - 2b - 6 + 12b + 3 + 4$ Remove parentheses for addition problem.
   $3a - 2b + 12b - 6 + 3 + 4$ Commute to group like terms.
   $3a + 10b + 1$
4. $(x + 2y) + (-2x - 3y - 1)$
   $x + 2y - 2x - 3y - 1$
   $x - 2x + 2y - 3y - 1$
   $-x - y - 1$
5. $(x + 1) + (x - 3) - (-x + 4)$
   $(x + 1) + (x - 3) + (x - 4)$ Change signs on subtraction portion first.
   $x + 1 + x - 3 + x - 4$ Remove parentheses.
   $3x - 6$
6. Subtract $(-4x^2 - 2y^2 - 7)$ from $(2x^2 + 5y^2 + 7)$.
   $(2x^2 + 5y^2 + 7) - (-4x^2 - 2y^2 - 7)$ Change signs.
   $(2x^2 + 5y^2 + 7) + (4x^2 + 2y^2 + 7)$
   $2x^2 + 5y^2 + 7 + 4x^2 + 2y^2 + 7$ Remove parentheses.
   $2x^2 + 4x^2 + 5y^2 + 2y^2 + 7 + 7$ Regroup.
   $6x^2 + 7y^2 + 14$

Notice that 0 is used as a coefficient for x^2 to hold the place.

Multiplying Polynomials

For all numbers x and all integers m and n, $x^m \cdot x^n = x^{m+n}$.

Example 1: $x^2 \cdot x^3 = x^5$. In expanded form, this means

$$(x \cdot x) \cdot (x \cdot x \cdot x) = (x \cdot x \cdot x \cdot x \cdot x) = x^5.$$

Example 2: $2^2(2^3)(2^4) = 2^{(2+3+4)} = 2^9$

Example 3: $3x^2(2x^3) = 6x^5$

If you are multiplying and the base variables are the same, the exponents should be added.

Example 4: $(a^2b^3)(a^5b^2) = a^7b^5$

Example 5: $-3q^3r^2(5q^4)(6r) = -90q^7r^3$

Example 6: $3x(x^2 + 2x - 1) = 3x^3 + 6x^2 - 3x$ Multiply $3x$ by every term inside parentheses.

Sample Questions

Simplify each problem:

1. $5x^2(10x^5) =$
2. $(a^3b^4c^2)(a^5c^4) =$
3. $2x^2(x - 1)$
4. $(-2x)^3$
5. $(4a^2)^3$

Answers

1. $50x^7$
2. $a^8\, b^4c^6$
3. $2x^3 - 2x^2$
4. $-8x^3$ — $(-2x)^3$ in expanded form is $(-2x)\,(-2x)\,(-2x) = (-2)(-2)(-2)x^3$.
5. $64a^6$ — $(4a^2)^3$ in expanded form is $(4a^2)\,(4a^2)\,(4a^2) = (4)(4)(4)a^6$.

DIVIDE POLYNOMIALS BY MONOMIALS

Divide each term of the polynomial by the monomial by dividing the numbers and subtracting the exponents.

When dividing polynomials, you can subtract the exponents.

Note: $4x^2 \div 4x^2 = 1$, they do not cancel.

Example 1: $\dfrac{10x^5}{2x^2} = 5x^3$ Step 1: Divide 10 by 2. Step 2: x^5 divided by $x^2 = x^{(5-2)} = x^3$.

Example 2: $\dfrac{12x^4 + 8x^3 + 4x^2}{4x^2} = 3x^2 + 2x + 1$

Sample Questions

Simplify:

1. $\dfrac{ab - b}{b} =$
2. $(10q^5 - 5q^4) \div 5q^3 =$
3. $\dfrac{20y^7 - 15y^5 + 5y^3}{5y^3} =$

Answers

1. $a - 1$
2. $2q^2 - q$
3. $4y^4 - 3y^2 - 1$

WORD PROBLEMS FOR POLYNOMIALS

Sample Questions

1. The sides of a triangle are represented by the expressions $2x + 5$, $x^2 - 1$, and $2x^2 - 2x$. What is the simplest expression for the perimeter of that triangle?

2. A rectangular park's area can be represented by $10x^2 - 5x$. Two paved areas have areas of $3x^2$ and $9x$, respectively. Write a simplified expression of the park's remaining grassy area after the paved areas have been removed.

It may be helpful to draw a picture.

3. The hare and the tortoise are in a race. After 30 minutes, the hare's distance from the starting line can be represented by $12d - 5$ and the tortoise's distance can be represented by $10d + 4$. How far apart are they?

Answers

1. $3x^2 + 4$. To find perimeter add all sides:
$(2x + 5) + (x^2 - 1) + (2x^2 - 2x) = 2x^2 + x^2 + 2x - 2x + 5 - 1 = 3x^2 + 4$.
2. $7x^2 - 14x$. To find the remaining grassy area subtract the two paved areas and then simplify: $(10x^2 - 5x) - 3x^2 - 9x$.
3. $2d - 9$. To find the distance between them, subtract: $(12d - 5) - (10d + 4) = (12d - 5) + (-10d - 4) = 12d - 10d - 5 - 4 = 2d - 9$.

SECTION 5: QUADRATICS

Some equations form graphs that are nonlinear. This means that the graph of the line is not straight. There are many different nonlinear graphs, but the one shown here is called a *parabola*. Parabolas are formed when the graph is in the form of $y = ax^2$, $y = ax^2 + b$, or $y = ax^2 + bx + c$. Notice that in all forms of this equation, the x is raised to the second power. Linear equations form straight lines. They are first-degree equations because the x is raised to the first power.

This graph represents the equation $y = x^2$. On this graph the vertex is the lowest point and is located at (0, 0). The other marked points on the graph are (1, 1), (–1, 1), (2, 4), and (–2, 4).

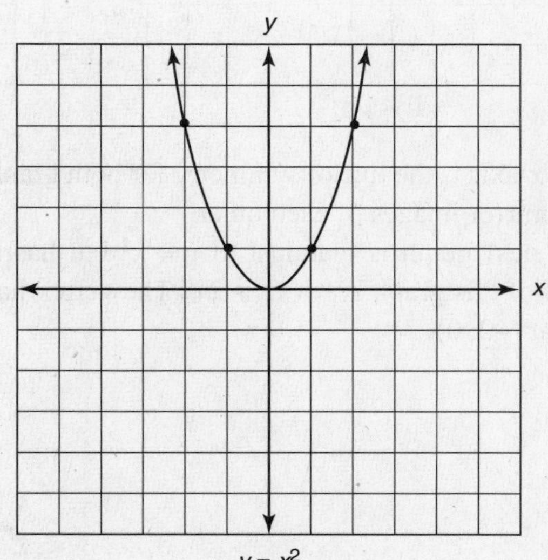

$y = x^2$

Parabolas are bowl-shaped and symmetrical.

The following graph represents the equation $y = -x^2$. Notice that the graph is inverted. This is due to the negative sign. The vertex, in this case the highest point, is also located at (0, 0). The other marked points on the graph are (1, −1), (−1, −1), (2, −4), and (−2, −4).

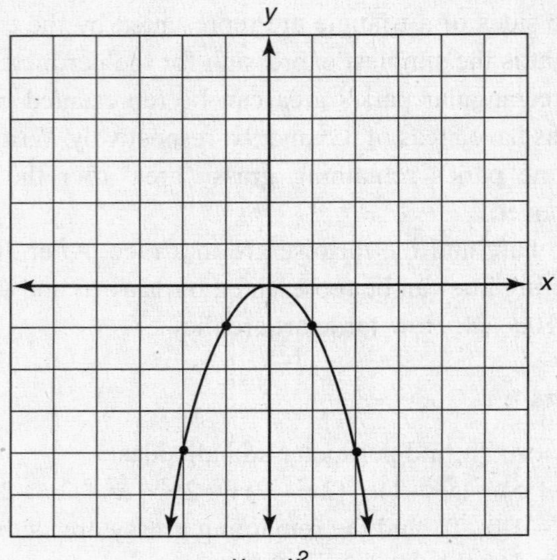

$y = -x^2$

A math constant is a real number.

Adding a constant to either of these equations results in the entire graph being shifted up or down. Positive constants shift the graph up, and negative constants shift the graph down.

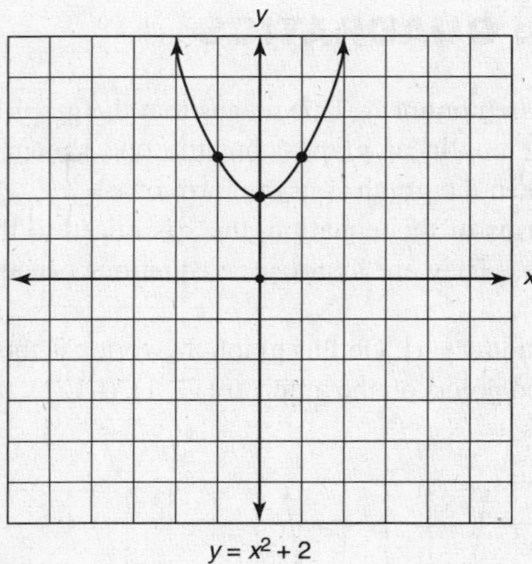

$y = x^2 + 2$

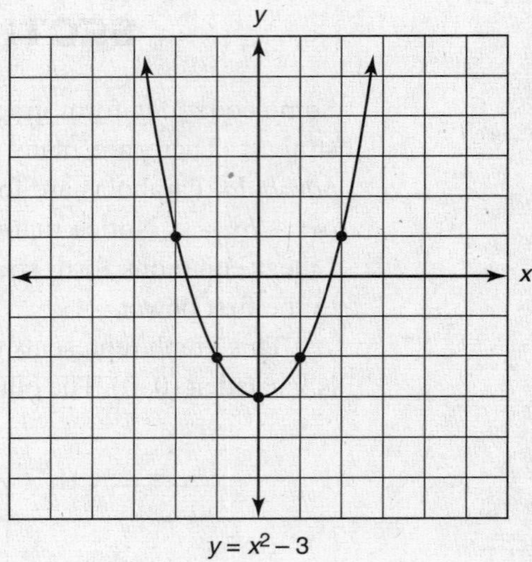

$y = x^2 - 3$

The x-axis is the line of symmetry for both graphs. It divides the parabola into two halves that are mirror images of each other.

The next graph is identical to $y = x^2$ but has been shifted two spaces to the left. The equation of this graph is $y = (x + 2)^2$. The vertex has also shifted two spaces left and is now located at (−2, 0).

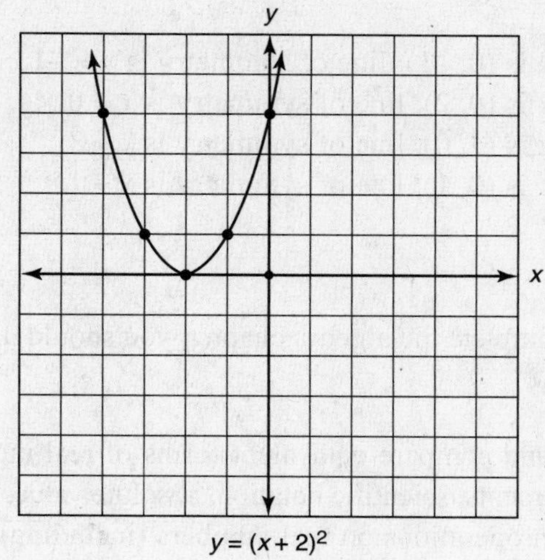

Sample Questions

Identify the vertex and line of symmetry on the graphs shown here.

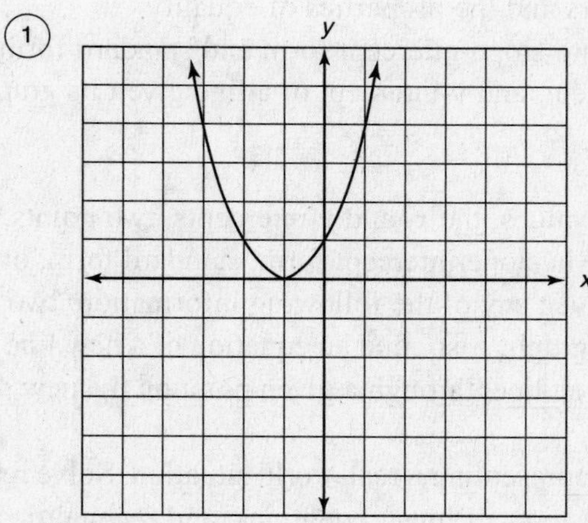

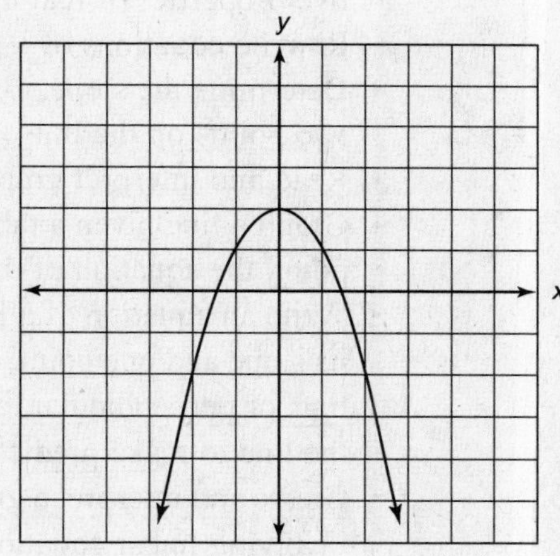

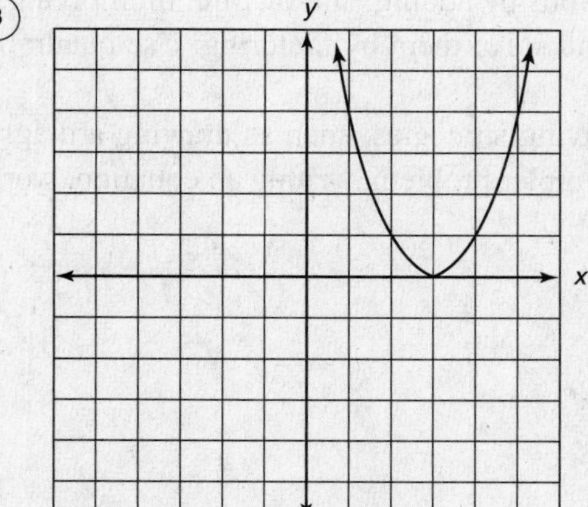

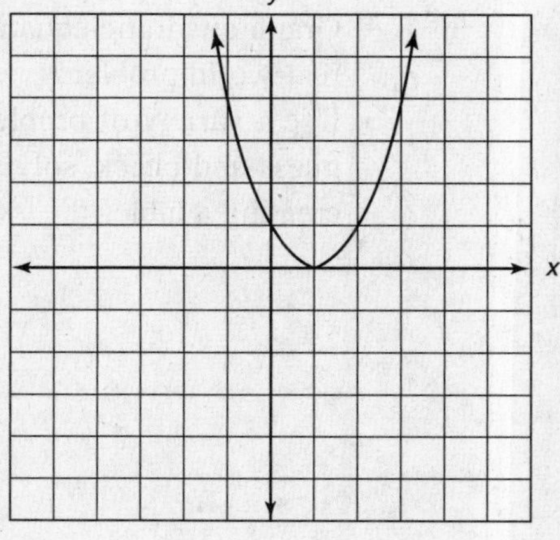

Answers

1. Vertex is (0, −1); line of symmetry is $x = -1$.
2. Vertex is (0, 2); line of symmetry is $x = 0$.
3. Vertex is (3, 0); line of symmetry is $x = 3$.
4. Vertex is (0, 1); line of symmetry is $x = 1$.

SUMMARY

As you complete the algebra chapter, you should understand or be able to do most of the following:

- Know and compare equivalent forms of real numbers (including integer exponents, radicals, percents, scientific notation, absolute value, rational numbers, and irrational numbers).
- Perform operations on real numbers (including integer exponents, radicals, percents, scientific notation, absolute value, rational numbers and irrational numbers) using multistep and real-world problems.
- Determine the domain and range of a relation.
- Simplify algebraic expressions and solve linear equations and inequalities by using order of operations and by identifying and applying the distributive, associative, and commutative properties of real numbers and the properties of equality.
- Rewrite equations of a line into slope-intercept form and standard form.
- Determine the slope, x-intercept, and y-intercept of a line given its graph, its equation, or two points on the line.
- Read and interpret graphs.
- Graph a line given a table of values, the x- and y-intercepts, two points, the slope and one point, the equation of the line in slope-intercept form, standard form, or point-slope form.
- Write an equation of a line given any of the following information: two points on the line, its slope and one point, or its graph. Also, find an equation of a new line parallel to a given line, or perpendicular to a given line, through a given point on the new line. Be able to use the line to make predictions.
- Create and interpret a graph representing a real-world situation. Solve real-world problems involving linear equations, systems of linear equations, and inequalities.
- Simplify algebraic rations and solve algebraic proportions.
- Simplify square root expressions by adding, subtracting, multiplying, and dividing.
- Graph quadratic equations and solve them by factoring. Use quadratic equations to solve real-world problems.
- Use a variety of problem-solving strategies, such as drawing a diagram, making a chart, guess-and-check, solving a simpler problem, writing an equation, working backwards, and creating a table.

PRACTICE PROBLEMS

1. Which of the following is equivalent to $3(2x - 4) + 6$?

   A. $6x - 12$
   B. $6x - 6$
   C. $6x + 6$
   D. $3x - 12$

   1. ________

2. A school fundraiser is selling $\frac{1}{4}$-pound bags of candy. Each bag will sell for \$.99. If 55 pounds of candy has been donated for the fundraiser, to the nearest whole dollar, how much money will be raised for the sale?

   2.

3. If $2a = b$, $2b = 12$, $3c = 4a$, and $d = c$, find the value of d. Show your work.

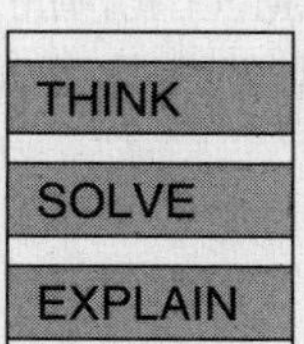

   3. ________

4. Simplify $4(x - 2) + 3(2 + x) - (x + 9)$. Show how you arrived at your answer.

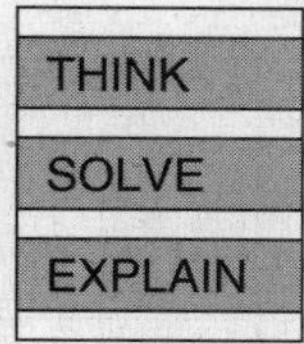

   4. ________

5. A toll booth in Orlando costs \$.50. If Harry goes though the toll gate twice a day, five times a week, how much will it cost him after 4 weeks?

   5.

6. Newton's second law of motion says that $F = Ma$, where F stands for force, M stands for motion, and a stands for acceleration. If a force of 250 newtons acts on a mass of 20 grams, what will the acceleration be?

6.

---

7. If Mary can paint a room in 3 hours and Nick can paint a room of the same size in $2\frac{1}{2}$ hours, approximately how long will it take them to paint the room if they work together?

F. 2 hours, 45 minutes
G. 1 hour, 45 minutes
H. $5\frac{1}{2}$ hours
I. 1 hour, 25 minutes

7. ________

---

8. The pressure exerted on the floor by a person's heel can be calculated using the formula $P = 1.2W/H^2$, where W stands for the weight of the person (in pounds) and H stands for the width of the heel, (in inches). A 110-pound woman wearing $\frac{1}{4}$-inch spiked heels exerts how much pressure per square inch on the floor?

A. 2112 pounds
B. 1760 pounds
C. 440 pounds
D. 44 pounds

8. ________

---

9. Simplify $3[4a + 2(7 - a)] - 22$. Show your work.

THINK

SOLVE

EXPLAIN

9. ________

---

10. Evaluate $(4b + 6)(4b - 6)$ when $b = -5$.

10.

---

11. Simplify $\frac{7c - 3}{3c} - 2$ expression when $c = 3$.

F. 3
G. 2
H. 1
I. 0

11. ________

---

12. The School Board budgets 84% of its entire budget for salaries. If their budget for the school year is \$60 million, how much is left for other costs?

A. \$9,600,000
B. \$50,040,000
C. \$50,400,000
D. \$96,000,000

12. ________

13. Frankie is 20 years younger than Joannie who is twice Amy's age. If Amy is 36, how old is Frankie?

F. 4
G. 16
H. 52
I. 72

13. ________

14. The Stock Exchange reported that Dow Jones Industrials on Monday were 195 points below Friday's close. On Tuesday, the Dow gained 188 points, and on Wednesday there was an additional gain of 169.59 points. What was the overall gain or loss over this time period? Use a negative number if there was a loss.

A. 552.59
B. 357.59
C. 162.59
D. −162.59

14. ________

15. General Electric Company said their third-quarter earnings rose 3% to \$3.28 billion. Which of the following could represent the earnings in the second quarter?

F. \$2.98 billion
G. \$3.25 billion
H. \$3.18 billion
I. \$3.35 billion

15. ________

16. Simplify $4\sqrt{3} + 5\sqrt{3}$.

A. $20\sqrt{3}$
B. $14\sqrt{3}$
C. $9\sqrt{3}$
D. $9\sqrt{9}$

16. ________

17. Nancy is participating in a Red Cross walkathon that takes place over a week. On Monday she walks 15 miles, and on Tuesday she walks $\frac{1}{3}$ as far. On Wednesday, Thursday, and Friday, she walks twice as far each day as she did on Tuesday. How many miles did she walk in all?

17.

	⊘	⊘	⊘	
⊙	⊙	⊙	⊙	⊙
0	0	0	0	0
1	1	1	1	1
2	2	2	2	2
3	3	3	3	3
4	4	4	4	4
5	5	5	5	5
6	6	6	6	6
7	7	7	7	7
8	8	8	8	8
9	9	9	9	9

18. Look at the accompanying table showing U.S. trade with selected countries and select the statement that is the most accurate.

Country	Total Trade With U.S. (millions of dollars)
Canada	365,311.1
Mexico	196,629.5
Japan	188,329.6
China	94,899.2
Germany	82,029.6

F. Canada's trade is twice that of Mexico.
G. Germany's trade is one-fourth that of Canada.
H. China's trade is one-fourth that of Canada.
I. Japan's trade is twice that of China.

18. ________

19. A jiffy is an actual time unit of $\frac{1}{100}$ second. Which of the following could *not* be a jiffy?

A. 1×10^{-1} second
B. $\sqrt{.0001}$ second
C. .01 second
D. 10^{-2} second

19. ______

20. The distance between bases on a baseball diamond is 90 feet. Dan used the Pythagorean Theorem and found the distance between first and third bases could be expressed as $\sqrt{16,200}$. This number $\sqrt{16,200}$ is between what two integers?

F. 126 and 127 feet
G. 127.2 and 127.3 feet
H. 127 and 128 feet
I. 8100 feet

20. ______

21. The following are some common wrench measurements. Arrange them in ascending (smallest to largest) order.

A. $\frac{3}{8}$ $\frac{5}{8}$ $\frac{3}{16}$ $\frac{1}{4}$
B. $\frac{5}{8}$ $\frac{3}{8}$ $\frac{1}{4}$ $\frac{3}{16}$
C. $\frac{3}{16}$ $\frac{1}{4}$ $\frac{3}{8}$ $\frac{5}{8}$
D. $\frac{5}{8}$ $\frac{3}{16}$ $\frac{1}{4}$ $\frac{3}{8}$

21. ______

22. Which of the following numbers is *not* irrational?

F. $.\overline{3}$
G. $\sqrt{122}$
H. π
I. .121122111 . . .

22. ______

23. Which of the following is arranged in descending (smallest to largest) order?

A. $2\sqrt{3},\ \sqrt{10},\ \pi,\ \frac{23}{7},\ |-3.5|$
B. $|-3.5|,\ \pi,\ \sqrt{10},\ \frac{23}{7},\ 2\sqrt{3}$
C. $\pi,\ \sqrt{10},\ \frac{23}{7},\ 2\sqrt{3},\ |-3.5|$
D. $\sqrt{10},\ \pi,\ \frac{23}{7},\ 2\sqrt{3},\ |-3.5|$

23. ______

24. The distance from the earth to the sun is approximately 93,000,000 miles. What is this number expressed in scientific notation?

F. $.93 \times 10^{8}$
G. 9×10^{6}
H. 9.3×10^{6}
I. 9.3×10^{7}

24. ______

25. An erg is a unit of energy. It is equal to 0.0000000239 Calories. What is this number expressed in scientific notation?

A. 2.39×10^{-8}
B. 2.39×10^{8}
C. 2.39×10^{-7}
D. $.239 \times 10^{-10}$

25. ______

26. The number 4.56×10^{-3} lies between which of these two integers?

F. −5 and −4
G. −1 and 0
H. −3 and −2
I. 0 and 1

26. ______

27. In the diagram of three towns shown here, the estimated distance from Freeport to Hollywood Beach could be which one of the following?

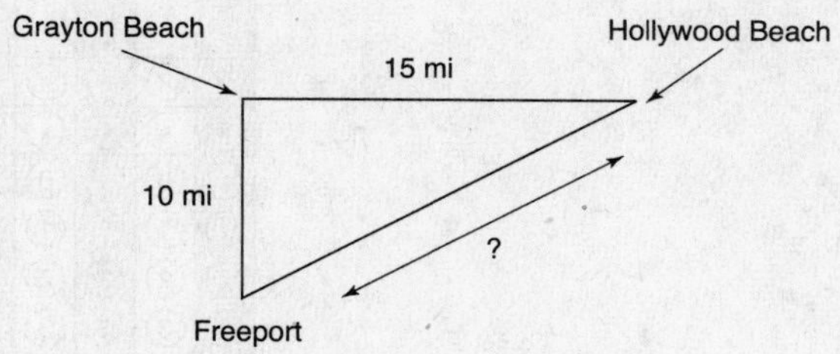

A. 14.9
B. $\sqrt{325}$
C. $\frac{52}{4}$
D. $|-15|$

27. ________

28. A 96-mile trip from Port Charlotte to Tampa took 1 hour, 47 minutes, and 30 seconds. Estimate the average number of miles per hour traveled for this trip.

THINK

SOLVE

EXPLAIN

28. ________

29. What is the value of x when y is 33?

x	y
41	17
47	21
53	25
59	29
	33

29.

	⊘	⊘	⊘	
•	•	•	•	•
0	0	0	0	0
1	1	1	1	1
2	2	2	2	2
3	3	3	3	3
4	4	4	4	4
5	5	5	5	5
6	6	6	6	6
7	7	7	7	7
8	8	8	8	8
9	9	9	9	9

30. To get from 10^{-4} to 10^5, how many times would you have to multiply by 10?

F. 1
G. 5
H. 9
I. 10

30. ________

31. A spider web extended from point A to point B as shown here. The distance from point A to point B was determined to be $5\sqrt{55}$. Which of the following is closest to this number?

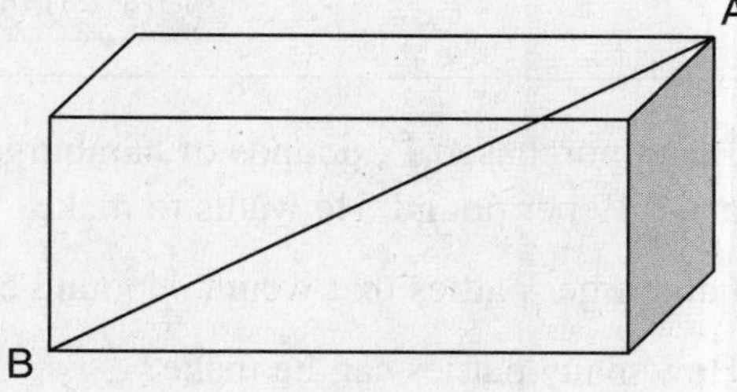

A. $37\frac{2}{25}$
B. $37\frac{4}{5}$
C. $37\frac{1}{50}$
D. $37\frac{1}{5}$

31. ________

32. The decimal .005 is equivalent to what percent?

F. 50%
G. 5%
H. .5%
I. .05%

32. ________

33. Write 4.1×10^{-2} in standard notation.

33.

34. Justin purchased 12 pounds of hamburger at \$1.49 per pound. He wants to make hamburger patties that weigh $\frac{1}{4}$ pound each. How many patties can he make?

34.

35. Select the appropriate algebraic expression from the verbal phrase 17 less than the product of 8 and c.

A. $17 - 8c$
B. $8c - 17$
C. $8 + c - 17$
D. $17 - (8 + c)$

35. ________

36. A bacterial culture starts out with 20 cells and doubles every hour. At the *end* of the 9th hour, how many bacteria are in the culture?

36.

37. If you receive a base salary of \$97.50 plus \$7.50 for x hours, how many hours will you have to work before you earn \$255.00?

F. 10
G. 15
H. 21
I. 33

37. ________

38. Shannon works as a real estate agent. She makes a 3.5% commission on all sales. If she received \$1250 in one month, to the nearest dollar, what was the amount of her sales for that month?

38.

39. A CD can be purchased for C dollars. Which of the following is an expression for the number of CDs you can purchase for $200?

A. $\frac{200}{C}$

B. $\frac{C}{200}$

C. $200C$

D. $200 - C$

39. ________

40. Betty bought a 2-carat gemstone for $379 per carat last year. This year gemstones cost 20% more. To the nearest $100, how much is her gemstone worth this year?

40.

41. A box of 30 chocolate peanut clusters costs $2.50. If the manufacturer increases the number of peanut clusters in a new box to 40, to the nearest half-dollar, how much should he charge?

41.

42. Tony works part-time at a restaurant as a busboy. He is paid an hourly wage of $5.65 for 20 hours per week. If he takes a job as a waiter, he will lose his hourly wage but will make tips. If he earns $200 per week in tips, how much more will he earn as a waiter?

42.

43. Which of the following is the solution to the equation $20 = -2y + 4$?

F. 20
G. 8
H. −8
I. −12

43. ________

44. Which of the following represents the solution to the inequality $-2y + 4 \geq 20$?

A. $y = -8$
B. $y \geq -8$
C. $y \leq -8$
D. $y \geq -12$

44. ________

45. Which algebraic expression represents the total number of yards in *y* yards and *f* feet?

F. $y + \frac{3}{f}$
G. $y + \frac{f}{3}$
H. $3y$
I. $\frac{y}{f}$

45. ________

46. Solve the equation $4(x - 2) = 3(x + 4)$ showing all steps.

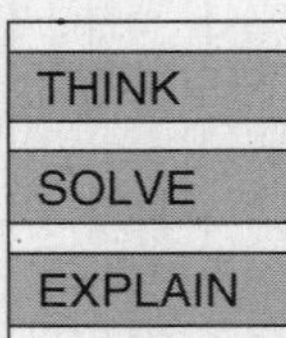

46. ________

47. According to the American Kennel Club, Shetland sheepdogs must be from 13 to 16 inches high at the shoulder. Using *h* for height, which of the following is an algebraic expression representing this restriction?

A. $13 < h < 16$
B. $13 > h > 16$
C. $13 \leq h \leq 16$
D. $h \leq 13$ and $h \geq 16$

47. ________

48. Find the solution to the system of equations by graphing on the same set of axes. Locate the solution on the graph and state as an ordered pair.

$y = x - 2$
$2y + 2x = 8$

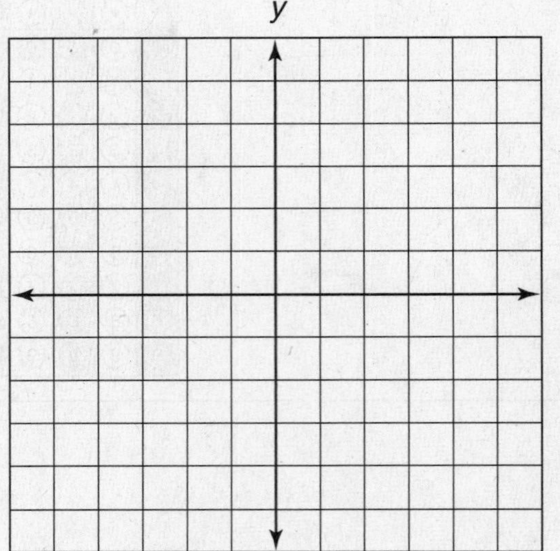

THINK
SOLVE
EXPLAIN

49. Solve the system of equations using the substitution method. Show all steps.

$x + y = 4$
$3x + 4y = 15$

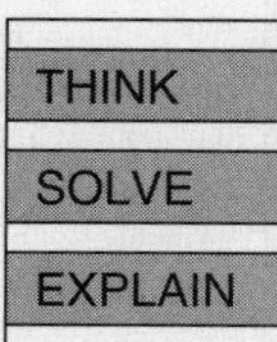

49. ______

50. Solve the system of equations using the linear combination method. Show all steps.

$-4x + 17 = 5y$
$2x + 14 = 2y$

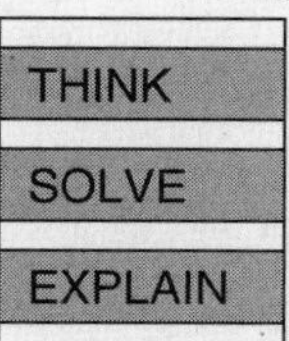

50. ______

51. Determine which area of the graph satisfies the inequality statements $y - x \leq -1$ and $x + y \leq 4$.

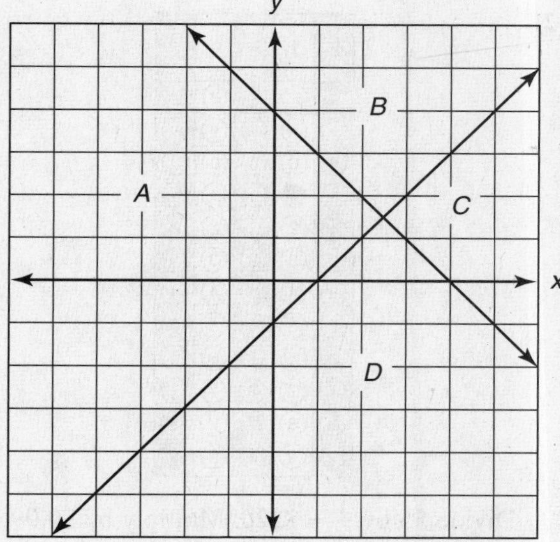

F. *A*
G. *B*
H. *C*
I. *D*

51. ______

Answers Explained

1. **B.** $3(2x - 4) + 6 = 6x - 12 + 6 = 6x - 6$

2.

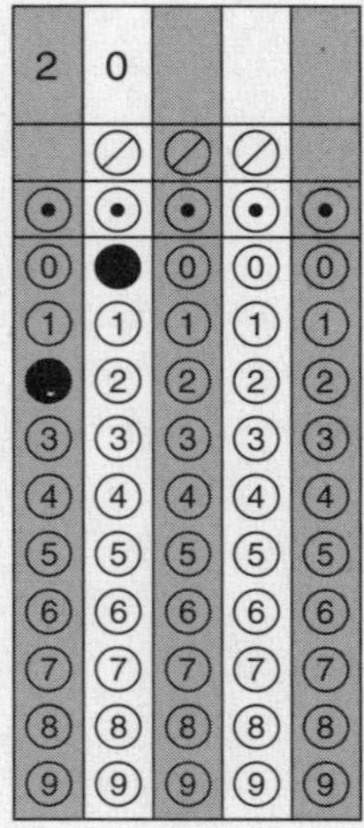

Divide 55 by $\frac{1}{4}$ = \$220. Multiply by \$0.99.
You can use 0.25 for $\frac{1}{4}$.

3. $2b = 12$, so $b = 6$.
$2a = b$ and $b = 6$, so $a = 3$, and $4a = 12$.
$4a = 3c$, $3c = 12$, and $c = 4$.
$d = c$, so if $c = 4$, then $d = 4$.

4. $4(x - 2) + 3(2 + x) - (x + 9)$
$4x - 8 + 6 + 3x - x - 9$
$4x + 3x - x - 8 + 6 - 9$
$6x - 11$

5.

$\$.50 \cdot 2 \cdot 5 \cdot 4 = \$20$

6.

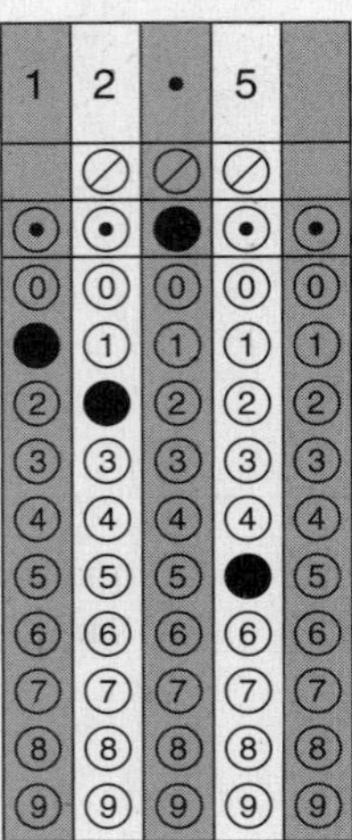

$F = Ma$	Substitute given values into the formula.
$250 = 20a$	Divide by 20; $F = 12.5$

7. **I.** Since this is only an estimate, you can assume Mary will paint half the room in 1 hour, 30 minutes and Nick will paint the other half in 1 hour, 15 minutes. Working together, they will finish in about 1 hour, 25 min.

8. **A.** Use 0.25 for $\frac{1}{4}$.

$$P = \frac{1.2W}{H^2} = \frac{1.2 \cdot 110}{0.25^2} = \frac{132}{0.0625} = 2112$$

9. $6a + 20$
Work from the inside out when you have brackets and parentheses.

$3[4a + 2(7 - a)] - 22$	Distribute the 2.
$3[4a + 14 - 2a] - 22$	Add like terms inside the brackets.
$3[2a + 14] - 22$	Distribute the 3.
$6a + 42 - 22$	Add like terms.
$6a + 20$	

10.

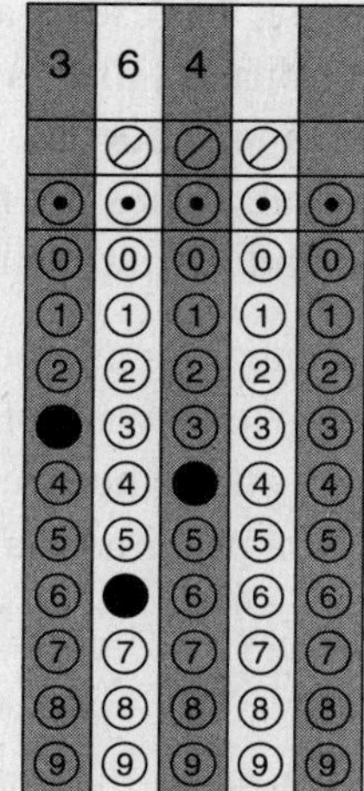

$(4b + 6)(4b - 6)$

Substitute −5 in the expression.

$[4(-5) + 6]\ [4(-5) - 6]$

Perform all inside multiplications.

$(-20 + 6)(-20 - 6)$

Do all the addition inside the parentheses.

$-14 \bullet -26 = 364$ Multiply.

11. I.

$$\frac{7(3) - 3}{3(3)} - 2 = \frac{21 - 3}{9} - 2 = \frac{18}{9} - 2 = 2 - 2 = 0$$

12. A. Multiply: .84 × \$60 million = 50.4 million. \$60.0 million − 50.4 million = 9.6 million.

13. H. Work backward. Joannie is twice Amy's ages, or 72. Subtract 20 for Frankie's age: 72 − 20 = 52.

14. C. Add: −195 + 188 + 169.59 = 162.59.

15. H. One way to do this problem is to add 3% to each answer until you get \$3.28. Another method is to set up an equation: 103% of earnings = 3.28 billion (English version) or $1.03e$ = \$3.28 (algebraic equation). Divide \$3.28 by 1.03 to solve.

16. C. $4\sqrt{3} + 5\sqrt{3} = 9\sqrt{3}$. Add 4 + 5 and keep $\sqrt{3}$.

17.

5 0

Monday: 15 miles

Tuesday: 5 miles $\quad \frac{1}{3}$ of $15 = \frac{1}{3} \times 15 = 5$

Wednesday: 10 miles $\quad 5 \times 2 = 10$

Thursday: 10 miles $\quad 5 \times 2 = 10$

Friday: 10 miles $\quad 5 \times 2 = 10$

Total: 50 miles

18. I. 2 × 94,899.2 = 189,798.4. In **F**, 2 × 196,629.5 = 393,259. In **G** and **H**, $\frac{1}{4} \times 365311.1 = 364311.1 \div 4 = 91,327.8$.

19. A. In **B**, $\sqrt{.0001} = .01$, in **C** $.01 = \frac{1}{100}$, and in **D**, $10^{-2} = \frac{1}{100}$ or .01.

20. H. You do not need to know the Pythagorean Theorem to do this particular problem. **G** most closely matches the exact answer, but 127.2 and 127.3 are not integers. **I** might seem attractive if you divide by 2 instead of taking the square root, but your estimation skills should come into play here as you realize that this answer is much too large to be correct. The correct answer is **H** because $\sqrt{16,200} \approx 127.28$. (The symbol $\approx$ means "approximately.") 127.28 is slightly more than 127 but is still less than 128.

Note: Multiple-choice answers can be deceiving. Just because the answer you arrive at is one of the choices does not necessarily mean that it's the correct answer. Test writers know about potential mistakes and include tempting (but wrong) answers called *distracters* because they distract you from the real answer.

Test Tip: Eliminate answers that do not make sense. Choice **I**, 8100 feet, could have been eliminated as a possible answer because of the distance involved. 8100 feet is more than a mile, which does not make sense in terms of a baseball diamond. Be certain that you know the correct names of numbers such as whole numbers and integers.

21. C. One way to arrange these numbers is to change them all to fractions with the same denominator (Section 2). Using this technique, $\frac{5}{8} = \frac{10}{16}$ and $\frac{1}{4} = \frac{4}{16}$. Now, comparing only the numerators, $\frac{3}{16}$ is lowest, next comes $\frac{4}{16}\left(\frac{1}{4}\right)$, then $\frac{3}{8}\left(\frac{6}{16}\right)$, and then $\frac{5}{8}\left(\frac{10}{16}\right)$.

Test Tip: If you are having trouble following the technique above, try using your calculator to change each fraction to a decimal by dividing the numerator by the denominator and then compare the decimals. $\frac{3}{8} = 0.375$, $\frac{1}{4} = 0.25$, $\frac{5}{8} = 0.625$, and $\frac{3}{16} = 0.1875$.

22. F. This number represents a repeating decimal and is rational. **G** also is nonterminating, nonrepeating, and irrational. **H** is pi, which is always irrational. **I** does not repeat in the same pattern and is irrational.

23. C. When comparing numbers, try to change all the numbers to the same form. In this case, changing all the numbers to a decimal equivalent (or approximation) is the easiest way: $\pi \approx 3.143$; $\sqrt{10} \approx 3.16$; $\frac{23}{7} \approx 3.29$; $2\sqrt{3} \approx 3.46$; $|-3.5| = 3.5$.

24. I. Numbers in scientific notation must have a decimal point after the first nonzero digit. This eliminates **F.** If you selected **H**, you should know that a common error is to count the number of zeros in the standard version of the number and use that number as the exponent. Be sure to count the decimal places instead.

25. A. A number expressed in scientific notation must have a decimal point *after* the first significant digit and be multiplied by a power of 10. You can find the first significant digit by moving from left to right until you come to a nonzero number. The 2 is the first significant digit. This rule eliminates **D.** Standard numbers less than 1 always have a negative exponent. **C** incorrectly moves the decimal point in a positive direction, making the number larger than 1 in its expanded form.

26. I. $4.56 \times 10^{-3} = .00456$, which is less than 1 but more than 0.

27. B. Although knowing the Pythagorean Theorem would be helpful here, it's not really necessary. Change all the answers to the same form. Notice that the distance from Freeport to Hollywood Beach is farther than the distance between the other two cities. **B** is the only answer that provides a reasonable distance.

28. Estimation is a form of rounding. The main difference between estimating and rounding is that when you estimate, you round *before* you begin calculations. The more you round, however, the more approximate your answer.

In this problem if you rounded the mileage to 100 miles and the time to 2 hours, you would have an average speed of 50 miles per hour (miles divided by hours). However, if you require a closer estimate, you might round mileage to 95 miles and time to 1 hour, 45 minutes. To calculate this divide 95 miles by 1.75 hours.

95 divided by 1.75 is approximately 54 miles per hour. This is a much more accurate estimate than 100 miles divided by 2 hours and demonstrates your knowledge much more effectively to the FCAT scorer. Remember that think-solve-explain problems are worth 2 or 4 points.

Note: If you thought 1 hour, 45 minutes should be 1.45 hours, read the following: You should express 45 minutes as a part of an hour. Since there are 60 minutes in an hour, the fractional part of the hour is represented as $\frac{45}{60}$ or .75. So 1 hour, 45 minutes is then expressed as 1.75 hours.

29.

6	5			

Each number under x is increasing by 6 each time.

30. H. To get from 10^{-4} to 10^5 you have to multiply by 10 nine times. Each time you multiply by 10 you move in a positive direction toward 10^5. To see this more clearly, start on a number line at –4 and move toward the 5 counting every time you move.

31. A. $5 \times \sqrt{55} \approx 37.08$, and $.08 = \frac{8}{100} = \frac{2}{25}$.

Be sure to take the square root first before multiplying on your FCAT calculator.

32. H. To change a decimal to a percent, move the decimal to the right two places and add the percent sign.

33.

•	0	4	1	

The –2 power indicates that you are to move the decimal point two places to the left.

34.

4	8			

Justin can make $12 \div \frac{1}{4} = \frac{12}{1} \bullet \frac{4}{1} = 48$.

The cost per pound is not needed to solve the problem.

35. B. The product of 8 and b is $8b$. 7 less than $8b = 8b - 7$.

36.

1 0 2 4 0

Hour	No. of Bacteria
0	20
1	40
2	80
3	160
4	320
5	640
6	1,280
7	2,560
8	5,120
9	10,240

20×2^9 bacteria.

37. H. $7.50x + 97.50 = 255$
$(255 - 97.50) \div 7.50$.

38.

3 5 7 1 4

$1250 = .035x$. Divide by .035.

39. A.

40.

9 0 0

Multiply: $2 \times 379 \times 1.2 = \$909.60$. Round to the nearest hundred.

41.

3 . 5 0

You can solve with a proportion: $\frac{30}{40} = \frac{2.50}{x}$;

$30x = 100$; $x = \$3.33$. Round to the nearest \$.50.

42.

8 7

$\$200 - 5.65 \times 20. = \$200 - \$113 = \$87$.

43. H. Subtract 4 from both sides and divide by -2.

44. C. Subtract 4 from both sides and divide by -2. Remember to flip the inequality for this problem.

45. G. For example, 6 yards, 2 feet = $6\frac{2}{3}$ yards, and 4 yards, 1 foot = $4\frac{1}{3}$ yards.

46.

$4(x-2) = 3(x+4)$	Distribute the 4 and the 3.
$4x - 8 = 3x + 12$	Subtract $3x$ from both sides.
$4x - 3x = 12 + 8$	Add 8 to both sides.
$x = 20$	

47. C. The statement allows for the dogs to be exactly 13 inches and 16 inches in addition to any height in between.

48. For ease in graphing, change $2y + 2x = 8$ to the slope-intercept form:

$2y + 2x = 8$
$2y = -2x + 8$
$y = -x + 4$

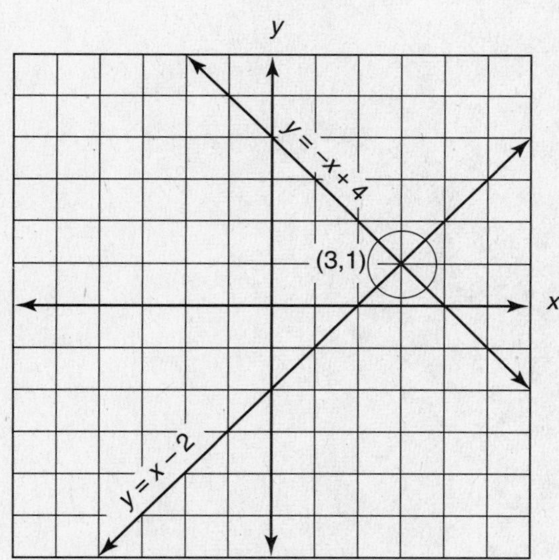

Solution

49. (1, 3). Substitute $x = 4 - y$ for x in $3x + 4y = 15$.

$3(4 - y) + 4y = 15$
$12 - 3y + 4y = 15$
$12 + y = 15$
$y = 3$ Substitute into $x + y = 4$, and $x = 1$.

50. (−2, 5). Multiply the second equation by 2. Then, add the equations.

$$\begin{aligned} -4x + 17 &= 5y \\ 4x + 28 &= 4y \\ \hline 45 &= 9y \end{aligned}$$

Substitute 5 for y in one of the original equations:

$2x + 14 = 2(5)$
$2x + 14 = 10$
$2x = -4$
$x = -2$

51. I.

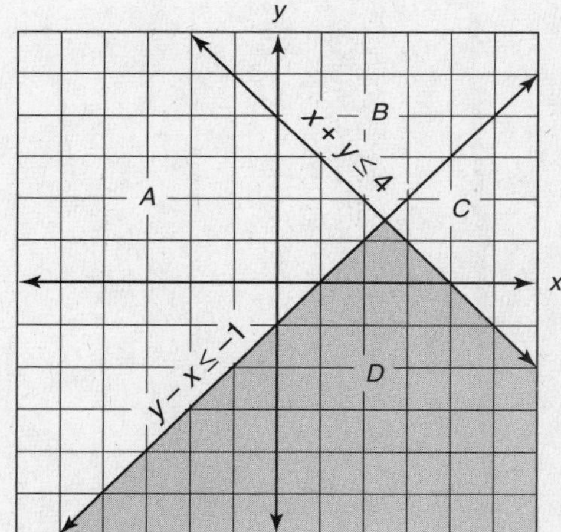

Chapter 2 Geometry

KEY TERMS/CONCEPTS

- Points, lines, angles, and planes
- Parallel and perpendicular lines
- Triangles
- Quadrilaterals
- Changing and moving shapes
- Proportion, scale, and coordinate geometry
- Perimeter, area, surface area, and volume

SECTION 1: POINTS, LINES, ANGLES, AND PLANES

Geometry is built on a foundation of the *point*, the *line*, and the *plane*. In order to understand other terms used in geometry, it is important to understand these three basic terms. Each term will be defined as we come to it in this section.

POINTS

A *point* is a location in space. Because it is simply a location, you cannot actually see it. We generally use a dot (.) and a capital letter (for example, *P*) to represent the place where a point is located. A point has no dimensions; that is, it has no length, no width, and no height. To identify its location we use *coordinates*, such as (5, 6) to indicate where it can be found.

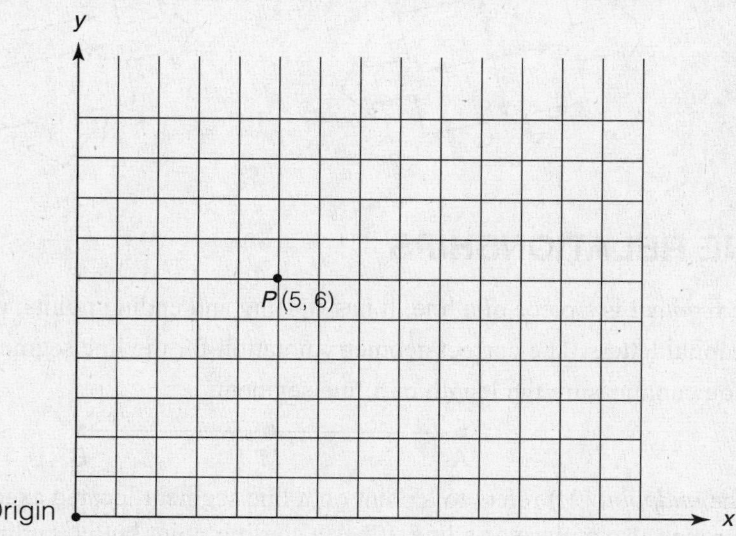

P (5, 6) means 5 spaces to the right of the origin and 6 spaces up.

Collinear points lie on the same line. *Coplanar points* lie on the same plane.

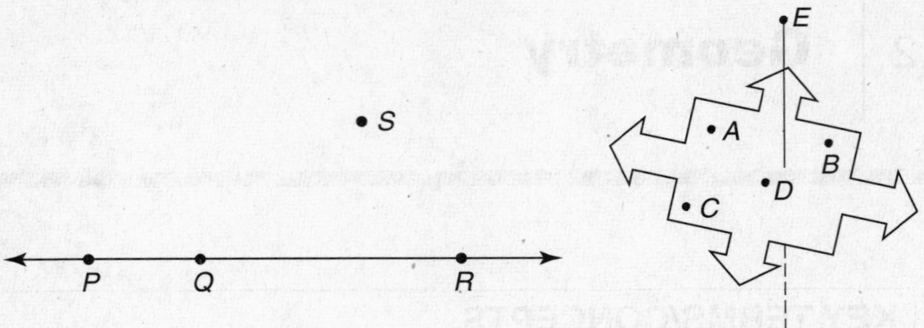

Points P, Q, and R are collinear. Point S is not collinear.

Points A, B, C, and D are coplanar. Point E is not coplanar.

LINES

A *line* is defined by a minimum of two points. A line has no beginning and no end; thus, when we draw a line, we use a double-headed arrow. We can label at least two points on the line, or the entire line can be labeled by a small letter, for example, *l*. The length of a line cannot be measured because it continues infinitely in two directions. The correct geometry notation for the line shown below is $\overleftrightarrow{AB}$ or $\overleftrightarrow{BA}$.

PLANES

A *plane* is one of the three foundations of geometry mentioned at the beginning of this section. Planes are determined by a line and a point not on the line. Recall that a line extends infinitely in two directions. A plane extends infinitely in all directions. It actually does not have boundaries, but so that we can use it and identify it, we draw boundaries to help us visualize it. Planes can be determined by intersecting or parallel lines.

It helps to think of a plane as a flat surface.

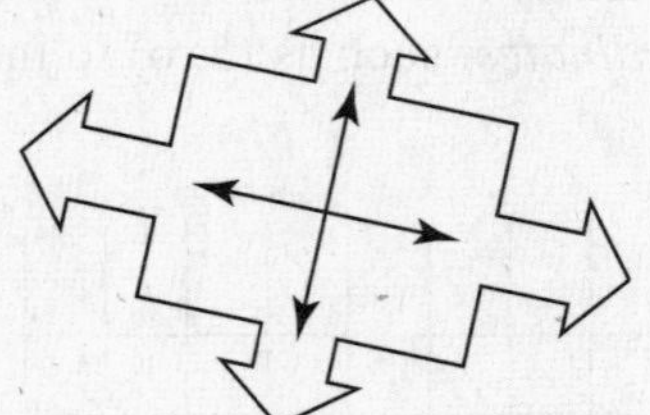

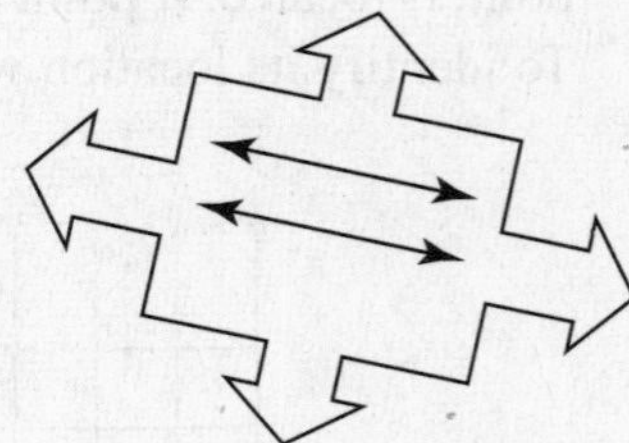

LINES AND LINE RELATIONSHIPS

A *line segment* is a piece of a line. It has starting and ending points, which should be labeled with capital letters. The correct geometry notation for the line segment shown here is $\overleftrightarrow{AB}$ or $\overleftrightarrow{BA}$. You can measure the length of a line segment.

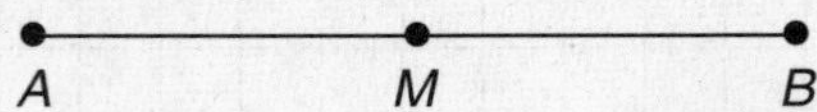

The *midpoint* (M) refers to a point on a line segment located exactly in the middle.

A *ray* is also a piece of a line. It has a starting point but no ending point. As with a line, you cannot measure the length of a ray because although it does have a starting point (A), it has no ending point. The correct geometry notation for both the rays shown here is $\overrightarrow{AB}$. *A* represents the *starting* point for the ray, regardless of the direction in which the ray is pointing.

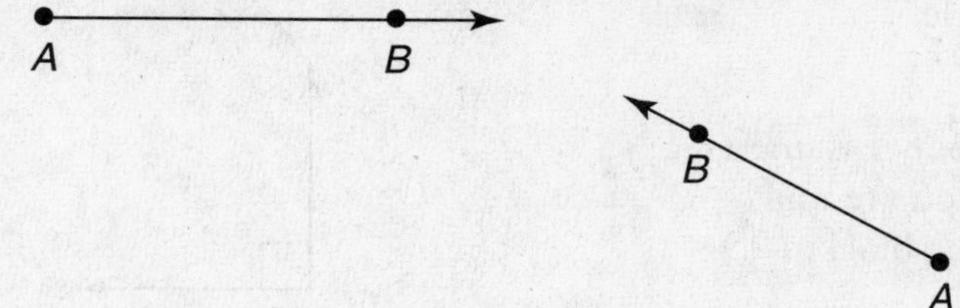

Line Relationships		
Diagram	**Relationship**	**Definition**
	Parallel	Lines in the same plane that do not intersect: $\overleftrightarrow{AB} \parallel \overleftrightarrow{CD}$
	Perpendicular	Lines that intersect at a 90° angle: $AB \perp \overleftrightarrow{CD}$
	Intersecting	Lines that meet at a point
	Skew	Lines in different planes that do not intersect
	Concurrent	Three or more lines that meet at the same point

ANGLES

Angles are defined as two rays that are joined at their endpoints. This shared endpoint is called the *vertex*, and the rays form the sides of the angle. Angles can be named in three ways: using three letters with the central letter at the vertex, using one letter at the vertex, or using a number or symbol located inside the angle.

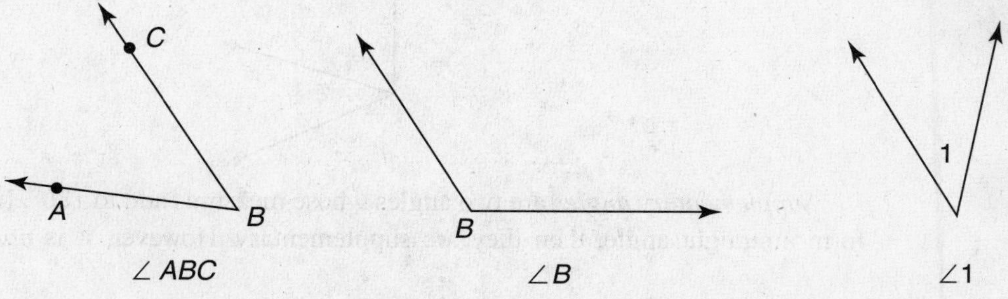

Classification of Angles

Right angles measure exactly 90°. They are generally denoted by a small block located in the angle.

> ***Test Tip:*** You can only assume an angle is 90° if you are told it is or if you see the block.

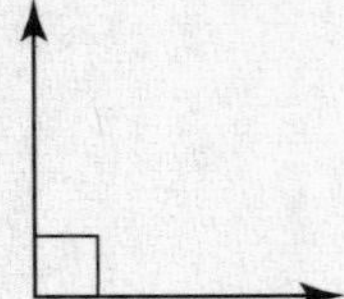

Acute angles have measurements of less than 90°.

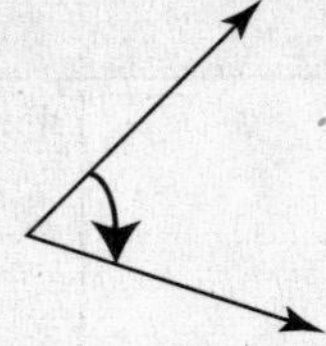

Obtuse angles have measurements of more than 90° but less than 180°.

> A line and a straight angle are the same thing.

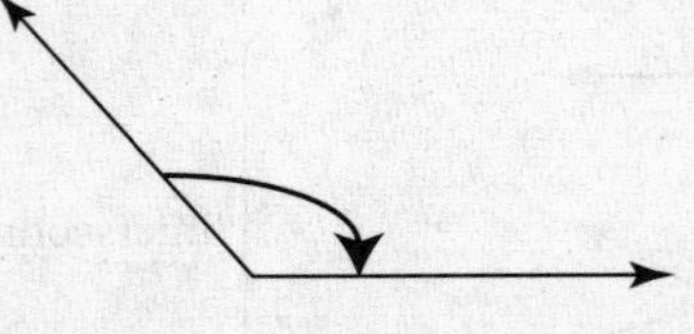

Straight angles measure exactly 180°.

Angle Relationships

Congruent angles are angles with the same measure. If $\angle ABC$ is congruent to $\angle CDE$, we use the notation $\angle ABC \cong \angle CDE$. The symbol $\cong$ is used to establish that two figures have the same shape and measure. It is not necessary, however, that the two figures be in the same position. In the diagram here, $\angle B \cong \angle C$.

Adjacent angles are any two nonoverlapping angles that share a vertex and a side. In this diagram, angles 1 and 2 are adjacent.

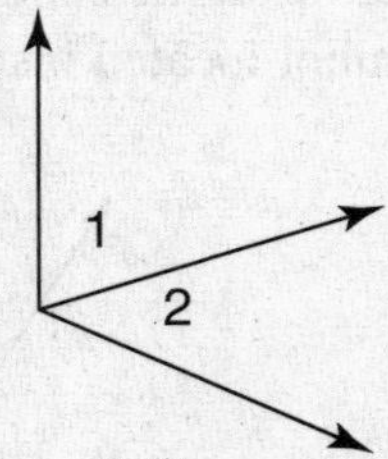

Supplementary angles are two angles whose measures add to 180°. If two adjacent angles form a straight angle, then they are supplementary. However, it is *not* necessary that two angles be adjacent, only that the sum of their measures equal 180°. The angles in the diagram here are supplementary.

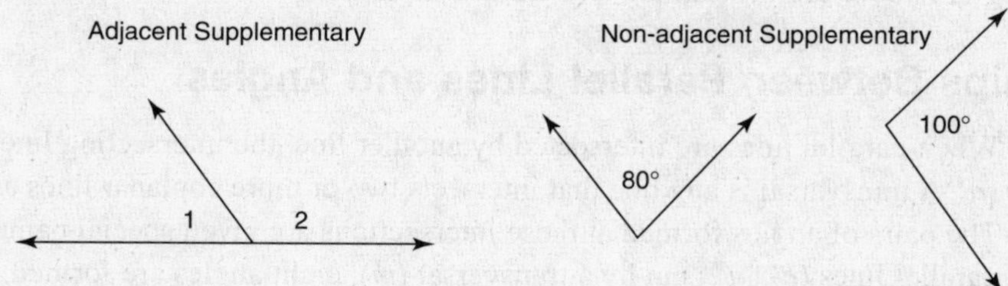

Complementary angles are two angles whose measures add to 90°. If two adjacent angles form a 90° angle, then they are complementary. As with supplementary angles, it is not necessary that two angles be adjacent, only that the sum of their measures be 90°. The angles in the diagram shown here are supplementary.

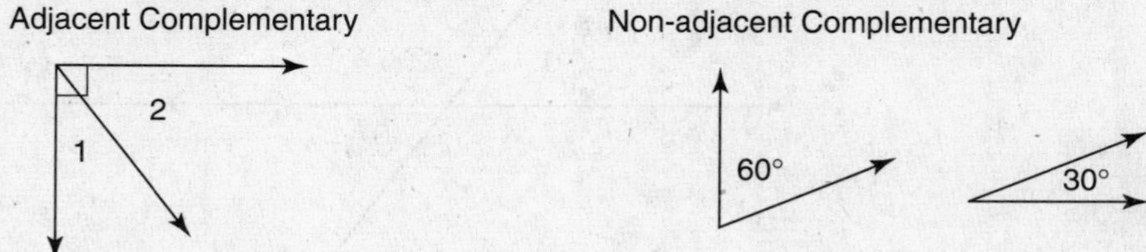

Angle bisectors divide an angle into two equal parts. In the diagram here, $\overrightarrow{QR}$ is the angle bisector of $\angle AQB$. This means that $\angle AQR$ is congruent to (≅) $\angle RQB$.

In geometry, ≅ means congruent or of equal measure.

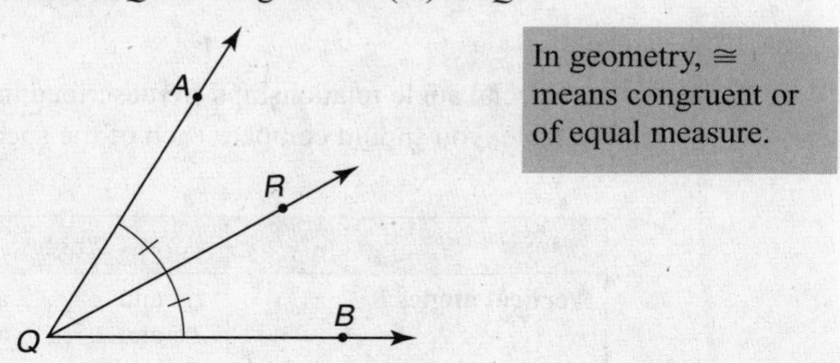

Relationships Between Intersecting Lines and Angles

When two lines intersect, the angles formed have special relationships.

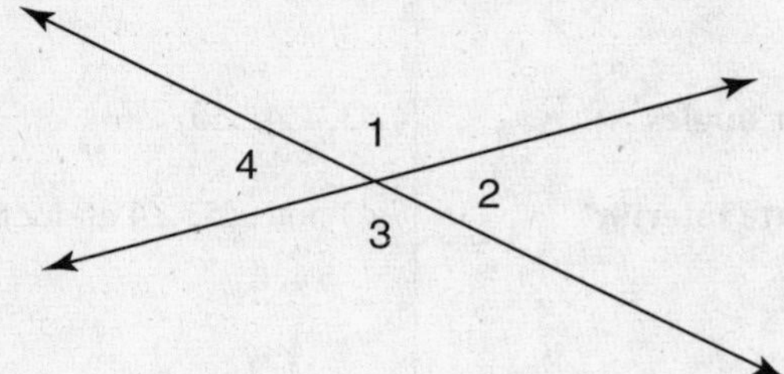

Adjacent supplementary angles share a side and a vertex. The sum of their measures equals 180°. The adjacent supplementary angle pairs in the diagram shown are $\angle 1$ and $\angle 2$, $\angle 2$ and $\angle 3$, $\angle 3$ and $\angle 4$, and $\angle 4$ and $\angle 1$.

Vertical angles are also formed when two lines intersect. Vertical angles are opposite pairs of congruent angles. They are nonadjacent and share a vertex. The vertical angles in the diagram above are $\angle 1$ and $\angle 3$ and $\angle 2$ and $\angle 4$.

Relationships Between Parallel Lines and Angles

When parallel lines are intersected by another line, the intersecting line is called a *transversal*. A transversal is any line that intersects two or more coplanar lines at two or more points. The pairs of angles formed at these intersections are given special names. In the case of two parallel lines ($\ell_1 \parallel \ell_1$) cut by a transversal (m), eight angles are formed. Notice the additional arrows between the ends on the parallel lines shown here. When you see these arrows, you can rely on them as a signal the lines are parallel.

Line *m* is the transversal.

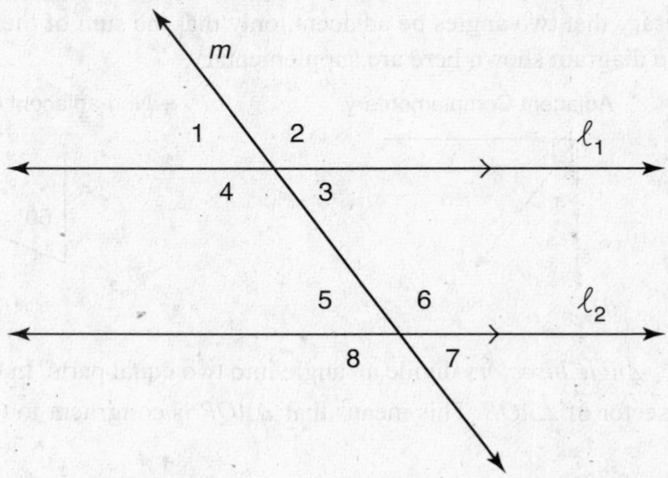

Special angle relationships are described in the accompanying table. As you read through this table, you should compare each of the special angle relationships to the diagram above.

Name	Angles	Description
Vertical angles	$\angle 1$ and $\angle 3$, $\angle 2$ and $\angle 4$, $\angle 5$ and $\angle 7$, $\angle 6$ and $\angle 8$	Congruent pairs. Pairs share a vertex and are nonadjacent.
Corresponding angles	$\angle 1$ and $\angle 5$, $\angle 2$ and $\angle 6$, $\angle 4$ and $\angle 8$, $\angle 3$ and $\angle 7$	Congruent pairs. Pairs are located on same side of the transversal and are in the same position above or below their respective lines (ℓ_1 and ℓ_2).
Interior angles	$\angle 3$, $\angle 4$, $\angle 5$, $\angle 6$	Located on the inside of parallel lines.
Alternate interior angles	$\angle 3$ and $\angle 5$, $\angle 4$ and $\angle 6$	Congruent pairs. Pairs are interior angles on opposite sides of the transversal.
Exterior angles	$\angle 1$, $\angle 2$, $\angle 7$, $\angle 8$	Located on the outside of parallel lines.
Alternate exterior angles	$\angle 1$ and $\angle 7$, $\angle 2$ and $\angle 8$	Congruent pairs. Pairs are exterior angles on opposite sides of the transversal.

Sample Questions

1. $\overrightarrow{BD}$ is the angle bisector for $\angle ABC$. If m$\angle ABC = 150°$, find m$\angle DBC$.

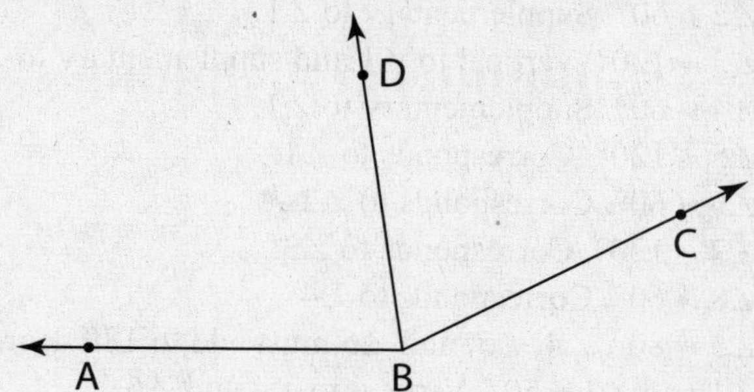

2. In this diagram two parallel lines are cut by a transversal. m$\angle 1 = 120°$. Find the measures of all the other angles. Give a reason for each.

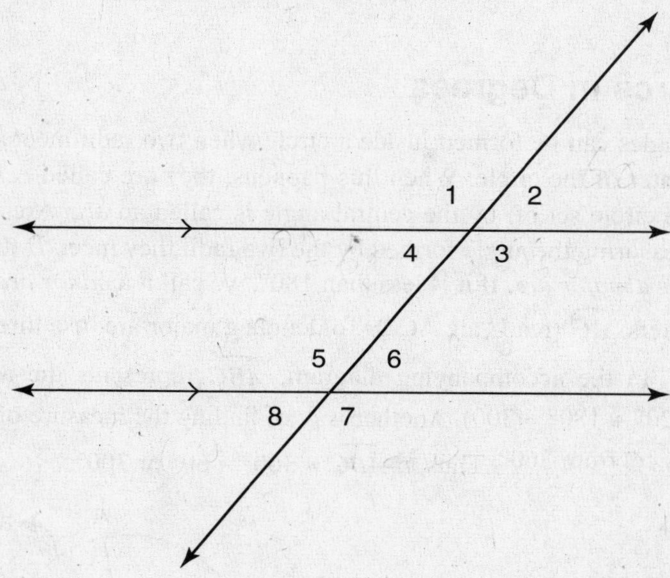

3. In this diagram m$\angle 6 = 50°$ and $\angle 6 \cong \angle 4$. Find m$\angle 1$, $\angle 2$, $\angle 3$, and $\angle 5$.

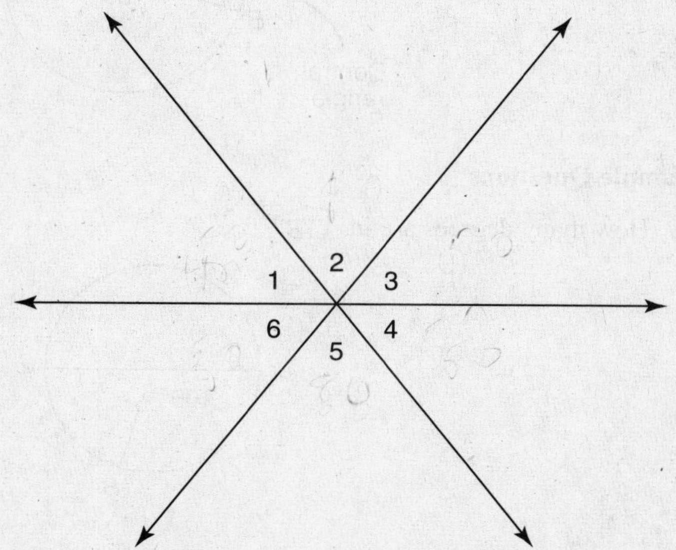

Answers

1. $m\angle DBC = 75°$. An angle bisector cuts an angle in half.
2. $m\angle 2 = 60°$. Supplementary to $\angle 1$.
   $m\angle 3 = 120°$. Vertical to $\angle 1$ and supplementary to $\angle 2$.
   $m\angle 4 = 60°$. Supplementary to $\angle 1$.
   $m\angle 5 = 120°$. Corresponds to $\angle 1$.
   $m\angle 6 = 60°$. Corresponds to $\angle 2$.
   $m\angle 7 = 120°$. Corresponds to $\angle 3$.
   $m\angle 8 = 60°$. Corresponds to $\angle 4$.
3. $m\angle 5 = 80°$. $\angle 4$, $\angle 5$, and $\angle 6$ must add to 180 degrees.
   $m\angle 1$ and $\angle 3 = 50°$. Vertical to $\angle 4$ and $\angle 6$.
   $m\angle 2 = 80°$. Vertical to $\angle 5$.

ARCS

Measuring Arcs in Degrees

Angles can be formed inside a circle when two radii meet and their vertex is located at the center of the circle. When this happens, they are called *central angles*. The curved part of the circle set off by the central angle is called an *arc*. Arcs can be measured in degrees by measuring the angle formed by the two radii they meet. If the arc is larger than 180°, we say it is a *major arc*. If it is less than 180°, we call it a *minor arc*. To denote a minor arc, use two letters, $\overset{\frown}{AC}$ (read "arc AC"). To denote a major arc, use three letters, $\overset{\frown}{ABC}$.

In the accompanying diagram, $\overset{\frown}{ABC}$ represents the arc found by adding $\overset{\frown}{AB}$ to $\overset{\frown}{BC}$ (120° + 180° = 300). Another way of finding the measure of $\overset{\frown}{ABC}$ is to subtract the measure of $\overset{\frown}{AC}$ from 360°. Thus, $m\overset{\frown}{ABC} = 360° - 60°$ or 300°.

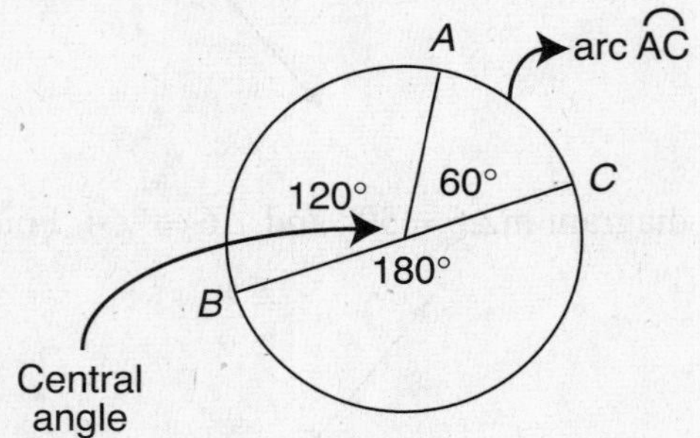

Sample Questions

1. How many degrees are in $\overset{\frown}{XYB}$?

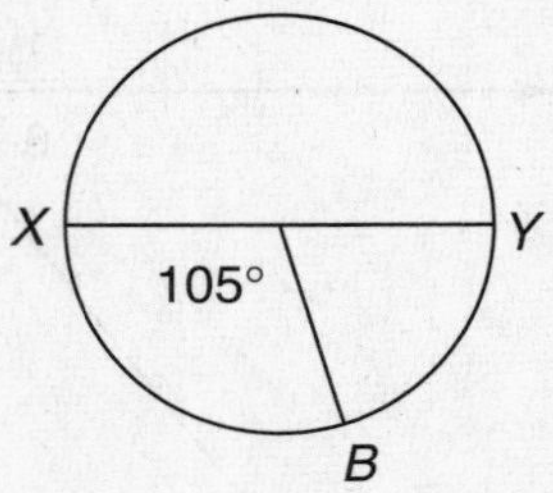

2. Find the measure of arc K in the diagram.

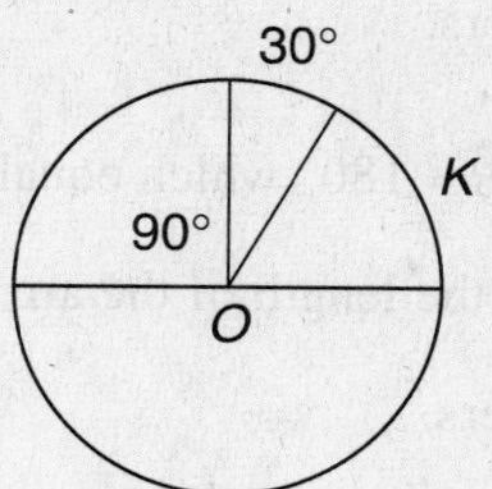

3. Find the measure of $\widehat{DE}$ in the diagram.

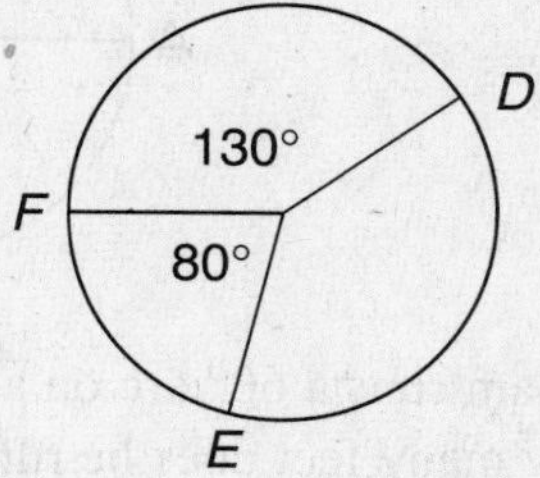

Answers

1. Subtract: 360° – 105° = 255°.
2. Subtract: 180° – 90° – 30° = 60°.
3. Subtract: 360° – 130° – 80° = 150°

Measuring the Length of an Arc

The length of an arc is a portion or fraction of the circumference. It's as if you made a circle out of string and then cut off a piece of the string. The piece you cut off is an *arc*. The cut piece of string (the arc) represents a portion, or fraction of the entire length of string (the circumference). You can use the information from measuring the degrees in an arc to find the length of the arc.

A 180° arc represents 180°/360° or $\frac{1}{2}$ of a circle. If you know the distance around the circle (circumference), you can multiply it by $\frac{1}{2}$ to find the *length* of the 180° arc.

$$\text{arc length} = \frac{\text{degrees in central angle}}{360} \bullet \text{circumference}$$

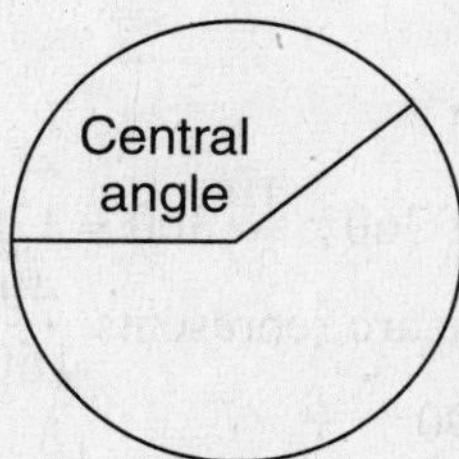

Example 1: Find the length of a 180° arc in a circle with a circumference of 18.84 meters.

$m\widehat{AB} = 180°$ which equals $\frac{180}{360}$ or $\frac{1}{2}$ of the circle. Use the circumference to find the length of the arc: $18.84 \cdot \frac{1}{2} = 9.42$. The length of the arc is 9.42 meters.

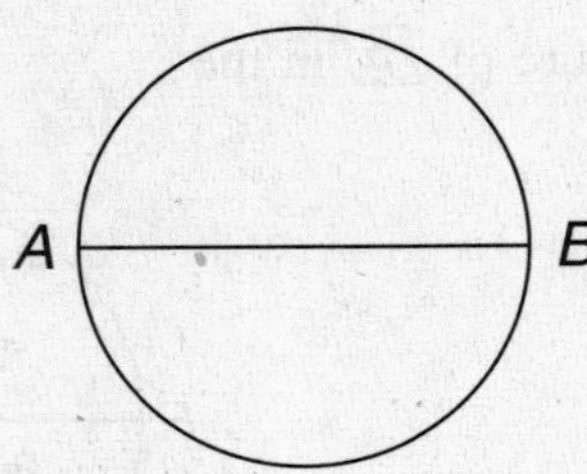

Example 2: If Sam runs a 60° arc on a circular track with a circumference of 1 mile, how many feet does he run?

1 mile = 5280 feet

$\frac{60°}{360°} \cdot 5280 = 880$ feet

Sample Questions

1. Freida is making a pie chart using information she collected during a survey. She needs to calculate the central angle measurement required if one-third of those surveyed like chocolate ice cream. What angle measurement does she need?
2. The length of $\widehat{PQ}$ measures 40 centimeters. The circumference of the circle containing the arc is 160 centimeters.
   Find the measure of the angle intercepting $\widehat{PQ}$.
3. Find the length of the arc intercepted by angle **O**.

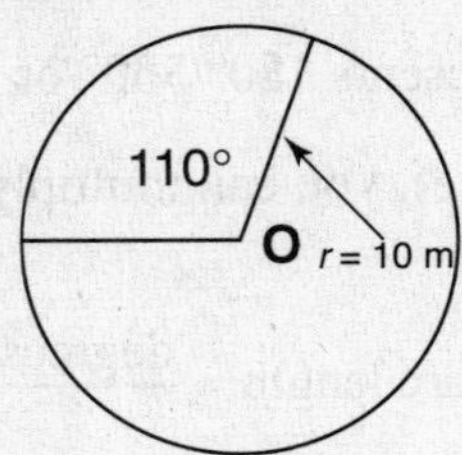

Answers

1. Find one-third of 360°: $\frac{1}{3} \cdot 360 = 120°$.
2. The length of the arc represents $\frac{40}{160}$ or $\frac{1}{4}$ of the circle. Multiply the degrees in a circle by $\frac{1}{4}$: $\frac{1}{4} \cdot 360 = 90°$.
3. 110° represents $\frac{110}{360}$ of the circumference.
   $\frac{110}{360} \cdot 2\pi r = \frac{110}{360} \cdot 2 \cdot 3.14 \cdot 10 \approx 19.188 = 19.19$ meters.

SECTION 2: POLYGONS

POLYGONS

Before we begin working with perimeter, we should review polygons. A polygon is a closed plane (flat) figure made up of line segments. Polygons are generally classified by the number of sides they have.

Number of Sides	Name of Polygon	Example
3	Triangle	
4	Quadrilateral	
5	Pentagon	
6	Hexagon	
8	Octagon	
10	Decagon	
12	Dodecagon	
N	*N*-gon	*N* sides

Polygons can be *concave* or *convex*. The dotted lines in the figures represent diagonals. The diagonals of a convex polygon are all inside. In a concave polygon, one or more diagonals can be outside the polygon itself.

Example: **Example:**

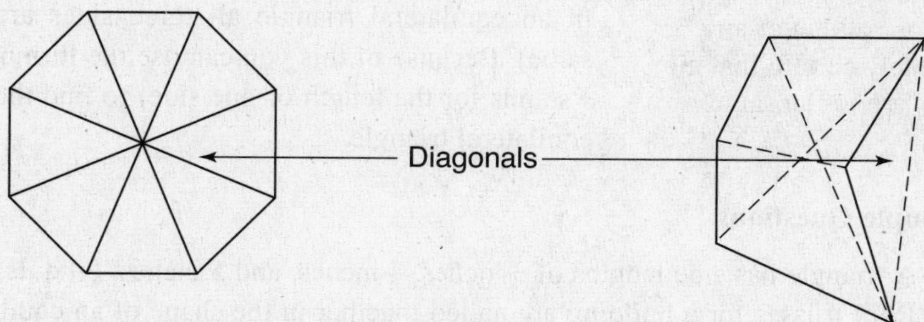

When the sides of a polygon are all the same length *and* all angles are the same, we say that the polygon is a *regular polygon*. The convex polygon in the diagram above is a regular polygon. The concave polygon is not regular.

To find the perimeter of a regular polygon, you can multiply the length of one side by the number of sides of that length: P = number of sides • length, or $P = 4s$ in the case of a square.

There are formulas on the FCAT reference sheet for finding the measures of the interior and exterior angles of these regular polygons.

The sum of the interior angles of a regular polygon: $180(n - 2)$, where n = the number of sides on the polygon.

The measure of one interior angle of a regular polygon: $\frac{180(n-2)}{n}$

The measure of one exterior angle of a regular polygon: $\frac{360}{n}$ or 180° – interior angle

Example: Find the measure of one exterior angle and one interior angle of a regular octagon.

***Test Tip*: In any polygon, an exterior angle plus its adjacent interior angle adds to 180°.**

$\frac{180(n-2)}{n} = \frac{180(8-2)}{8} = 135°$ for one interior angle

Exterior angle = 180° – 135° = 45°

Example: Find the perimeter of the polygon shown here.

Perimeter = sum of all sides.

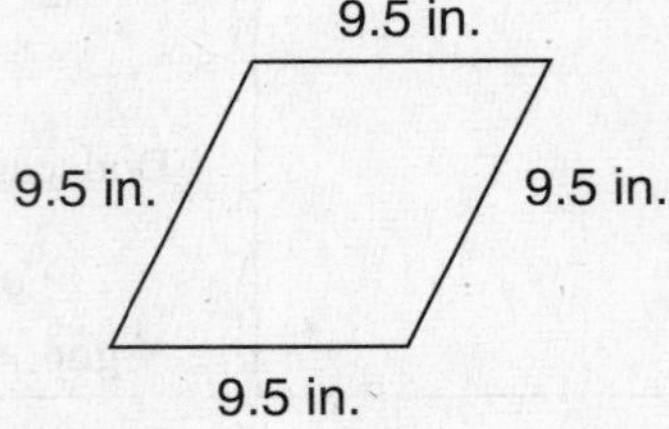

$P = 4s$ (4 times the length of the side s)
$4 \bullet 9.5$ in. = 38 in.

Triangles

***Test Tip:* Before finding the perimeter, area, surface area, or volume of any figure, make absolutely certain that all your measurements are in the same measurement units.**

To find the perimeter of a triangle, add the sides: $P = a + b + c$. In an equilateral triangle all three sides are congruent (the same). Because of this you can use the formula $P = 3s$ (where s stands for the length of one side) to find the perimeter of an equilateral triangle.

Sample Questions

1. A triangle has side lengths of 3 inches, 4 inches, and 5 inches. Find its perimeter.
2. Roof trusses for a building are nailed together in the shape of an equilateral triangle. If the board forming the sides of the triangle are each 12 feet, what is the total board length?

3. A triangle has sides with the following lengths: 4 feet, 6 inches; 6 feet, 5 inches; and 3 feet, 10 inches. Which of the following expresses its perimeter in feet?

A. $14\frac{3}{4}$ ft

B. $13\frac{11}{12}$ ft

C. $14\frac{1}{2}$ ft

D. $13\frac{3}{4}$ ft

Answers

1. $P = a + b + c = 3 + 4 + 5 = 12$ inches
2. $P = 3s = 3 \times 12 = 36$ feet long
3. **A.** $4\frac{6}{12} + 6\frac{5}{12} + 3\frac{10}{12} = 13\frac{21}{12} = 14\frac{9}{12} = 14\frac{3}{4}$

THE PYTHAGOREAN THEOREM

The Pythagorean Theorem is used to find the length of one side of a right triangle when the lengths of the other two sides are known. The theorem says that in any right triangle, the square of the length of the hypotenuse equals the sum of the squares of the other two sides. Written algebraically, $c^2 = a^2 + b^2$. You will find the Pythagorean Theorem extremely useful on the FCAT.

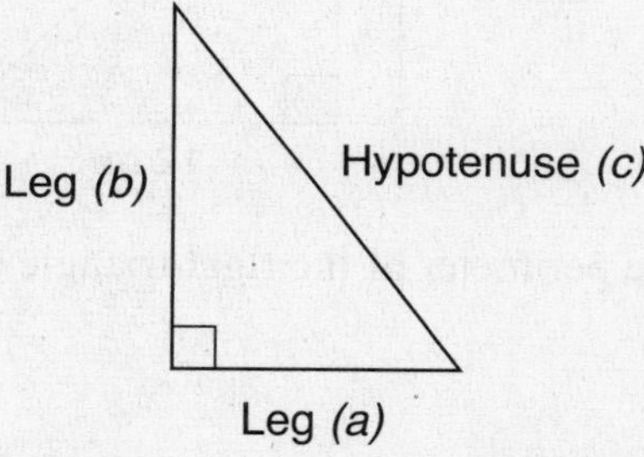

The *hypotenuse* is the longest side of a right triangle. It can be located easily because it is opposite the 90° angle. The *c* in the Pythagorean Theorem is always the hypotenuse, and the *a* and *b* represent the legs.

When solving for the hypotenuse using the Pythagorean Theorem ($c^2 = a^2 + b^2$), remember that you have to take the square root of c^2 to find *c*.

Example: Find the length of the hypotenuse in the triangle shown here.

$c^2 = a^2 + b^2$
$c^2 = 3^2 + 4^2$
$c^2 = 9 + 16$
$c^2 = 25$
$c = \sqrt{25} = 5$

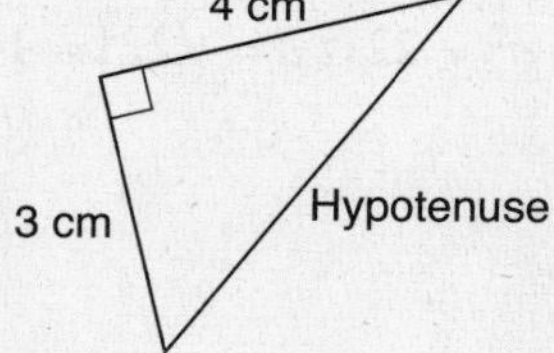

Test Tip: To find the hypotenuse, square and add. To find the length of a leg, square and subtract. Last, take the square root.

Sometimes you are asked to find the length of one of the legs rather than the hypotenuse.

Example: Find the length of the missing side.

$c^2 = a^2 + b^2$
$13^2 = 9^2 + b^2$
$13^2 - 9^2 = b^2$
$169 - 81 = b^2$
$88 = b^2$
$\sqrt{88} = b$
$9.38 \approx b$

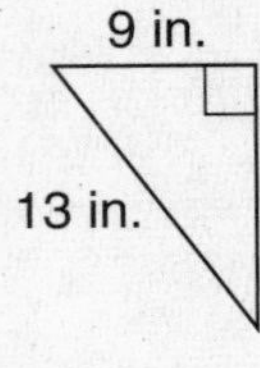

Sample Questions

1. Find the length of c if $a = 6$ and $b = 8$.
2. Find the length of a if $c = 15$ and $b = 2$. Round your answer to the hundredths position.
3. A 12-foot ladder leans against a building. The ladder is placed 5 feet from the base of the building. To the nearest foot, how many feet high above the ground does the ladder reach?
4. Find the missing length k. Show your work.

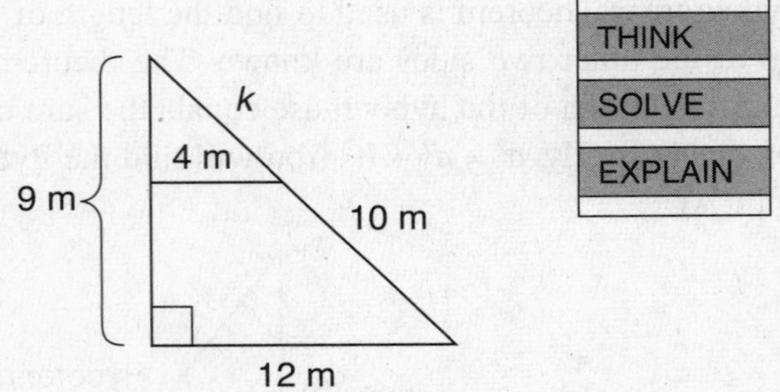

5. What is the perimeter of the right triangle shown here?

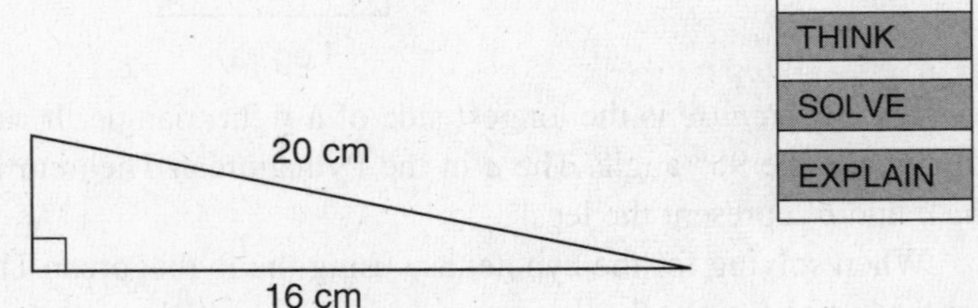

Answers

1. $c^2 = 6^2 + 8^2$; $c^2 = 100$; $c = 10$.
2. $a^2 = 15^2 - 2^2$; $a^2 = 221$; $a = \sqrt{221} = 14.46$.
3. $c^2 = a^2 + b^2$
   $12^2 = a^2 + 5^2$
   $144 = a^2 + 25$
   $144 - 25 = a^2$
   $119 = a^2$
   $\sqrt{119} = a^2 = 10.9087 = 11$ feet

12 ft
5 ft

4. First, find the hypotenuse of the larger triangle:
$c^2 = a^2 + b^2$
$c^2 = 9^2 + 12^2$
$c^2 = 81 + 144$
$c^2 = 225$
$c = \sqrt{225} = 15$
Subtract 10 m from 15 m to find $k = 5$ m.
5. Use the Pythagorean Theorem to find the length of the short side:
$c^2 = a^2 + b^2$
$20^2 = a^2 + 16^2$
$400 = a^2 + 256$
$400 - 256 = a^2$
$114 = a^2$
$\sqrt{114} = a = 12$
The perimeter = 12 + 16 + 20 = 48 centimeters

PERIMETER

The *perimeter* of a figure simply means the distance around the outside. To find the perimeter of any polygon, add the sides. In the regular hexagon shown below, the side lengths are 8 millimeters. The perimeter could be found by adding all six sides, or by multiplying 8 by 6, which equals 48 millimeters.

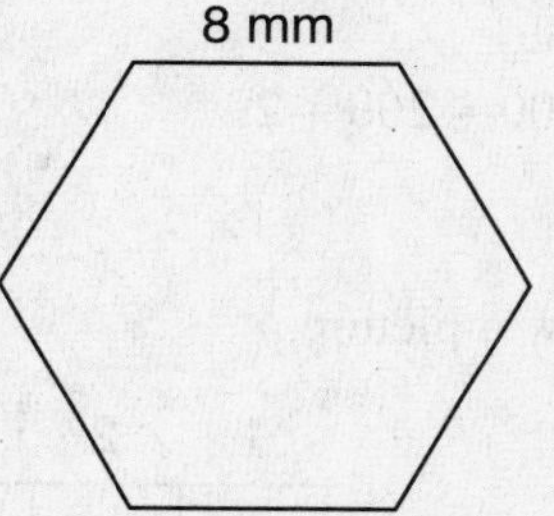

Perimeter of Quadrilaterals

In a rectangle, opposite sides are the same length. Therefore, you can use the formula $P = 2l + 2w$, where l stands for the length and w stands for the width.

Example: Find the perimeter of a rectangle with length 10 meters and width 5 meters.

$P = 2l + 2w$, where l = length and w = width
$= 2 \cdot 10 + 2 \cdot 5$
$= 20 + 10 = 30$ meters

Squares and Rhombuses

Squares and rhombuses are equal-sided quadrilaterals. Therefore, the formula for their perimeter is $P = 4s$, where s represents the length of the side.

Example: A baseball diamond is really a square. The distance between the bases is 90 feet. How far would a hitter have to run to make it all the way around the bases and back to home plate?

The distance around the outside of a polygon is the perimeter.

$P = 4s$
$= 4 \cdot 90$
$= 360$ feet

Sample Questions

1. How many inches of lace would it take to trim an 84-inch × 104-inch rectangular tablecloth?
2. Find the width of a rectangle that has a perimeter of 250 millimeters and a length of 100 millimeters.
3. If a square table seats 4 people, how many people can be seated if 6 tables are pushed together in a long row?

Answers

1. $P = 2l + 2w$
   $= 2 \cdot 84 + 2 \cdot 104$
   $= 168 + 208 = 376$ inches
2. $P = 2l + 2w$
   $250 = 2 \cdot 100 + 2w$
   $250 = 200 + 2w$
   $250 - 200 = 200 - 200 + 2w$
   $50 = 2w$
   $25 = w$
3. It helps to draw a picture.

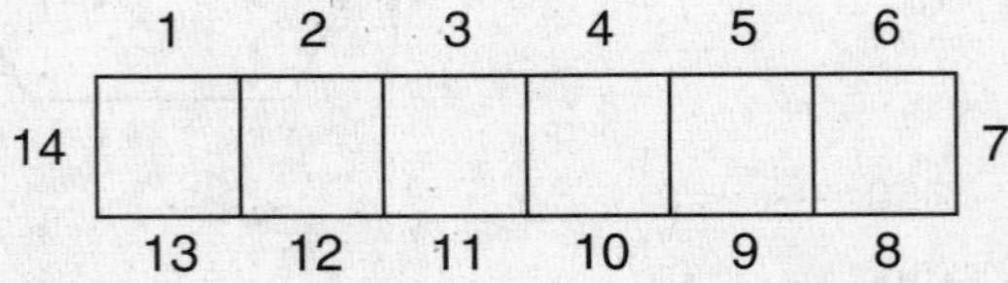

One person can be seated at each place, but one place is lost from each end table and two places are lost from the other tables because they are pushed together. Only 14 people can be seated with this arrangement.

CIRCUMFERENCE

$C = 2\pi r$
or
$C = \pi d$

Circumference is simply the perimeter of a circle. It is the distance all the way around the outer edge of the circle. To be able to use formulas relating to a circle, you must first be familiar with the parts of a circle. A circle is formed by a set of points that are equidistant from a point inside the circle (the *center*). The circle's *radius* is a line segment that extends from the center to an outside edge of the circle. It is exactly half the length of the *diameter*, which is a line segment that passes through the center of the circle and extends in both directions all the way across the circle.

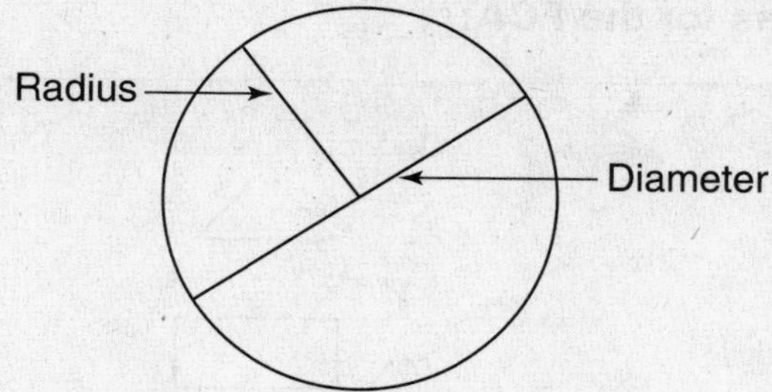

The following are formulas for the circumference of a circle.

$C = 2\pi r$ Use this formula when you have a radius.
$C = \pi d$ Use this formula when you are given a diameter.

Recall that π is an irrational number. Frequently, especially on tests like the FCAT, you will use the rational number 3.14 or $\frac{22}{7}$ instead of π. These numbers are close approximations of pi.

On the FCAT you will usually use 3.14 for π.

Example 1: Find the circumference of a circle whose radius is 4 centimeters. Use 3.14 for π.

$$
\begin{aligned}
C &= 2\pi r \\
&\approx 2 \cdot 3.14 \cdot 4 \\
&\approx 25.12 \text{ centimeters}
\end{aligned}
$$

Example 2: Find the circumference of a circle with a diameter of 10 inches.

$$
\begin{aligned}
C &= \pi d \\
&= 3.14 \cdot 10 \\
&= 31.4 \text{ inches}
\end{aligned}
$$

Sample Question

A bicycle wheel with a diameter of 20 inches travels approximately how far with each complete turn of the wheel?

Answer

$$
\begin{aligned}
C &= \pi d \\
&\approx 3.14 \cdot 20 \\
&\approx 62.8 \text{ inches}
\end{aligned}
$$

AREA AND VOLUME

Next we will look at calculating area and volume. On the FCAT you are not required to memorize formulas; they are provided for you. We will review examples of problems using these formulas.

Formulas for the FCAT*

Figure	Formula
Triangle	Area $= \frac{1}{2}bh$
Rectangle	Area $= lw$
Trapezoid	Area $= \frac{1}{2}(b_1 + b_2)h$
Parallelogram	Area $= bh$
Circle	Area $= \pi r^2$ $C = \pi d$ or $C = 2\pi r$
Right circular cylinder	Volume $= \pi r^2 h$ or volume $= Bh$ Surface area $= 2\pi rh + 2\pi r^2$
Rectangular solid	Volume $= lwh$ or volume $= Bh$ Surface area $= 2(lw) + 2(hw) + 2(lh)$
Right circular cone	Volume $= \frac{1}{3}\pi r^2 h$ or volume $= \frac{1}{3}Bh$
Square pyramid	Volume $= \frac{1}{3}lwh$ or volume $= \frac{1}{3}Bh$
Sphere	Volume $= \frac{4}{3}\pi r^3$ Surface area $= 4\pi r^2$

* b = base; h = height; l = length; w = width; r = radius; d = diameter; C = circumference; ℓ = slant height; $\pi = 3.14$ or $\frac{22}{7}$

AREA

Area Conversions

Area measurement units are expressed as *square units*. They are referred to as *square units* because the measurement unit can vary. For example, when calculating the area of a 10-inch × 5-inch rectangle, the measurement unit is expressed in square inches (in.^{2}). The accompanying figure is a square inch because it measures 1 inch on each side and 1 inch • 1 inch = 1 square inch or 1 in.^{2}.

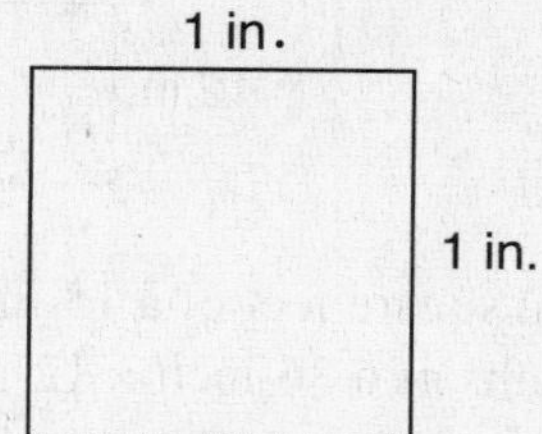

One square foot = 144 square inches because 12 inch • 12 inch = 144 square inches.

One square yard = 9 square feet because 3 feet • 3 feet = 9 square feet.

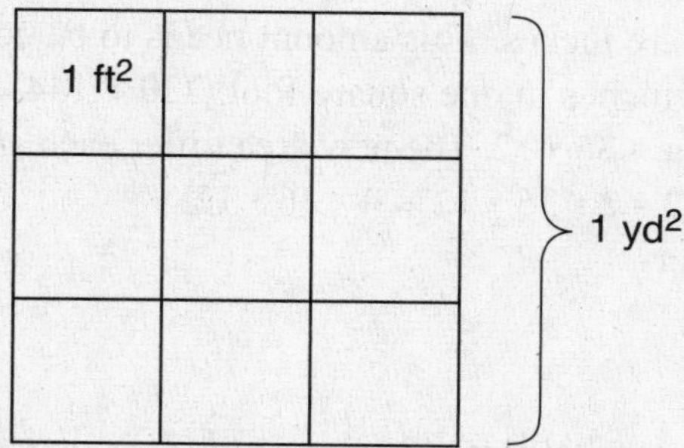

Note: As before, when converting from a larger unit to a smaller unit, multiply by the conversion factor. When converting from a smaller unit to a larger unit, divide by the conversion factor.

Example 1: Convert 36 square feet to square yards.

$36 \text{ ft}^2 \div 9 = 4 \text{ yd}^2$

Example 2: Convert 6.4 square feet to square inches.

$6.4 \text{ ft}^2 \cdot 144 = 921.6 \text{ in.}^2$

Area of a Rectangle

The area of a rectangular figure can be found by counting or calculating the number of square units in the figure. In the grid shown let each square represent 1 square centimeter. There are 50 squares, so the entire rectangle represents 50 square centimeters. We say that its area is 50 cm^2.

$A = lw$

Area = lw = 10 cm • 5 cm = 50 cm^2

Note: Because you are multiplying centimeters by centimeters, cm • cm = cm^2.

Sample Questions

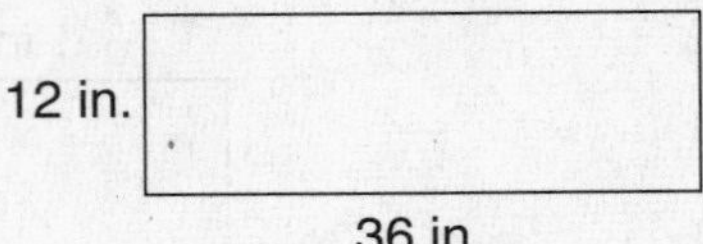

1. Find the area in square feet of a 15-inch × 48-inch rectangle.
2. If the side lengths of a 36-inch × 12-inch rectangle are doubled, how many times larger is the area of the new figure?
3. Find the area of a square if the side length is 6 cm.

Answers

1. $A = lw = 15 \cdot 48 = 720$ square inches. This amount needs to be converted to square feet. Since there are 144 square inches in one square foot, $720 \div 144 = 5$ square feet.
2. 4 times larger. Original area $= 36 \cdot 12$. The new area (after each side is doubled) $= 2 \cdot 36 \cdot 2 \cdot 12$. Commute the 2s: $2 \cdot 2 \cdot 36 \cdot 12 = 4 \cdot 36 \cdot 12$.
3. $6 \cdot 6 = 36$ square centimeters.

Area of a Parallelogram

The formula for the area of a parallelogram is area = *bh.* It is actually quite similar to the formula for the area of a rectangle. Cut a parallelogram along the dotted lines as shown here. Then slide one of the triangular pieces over and stack it on top of the other triangular piece. The resulting figure is a rectangle. The area of the figure doesn't change just because you rearrange the pieces, and so the formulas are similar. The base length of the parallelogram corresponds to the length of a rectangle, and the height of the parallelogram corresponds to the width of the rectangle.

$A = bh$

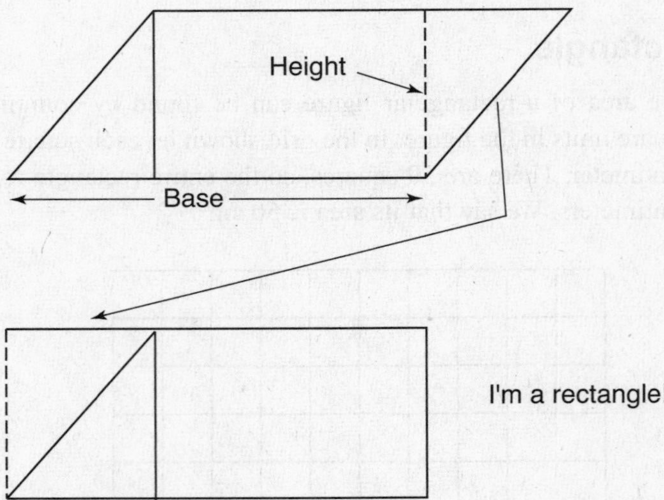

Example: Use area = *bh* to find the area of the parallelogram shown here.

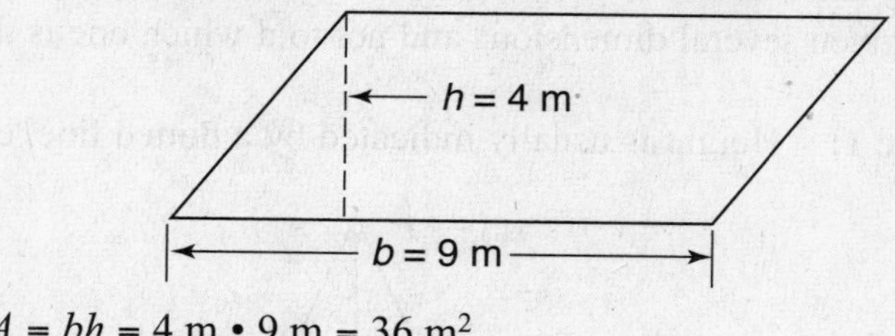

$$A = bh = 4\text{ m} \bullet 9\text{ m} = 36\text{ m}^2$$

Sample Questions

1. Find the area of the figure shown here.

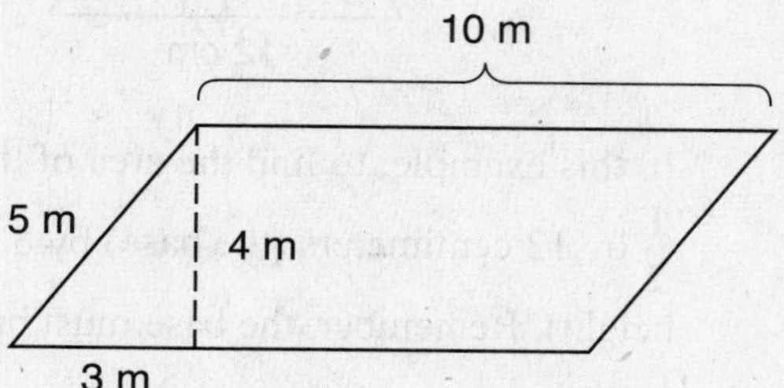

2. If a parallelogram has an area of 200 square inches and a height of 16 inches, what is the length of its base?

Answers

1. $A = bh = 10\text{ m} \bullet 4\text{ m} = 40\text{ m}^2$.
2. $A = bh$

$200 = bh$	Substitute 200 for A.
$200 = 16b$	Substitute 16 for h.
$12.5 = b$	Divide by 16 to solve.

Area of a Triangle

The formula for the area of a triangle is actually based on the formula for the area of a parallelogram.

$$\text{area} = \frac{1}{2}bh$$

$A = .5bh$

If you start with any parallelogram (including rectangles, rhombuses, and squares) and cut the figure in half diagonally, the resulting figures are triangles. Thus, you have cut the area of the parallelogram in half. The area of the triangle, then, is half that of the parallelogram.

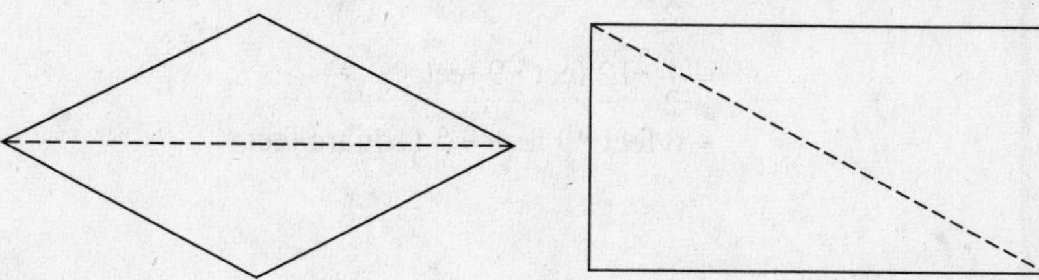

The altitude or height (h) of the triangle should always be perpendicular to the base (b). Knowing this can eliminate any problems with identifying which is which, particularly when you are given several dimensions and not told which one is the height or base.

Example 1: Height is usually indicated by a dotted line, except in a right triangle.

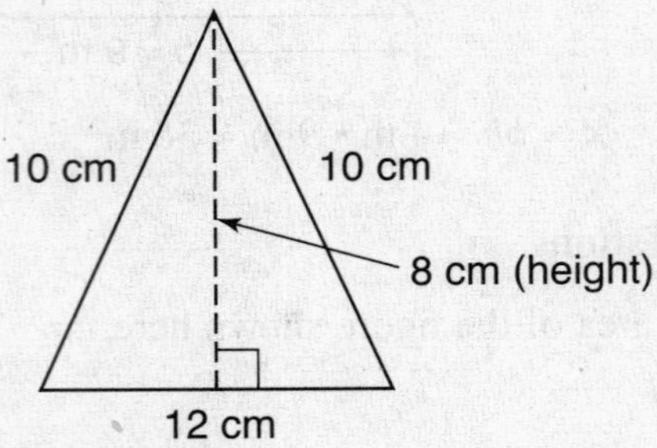

In this example, to find the area of the triangle, multiply $\frac{1}{2}$ by 12 centimeters (the base) by 8 centimeters (the height). Remember, the base must be perpendicular to the height.

$$A = \frac{1}{2}bh$$
$$= \frac{1}{2} \cdot 8 \text{ cm} \cdot 12 \text{ cm}$$
$$= 4 \text{ cm} \cdot 12 \text{ cm} = 48 \text{ cm}$$

Example 2:

The height and the base are always at a 90° angle to each other.

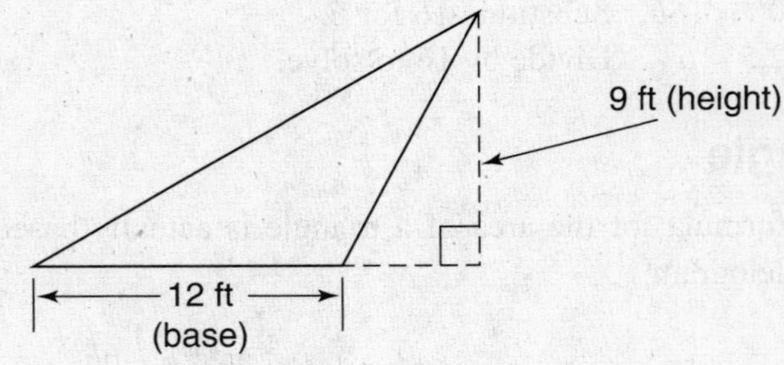

By examining this obtuse triangle carefully, you can see that the base (12 feet) is perpendicular to the dotted line representing the height (9 feet).

$$A = \frac{1}{2}bh$$
$$= \frac{1}{2} \cdot 12 \text{ feet} \cdot 9 \text{ feet}$$
$$= 6 \text{ feet} \cdot 9 \text{ feet} = 54 \text{ square feet}$$

Example 3: Find the area of the triangle shown here.

Note: The triangle measurements are given in both feet and inches. To use formulas, all measurements must be in the same unit. You can change feet to inches, or inches to feet, using one of the methods shown here.

Method 1: Change feet to inches

$2 \text{ ft} \cdot 12 = 24 \text{ in.}$

$$A = \frac{1}{2}bh$$
$$= .5 \cdot 12 \text{ in.} \cdot 24 \text{ in.}$$
$$= 180 \text{ in.}^2$$

Method 2: Change inches to feet

$15 \text{ in.} \div 12 = 1.25 \text{ ft}$

$$A = \frac{1}{2}bh$$
$$= .5 \cdot 1.25 \text{ ft} \cdot 2 \text{ ft}$$
$$= 1.25 \text{ ft}^2$$

Sample Questions

1. Find the area of the triangle in the diagram shown here.

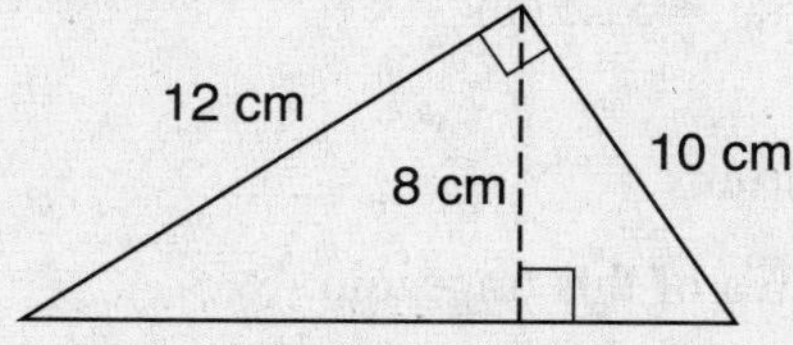

2. A triangle has an area of 90 centimeters and a base of 9 centimeters. Which of the following represents the height?

A. 120 centimeters
B. 60 centimeters
C. 48 centimeters
D. 20 centimeters

Answers

1. $A = \frac{1}{2}bh = \frac{1}{2} \bullet 10 \bullet 12 = 60$ square centimeters

2. **D.** $A = \frac{1}{2}bh$

$$90 = \frac{1}{2}9h$$
$$90 = 4.5h$$
$$20 = h$$

Area of a Trapezoid

The formula for the area of a trapezoid $\left[A = \frac{1}{2}(b_1 + b_2)h\right]$ is a combination formula. The bases of the trapezoid are the parallel sides of the trapezoid and are represented by b_1 and b_2. The equation combines the formulas for the area of a rectangle and the area of a triangle, and it is not important which base is b_1 or b_2. The height of the trapezoid is perpendicular to both bases. To use the formula to calculate the area of a trapezoid, look at the example given here.

Example: Find the area of the trapezoid shown here.

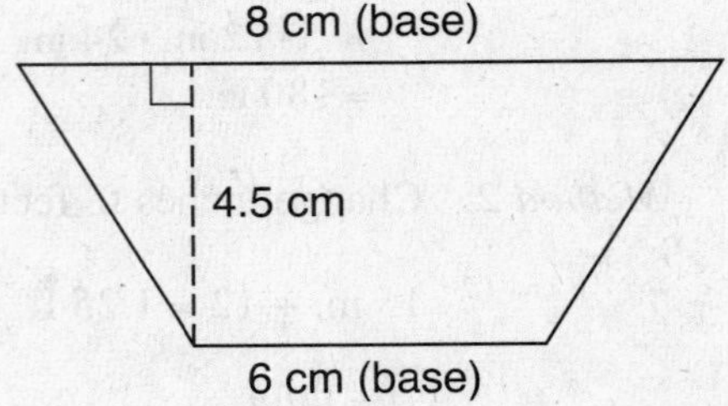

$$\begin{aligned} A &= \frac{1}{2}(b_1 + b_2)h \\ &= \frac{1}{2}(8 \text{ cm} + 6 \text{ cm})4.5 \text{ cm} \\ &= \frac{1}{2} \bullet 14 \text{ cm} \bullet 4.5 \text{ cm} = 7 \text{ cm} \bullet 4.5 \text{ cm} \\ &= 31.5 \text{ cm}^2 \end{aligned}$$

Sample Questions

1. Find the area of this trapezoid.

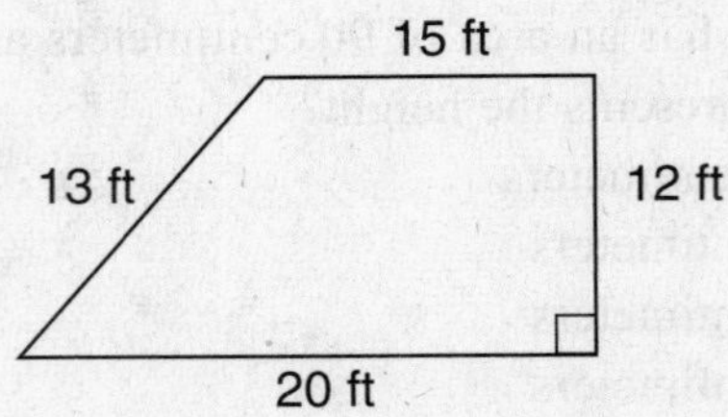

2. The area of the trapezoid shown here is 63 square millimeters. What is its height?

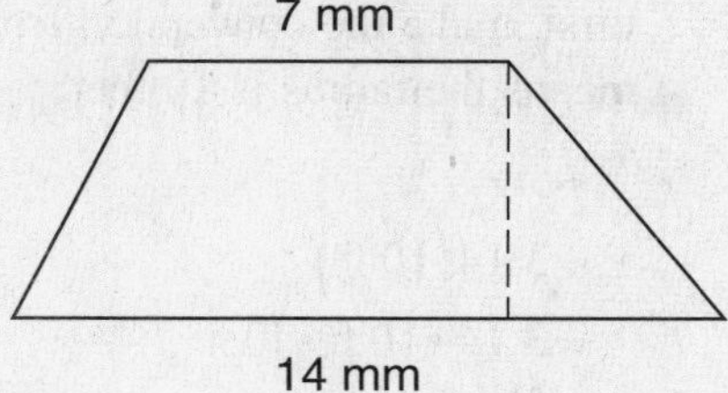

Answers

1. $$\begin{aligned} A &= \frac{1}{2}(b_1 + b_2)h \\ &= \frac{1}{2}(15\text{ ft} + 20\text{ ft}) \bullet 12\text{ ft} \\ &= \frac{1}{2} \bullet 35\text{ ft} \bullet 12\text{ ft} \\ &= 17.5\text{ ft} \bullet 12\text{ ft} \\ &= 210\text{ ft}^2 \end{aligned}$$

2. $$\begin{aligned} A &= \frac{1}{2}(b_1 + b_2)h \\ 63\text{ mm}^2 &= \frac{1}{2}(7\text{ mm} + 14\text{ mm})h \\ 63\text{ mm}^2 &= \frac{1}{2} \bullet 21\text{ mm} \bullet h \\ 63 &= 10.5h \\ 6\text{ mm} &= h \end{aligned}$$

Area of a Circle

To find the area of a circle, use the formula $A = \pi r^2$. Translated, this means pi times the squareof the radius. As is the case with circumference, either 3.14 or $\frac{22}{7}$ is generally substituted for π.

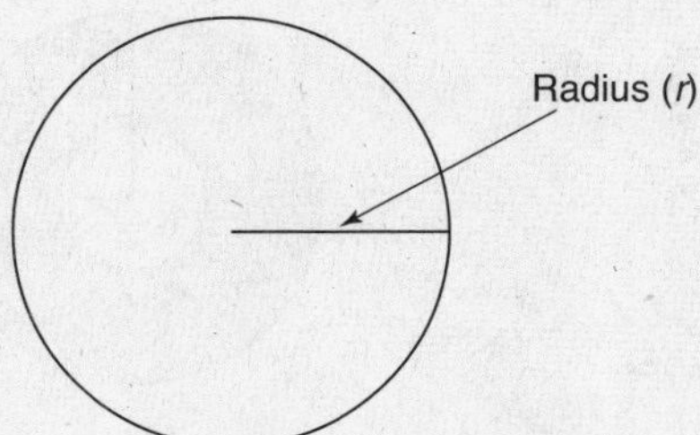

Example 1: Find the area of a circle with a radius of 28 millimeters. Use $\pi \approx \frac{22}{7}$.

$$\begin{aligned} A &= \pi r^2 \\ &\approx \frac{22}{7}(28\text{ mm})^2 \\ &\approx 22 \div 7 \bullet 28\text{ mm} \bullet 28\text{ mm} \\ &\approx 2464\text{ mm}^2 \end{aligned}$$

Example 2: Find the area of a circle with a diameter of 20 feet. Use $\pi \approx 3.14$.

First, make the conversion from diameter to radius. A 20-foot diameter means the radius is 10 feet.

$$\begin{aligned} A &\approx \pi r^2 \\ &\approx 3.14(10\text{ ft})^2 \\ &\approx 3.14 \cdot 10\text{ ft} \cdot 10\text{ ft} \\ &\approx 314\text{ ft}^2 \end{aligned}$$

Sample Questions

1. Find the circumference of a circle with a radius of 15 inches.
2. Find the circumference and area of a circle with a diameter of 7 feet.
3. A radio station in Tampa broadcasts its signal over a 100-mile radius. Use $\pi \approx 3.14$ to find out how many square miles this radio station serves.

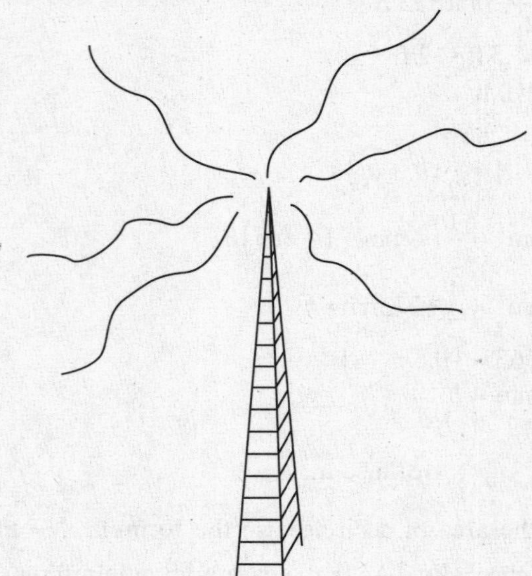

4. Find the area of the shaded region. Express it in terms of π.

Hint: Find the area of the large circle and then subtract the area of the small circle.

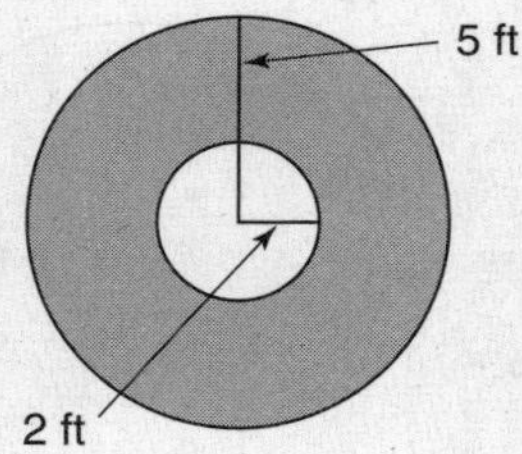

Answers

1. $C = 2\pi r$
   $= 2 \cdot 3.14 \cdot 15$
   $= 94.2$ inches
2. $C = \pi d$ $\qquad$ $A = \pi r^2$
   $= 3.14 \cdot 7$ $\qquad$ $= 3.14 \cdot 3.5 \cdot 3.5$
   $= 21.98$ feet $\qquad$ $= 38.47$ square feet

3. Try to picture the signal going out from the radio station. A radio signal goes out from the station to the same distance in all directions, forming a circle. The distance from the radio station to the outermost edge of the broadcast range is a radius.

$A = \pi r^2$
$\approx 3.14(100 \text{ miles})^2$
$\approx 3.14 \cdot 100 \text{ miles} \cdot 100 \text{ miles}$
$\approx 31{,}400 \text{ square miles}$

The radio station broadcasts over an area of approximately 31,400 square miles!

4. The area of the shaded region is the area of the large circle minus the area of the smaller circle. Use $A = \pi r^2$. Then, $\pi 5^2 - \pi 2^2 = 25\pi - 4\pi = 21\pi$.

Perimeter and Area of Composite Figures

A composite figure can be defined as a shape made up of two or more polygons or circles. To find the perimeter of a composite figure, simply find the sum of the sides.

Example: Jessica is buying a border for her garden as shown here. How many feet of border will she need to buy?

Perimeter = sum of all sides.

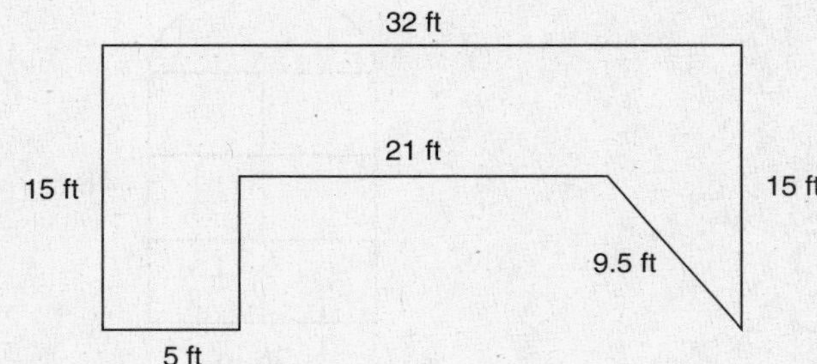

Because the border represents the perimeter of the figure, add the sides. You can begin anywhere but be sure to add all the sides.

Perimeter = 32 ft + 15 ft + 9.5 ft + 21 ft + 8 ft + 5 ft + 15 ft = 105.5 ft

Sample Questions

1. Harvey plans to fence a plot of land he has purchased next to his house to pasture his horses. If fencing costs $.85 per linear foot, how much will it cost Harvey to fence his land? Show how you arrived at your answer.

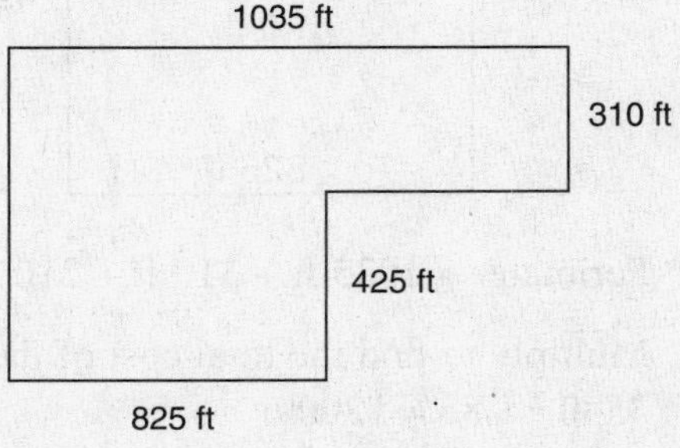

2. The accompanying floor plan represents the part of Jay's house that he is planning to tile. The shaded areas represent areas that will not be tiled. Find the number of square feet of tile that Jay will need to purchase. Show how you arrived at your answer.

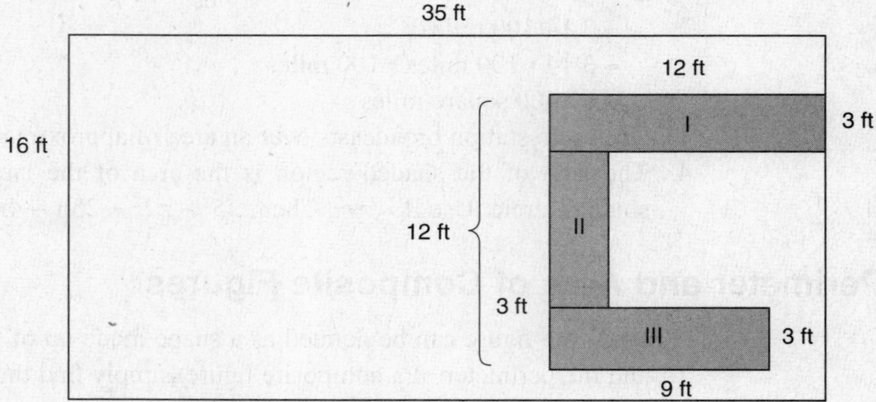

3. A decorator wants to trim the outside edge of a window with ribbon. To the nearest $\frac{1}{4}$ yard, how many yards of ribbon will it take?

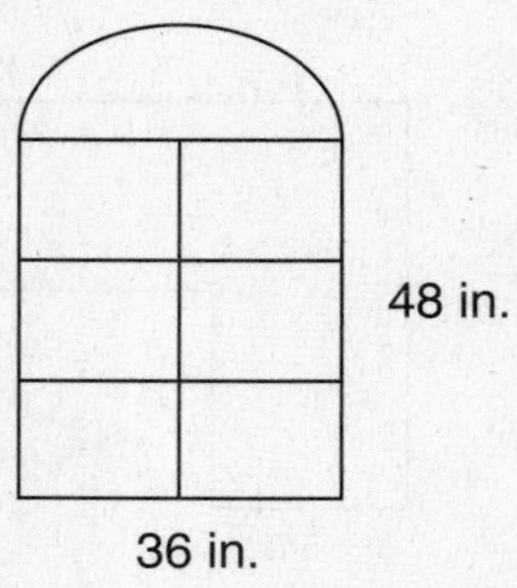

Answers

1. Step 1: Make sure you have all the sides labeled.

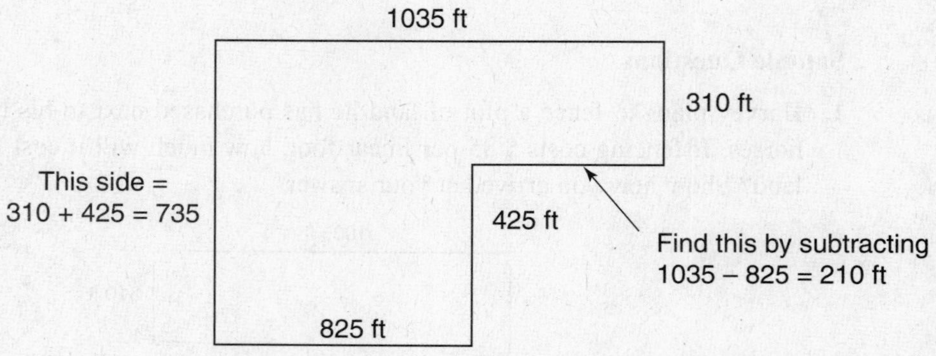

Step 2: Perimeter = 1035 ft + 310 ft + 210 ft + 425 ft + 825 ft + 735 ft = 3540 ft

Step 3: Multiply to find the total cost of the fencing.
3540 • \$.85 = \$3009

2. To find the area to be tiled, first find the area of the entire 35-feet × 16-feet rectangle and subtract the area of the shaded regions.
   Step 1: $A = lw = 35 \text{ ft} \cdot 16 \text{ ft} = 560 \text{ ft}^2$
   Step 2: There are three shaded regions:
   Part I is 12 ft × 3 ft. $A = 12 \text{ ft} \cdot 3 \text{ ft} = 36 \text{ ft}^2$
   Part II is 6 ft × 3 ft. $A = 6 \text{ ft} \cdot 3 \text{ ft} = 18 \text{ ft}^2$
   Part III is 9 ft × 3 ft. $A = 9 \text{ ft} \cdot 3 \text{ ft} = 27 \text{ ft}^2$
   Area of shaded region $= 81 \text{ ft}^2$
   Step 3: Total area $= 560 \text{ ft}^2 - 90 \text{ ft}^2 = 470 \text{ ft}^2$
3. To find the perimeter of the window, add all sides. First, find the circumference of the upper portion of the window. You will need to find the circumference of a circle with a 36-inch diameter and take half of the circumferences (since you are using only half a circle for the window): $C = \pi d = 3.14 \cdot 36 = 113.04$. $\frac{1}{2} \cdot 113.04 = 56.52$. Next, add the sides of the window: $56.52 + 36 + 48 + 48 = 118.52$ inches. Last, convert inches to yards by dividing by 36: $118.51 \div 36 = 5.23$, which is closest to $5\frac{1}{4}$ yards.

SURFACE AREA AND VOLUME

Surface Area and Volume of Rectangular Solids

A rectangular solid is a three-dimensional figure made up of six rectangles or squares called *faces*. The faces are joined together at *edges*. The *vertex* is the point where the edges join.

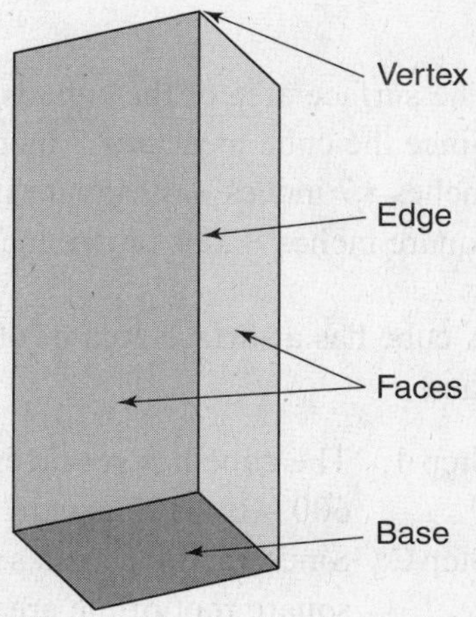

Surface Area of a Rectangular Solid

Surface area is simply the combined area of all the surfaces or faces of a three-dimensional figure. Surface area is expressed in square units.

CUBES

Example 1: Because rectangular solids are three-dimensional figures, they have *length, width,* and *height*. A cube is an example of a rectangular solid (remember that a square is also a rectangle). It is made up of six squares. If you were to take the cube apart and flatten it, it would look like this.

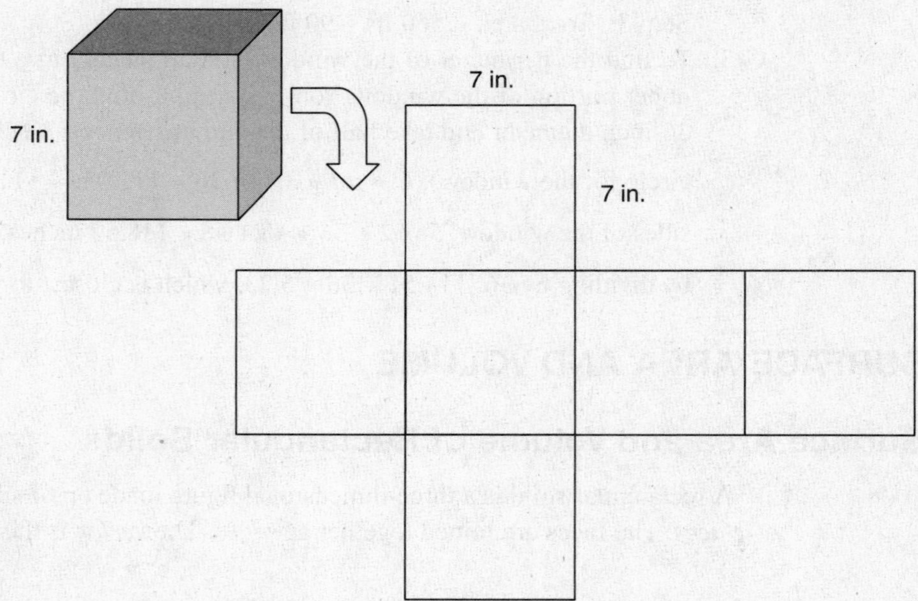

To find the **surface area**, find the area of each face of a solid and add them together.

The surface area of the cube is the sum of the areas of each of the faces. Since the cube measures 7 inches on an edge, each face has an area of 7 inches • 7 inches = 49 square inches. The sum of the areas is then 6 • 49 square inches = 294 square inches.

Example 2: A cube has a surface area of 600 square units. Find the edge length of one face.

Step 1: The cube has six faces, and so each face has an area of $600 \div 6 = 100$ square units.

Step 2: Since each face is a square, find the edge length by taking the square root of the area. $\sqrt{100} = 10$. The edge length is 10 units.

ANOTHER RECTANGULAR SOLID

For the cereal box shown here, a slightly different approach is needed to find the surface area. You can see that when the box is broken down, there are three pairs of rectangles representing the top and bottom, the two sides, and the front and back. To find the surface area, you must find the area of each of the faces and add them together.

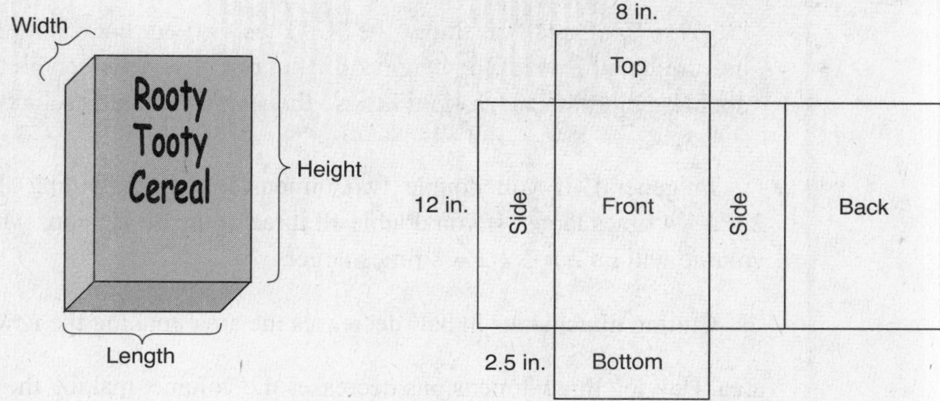

Area of top	8 in. • 2.5 in. =	20 in.^{2}
Area of top	8 in. • 2.5 in. =	20 in.^{2}
Area of side	12 in. • 2.5 in. =	30 in.^{2}
Area of side	12 in. • 2.5 in. =	30 in.^{2}
Area of front	12 in. • 8 in. =	96 in.^{2}
Area of back	12 in. • 8 in. =	96 in.^{2}
Surface area (sum)	=	292 in.^{2}

USING THE FORMULA FOR SURFACE AREA

The formula (which is provided for you on the FCAT) for the surface area of a rectangular solid compares to what has been done above.

$$\begin{aligned} SA &= 2(lw) + 2(hw) + 2(lh) \\ &= 2(8 \text{ in.} \cdot 2.5 \text{ in.}) + 2(12 \text{ in.} \cdot 2.5 \text{ in.}) + 2(12 \text{ in.} \cdot 8 \text{ in.}) \\ &= 2 \cdot 20 \text{ in.}^2 + 2 \cdot 30 \text{ in.}^2 + 2 \cdot 96 \text{ in.}^2 \\ &= 292 \text{ in.}^2 \end{aligned}$$

Test Tip: When working a surface area problem, it helps to make a drawing.

Sample Questions

1. The dimensions of a box are 16 inches × 10 inches × 2 inches. How many square inches of wrapping paper are needed to exactly cover the box?
2. A cube has a surface area of 486 square inches. What is the edge length of one face of the cube?

Answers

1. $SA = 2(lw) + 2(hw) + 2(lh)$
   $= 2(16 \cdot 10) + 2(2 \cdot 10) + 2(16 \cdot 2)$
   $= 320 + 40 + 64 = 424$ in.^{2} of wrapping paper
2. A cube has six sides, and so the area of each side is $486 \div 6 = 81$ square inches. Since the face of a cube is a square, take the square root of 81 to find the side length: $\sqrt{81} = 9$ inches.

Test Tip: One of the things the FCAT tests is your knowledge about the effect of changing the length, width, or height of a figure on its area or volume. You might think that doubling all dimensions would cause the area or volume to double, but this is not true.

In general, if you double two dimensions when finding the area, the area will be $2 \times 2 = 4$ times larger. If you double all three of the dimensions when finding the volume, the volume will be $2 \times 2 \times 2 = 8$ times larger.

Cutting dimensions in half decreases the area, making the new area $\frac{1}{2} \times \frac{1}{2} = \frac{1}{4}$ of the old area. Halving three dimensions decreases the volume, making the new volume $\frac{1}{2} \times \frac{1}{2} \times \frac{1}{2} = \frac{1}{8}$ as large.

Volume of a Rectangular Solid

$V = lwh$

Volume is a measurement unit used to describe how much space is inside an object. It refers to the *capacity* of the object. If the space is meant to be filled with liquid (for example, bottled water), the volume can be liters, milliliters (metric), ounces, quarts, or gallons (US customary).

Volume can also be expressed in cubic units. This refers to how many cubes, or parts of cubes, of a particular size would fit inside the solid. Cubic inches refers to cubes that are 1 inch × 1 inch × 1 inch. Similarly, a cubic centimeter is 1 cm × 1 cm × 1 cm. Once multiplied, the volume of one cubic centimeter is expressed as 1 cm^3.

Another formula for finding the volume of a rectangular solid is

$$V = lwh$$

Example 1: Find the volume of a rectangular solid with dimensions 4 cm, 5 cm, and 6 cm.

$V = lwh = 4 \bullet 5 \bullet 6 = 120 \text{ cm}^3$

Example 2: Find the volume of a cube with an edge length of 2 feet.

$V = lwh = 2 \bullet 2 \bullet 2 = 8$ cubic feet

The volume of a rectangular solid can be found by multiplying the area of the base (B) by the height (h): $V = Bh$.

Example 3: Find the volume of rectangular solid with a base area of 45 square feet and a height of 10 feet.

Using the formula $V = Bh$, multiply $45 \text{ ft}^2 \bullet 10 \text{ ft} = 450 \text{ ft}^3$.

Example 4: If a rectangular solid with a square base has a volume of 1584 cubic meters and a height of 11 meters, what is the side length of the base?

$V = Bh$

$1584 = B \bullet 11$ Substitute and divide both sides by 11.

$144 = B$

B represents the area of the base. Because the base is a square, $\sqrt{144} = 12$ is the side length.

Example 5: If you double the dimensions of a $3 \times 4 \times 5$ rectangular prism, the volume will

A. be twice as big
B. be 4 times as big
C. be 8 times as big
D. stay the same

C. The volume will be $2 \times 2 \times 2 = 8$ times larger. Try finding the volumes and comparing them: $3 \times 4 \times 5 = 60$ cubic units. After you double each dimension, the new volume will be $6 \times 8 \times 10 = 480$ cubic units. Since all three dimensions were affected, you need to multiply by 2 three times. $60 \times 2 \times 2 \times 2 = 480$.

Example 6: If you triple the length and width of a rectangular prism, the volume will be

A. 3 times as big
B. 6 times as big
C. 9 times as big
D. 27 times as big

C. The volume will be $3 \times 3 = 9$ times larger. Since only two dimensions were tripled, you need to multiply by 3 only two times. The old volume is lwh; the new volume is $3 \times 3 \times lwh = 9 \times lwh$.

Sample Questions

1. Find the volume of a rectangular prism with dimensions of 7 centimeters, 2 centimeters, and 11 centimeters.
2. Find the number of cubic inches in a rectangular prism with a base area of 50 inches and a height of 1 foot.
3. What is the height of a rectangular prism if the volume equals 450 cubic millimeters and the base area = 30 square millimeters?
4. Jeff has a custom-built aquarium with dimensions of 36 inches long by 16 inches wide, and 24 inches high. He filled the aquarium with water until the water level was 2 inches below the top. How many rocks can he add to the aquarium and still leave 1 inch of space at the top if each rock has an average displacement of 2.5 cubic inches? Show how you arrived at your answer.
5. An artificial rectangular holding pond is 150 feet long and 90 feet wide. It has an average depth of 4 feet. If one cubic foot is approximately 7.5 gallons, about how many gallons will the holding pond contain?
   A. 54,000 gallons
   B. 7200 gallons
   C. 405,000 gallons
   D. 202,500 gallons

Answers

1. $V = lwh$
   $= 2 \times 7 \times 11$
   $= 154 \text{ cm}^3$
2. $V = Bh$
   $= 50 \times 12$
   $= 600$ cubic inches. Convert 1 foot to 12 inches.
3. $V = Bh$
   $450 = 30\,h$ Divide by 30.
   $15 \text{ mm} = h$
4. $V = lwh = 36 \times 16 \times 1 = 576$ cubic inches of space allowable for rock displacement. $576 \div 2.5 = 230.4$. Jeff can put 230 rocks in the aquarium.
5. **C.** $V = 150 \cdot 90 \cdot 4 = 54{,}000$ cubic feet. Convert to gallons by multiplying by 7.5. Then, $54{,}000 \cdot 7.5 = 405{,}000$.

Surface Area and Volume of Right Cylinders

A right cylinder has two bases formed of circles. The rectangle that wraps around the circle to form its surface is called the *lateral* area.

Surface Area of a Right Cylinder

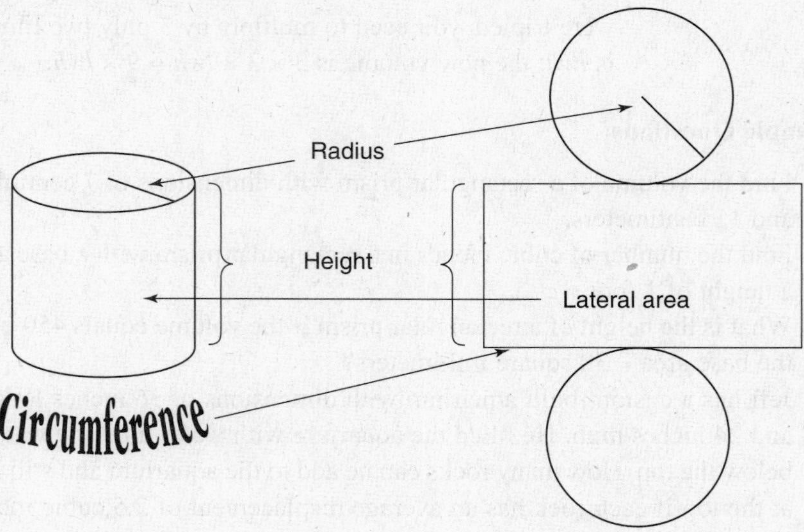

To find the surface area of a right cylinder, you need to add the area of the circular bases to the area of the lateral side (the "label" portion) of the can. The length of the label portion is the *circumference* of the can, and the width of the label portion is the height of the can.

To find the surface area, follow these steps:

> Notice that the formula for the surface area of a cylinder is $SA = 2\pi rh + 2\pi r^2$. It will be provided for you on the FCAT reference sheet when you take the test.

Step 1: Find the area of the two circular bases ($2\pi r^2$).

Step 2: Find the area of the lateral side ($2\pi rh$).

Step 3: Add the two areas together ($SA = 2\pi rh + 2\pi r^2$).

Example: Find the surface area of a right cylinder that has a radius of 5 inches and a height of 12 inches. Use $3.14 \approx \pi$.

Step 1: $SA = 2\pi r^2 + 2\pi rh$
First, $2\pi r^2 = 2 \cdot 3.14 \cdot 5^2 = 2 \cdot 78.5 = 157$ in.^{2} (area of both circular bases)

Step 2: Then, $2\pi rh = 2 \cdot 3.14 \cdot 5 \cdot 12 = 376.8$ in.^{2}

Step 3: Add 157 in^2 + 376.8 in^2 = 533.8 in.^{2} to obtain the total surface area.

Sample Questions

1. Find the surface area of a right cylinder if it has a radius of 3 inches and a height of 5 inches.
2. Find the surface area of a right cylinder if it has a diameter of 12 centimeters and a height of 24 centimeters.
3. Find the number of square inches required for the label of a can with a 4-inch radius and a height of 10 inches.

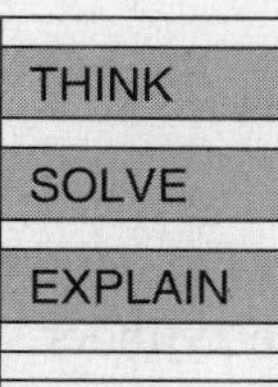

4. A soup can is $3\frac{7}{8}$ inches tall and has a diameter of $2\frac{5}{8}$ inches. How much paper is needed for the label if the label's height is $\frac{1}{4}$ inch shorter than the can and a $\frac{1}{4}$ inch overlap is required to glue the label? Use 3.14 for π.
   (*a*) Show what formula you need to find the amount of paper required for the label.
   (*b*) Draw the label and include its correct dimensions. Show how you calculated the dimensions of the label.
   (*c*) Calculate the amount of paper needed. Show the math you used to calculate your answer.

Answers

1. $SA_{cyl} = 2\pi r^2 + 2\pi rh$
   $2 \cdot 3.14 \cdot 3 \cdot 3 + 2 \cdot 3.14 \cdot 3 \cdot 5$
   $56.52 + 94.2 = 150.72$ square inches
2. $SA_{cyl} = 2\pi r^2 + 2\pi rh$
   $2 \cdot 3.14 \cdot 6 \cdot 6 + 2 \cdot 3.14 \cdot 6 \cdot 24$
   $226.08 + 904.32 = 1130.4$ square centimeters
3. $SA_{label} = 2\pi rh$
   $2 \cdot 3.14 \cdot 4 \cdot 10$
   251.2 square inches

4. (*a*) SA = $2\pi rh$.
(*b*) Step 1: Draw a rectangle.

Step 2: For the label, add $\frac{1}{4}$ inch to the length and subtract $\frac{1}{4}$ inch from the width. The length of the label is $2\pi r$, or πd. Add $\frac{1}{4}$ inch to the diameter: $2\frac{5}{8}+\frac{1}{4}$. Multiply by π: $3.14\left(2\frac{5}{8}+\frac{1}{4}\right) = 3.14 \bullet 2.875 = 9.03$ The width of the label (the height of the can) $= h = 3\frac{7}{8}-\frac{1}{4} = 3.625$.

$l = 2\pi r = 9.03$

$h = 3.625$

(*c*) $A = lw = 9.03 \bullet 3.625 = 32.7$ square inches

Volume of a Right Cylinder

To find the volume of a right cylinder, you can use the area of the base times the height ($V = Bh$) method by multiplying the area of the circular base by the height of the cylinder.

An alternate method is to use the formula for the volume of a right cylinder: $V = \pi r^2 h$. (Notice that the πr^2 portion of the formula is the formula for the area of a circle.)

Example: Find the volume of a cylinder with a base area of 314 mm^2 and a height of 15 mm. Use $3.14 \approx \pi$.

$V = Bh$
$= 314 \text{ mm}^2 \bullet 15 \text{ mm}$
$= 4710 \text{ mm}^3$

Example: Find the volume of a water tank that has a diameter of 30 feet and a height of 40 feet. Use $3.14 \approx \pi$.

$V = \pi r^2 h$
$= 3.14 \bullet 15^2 \bullet 40$ A diameter of 30 means a radius of 15.
$= 28{,}260$ cubic feet

Sample Questions

1. Find the volume of a right cylinder with a radius of 14 centimeters and a height of 18 centimeters.
2. Find the volume of a right cylinder with a diameter of 20 yards and a height of 30 yards.
3. Find the volume of a right cylinder with a base area of 45 square centimeters and a height of 12 centimeters.

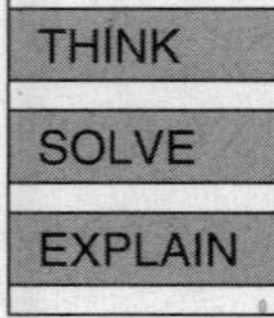

4. What is the volume of a cylinder with a height of $3\frac{3}{4}$ inches and a diameter of $2\frac{1}{2}$ inches? To the nearest half ounce, how many ounces of soup can it hold if 1 cubic inch = .6 ounce? Show how you arrived at your answer.

5. If the radius of a right cylinder is doubled, the volume of the cylinder will be
A. 2 times larger
B. 4 times larger
C. 3 times larger
D. 8 times larger

Answers

1. $V = \pi r^2 h$
$= 3.14 \cdot 14^2 \cdot 18$
$= 11{,}077.92 \text{ cm}^3$
2. $V = \pi r^2 h$
$= 3.14 \cdot 10^2 \cdot 30$
$= 9420 \text{ yd}^3$
3. $V = Bh$
$= 45 \text{ cm}^2 \cdot 12 \text{ cm}$
$= 540 \text{ cm}^3$
4. $V = \pi r^2 h$
$= 3.14 \cdot 1.25^2 \cdot 3.75 \cdot .6$
$= 11.03$ ounces
5. **B.** Use the $V = \pi r^2 h$. Double the radius: $V = \pi(2r)^2 h = \pi 4r^2 h = 4\pi r^2 h$. When the radius is doubled, squaring the radius causes the total volume to be multiplied by 4. $(2r)^2 = 4r^2$.

Surface Area and Volume of Cones

Surface Area of Cones

Unlike a cylinder, the cone has only one base. The surface of the cone curves up to a point called the *vertex*. The surface area of a cone is found by adding the area of the circular base to the area of the side or "cone" portion. To find the surface area of the cone, use the formula SA = $\pi r\ell + \pi r^2$. Notice that this formula uses ℓ, the lateral (slant) height, instead of the height of the cone, as shown in the diagram. Since the radius, the height of the cone, and the lateral height of the cone form a right triangle, the slant height can be found using the Pythagorean Theorem, or $a^2 + b^2 = c^2$, where a = radius, b = height, and c = slant height.

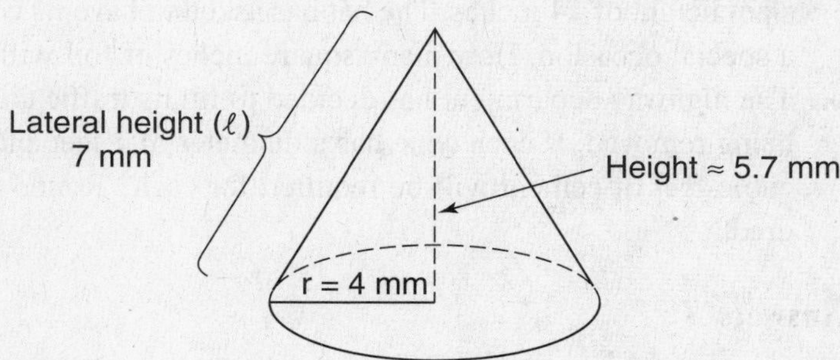

Example: Find the surface area of the cone shown here. Use $3.14 \approx \pi$.

To use the formula, follow these steps.

Step 1: Find the area of the side ($\pi r\ell$): $3.14 \cdot 4 \text{ mm} \cdot 7 \text{ mm} = 87.92 \text{ mm}^2$

Step 2: Find the area of the circular base (πr^2): $3.14 \cdot 4^2 = 50.24 \text{mm}^2$

Step 3: Add the areas together ($SA = \pi r\ell + \pi r^2$): $87.92 \text{mm}^2 + 50.24 \text{ mm}^2 = 138.16 \text{ mm}^2$

Volume of Cones

The volume of a cone represents the capacity of the cone. It can be found by using the formula $V = \frac{1}{3}\pi r^2 h$. If you already know the area of the base, use $V = \frac{1}{3}Bh$, where B represents the area of the circular base (πr^2).

Example: Find the volume of the cone shown here.

A cone has $\frac{1}{3}$ the volume of a cylinder of the same radius and height.

$$V = \frac{1}{3}\pi r^2 h$$
$$= \frac{1}{3} \cdot 3.14 \cdot 4^2 \cdot 5.7$$
$$\approx 95.5 \text{ mm}^3$$

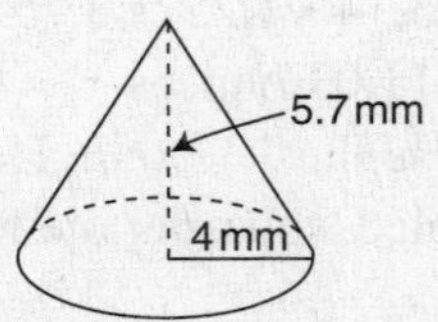

Sample Questions

1. Find the surface area of a cone with a radius of 2 meters and a slant height of 6 meters.
2. Which of the following represents the height of a cone with a volume of 33 mm^3 and a radius of 3 mm? Use $\frac{22}{7}$ for π.
   A. 3.5 mm
   B. 3.9 mm
   C. 9 mm
   D. 10.5 mm
3. A certain cone-shaped talking hat at a magician's school has a radius of 3.5 inches and a slant height of 14 inches. The hat has asked to have its cone portion covered with foil for a special occasion. How many square inches of foil will be needed to cover the hat?
4. The highway department has decided to fill its traffic cones with cement to prevent their being removed. If each cone has a diameter of 1 feet and a height of 2.5 feet, how many cubic feet of cement will be required for each? Round your answer to the nearest hundredth.

Answers

1. $SA = \pi r\ell + \pi r^2$
   $= 3.14 \cdot 2 \cdot 6 + 3.14 \cdot 2^2$
   $= 37.68 + 12.56 = 50.24 \text{ m}^2$

2. **A.**

$$V = \frac{1}{3}\pi r^2 h$$

$$33 = \frac{1}{3} \bullet \frac{22}{7} \bullet 3^2 \bullet h$$

$$33 = \frac{1}{3} \bullet \frac{22}{7} \bullet \frac{9}{1} \bullet h$$

$$33 = \frac{198}{21} h$$ Multiply both sides by $\frac{21}{198}$.

$$\frac{33}{1} \bullet \frac{21}{198} = \frac{21}{198} \bullet \frac{198}{21} h$$

$$h = 3.5$$

3. $SA = \pi r\ell + \pi r^2$

$$= 3.14 \bullet 3.5 \bullet 14 + 3.14 \bullet 3.5^2$$

$$= 153.86 + 38.465 = 192.3 \text{ in.}^2$$

4. $V = \frac{1}{3}\pi r^2 h$

$$= \frac{1}{3} \bullet 3.14 \bullet .5^2 \bullet 2.5 = .65 \text{ ft}^3$$ A diameter of 1 feet means a radius of .5 feet.

Surface Area and Volume of Pyramids

Pyramids are labeled very much the same way as cones. They have a height (h), a vertex, and a slant height (ℓ). They are named for the shape of their bases. The pyramid shown here is called a *square pyramid* because the base is a square. The sides of pyramids are made up of triangles. The base of a square pyramid has an edge length (l). Do not confuse this with the slant height (ℓ). The slant height can be found using the Pythagorean Theorem because the slant height represents the hypotenuse of a right triangle formed by the height of the triangle and an imaginary line drawn from the height to the outside edge of the square. (See the accompanying diagram).

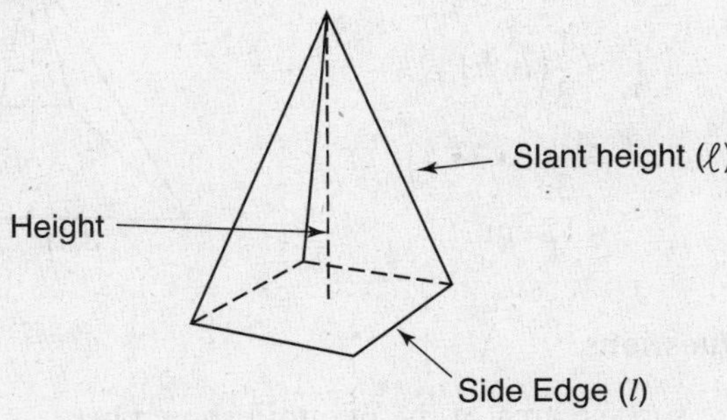

Surface Area of a Square Pyramid

The formula for the surface area of a pyramid is a combination of the formulas for the area of 4 triangles and a square:

$$\text{Surface area} = 4\left(\frac{1}{2} l\ell\right)(\text{area of four triangles}) + l^2$$

$$(\text{area of the square with side length } l)$$

Simplified, the formula looks like

$$SA = 2l\ell + l^2$$

Example: To find the surface area of the square pyramid shown here, use the formula:

$$SA = 2l\ell + l^2$$
$$= 2 \cdot 8 \cdot 10 + 8^2$$
$$= 224 \text{ cm}^2$$

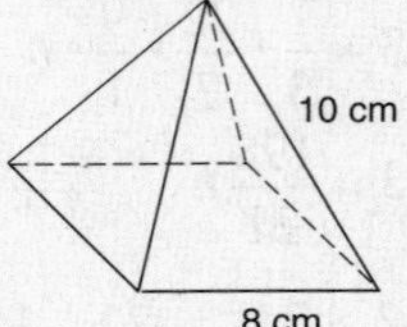

Volume of a Pyramid

The formula for the volume of a pyramid is based on the formula for a rectangular prism. A rectangular prism with the same base and height as a square pyramid has three times the volume as the pyramid. Since the formula for the volume of the rectangular prism is $V = lwh$, it follows that the formula for the volume of a pyramid is $V = \frac{1}{3}(lwh)$, where l is the base length, w is the base width, and h is the height of the pyramid.

A pyramid has $\frac{1}{3}$ the volume of a rectangular prism of the same dimensions.

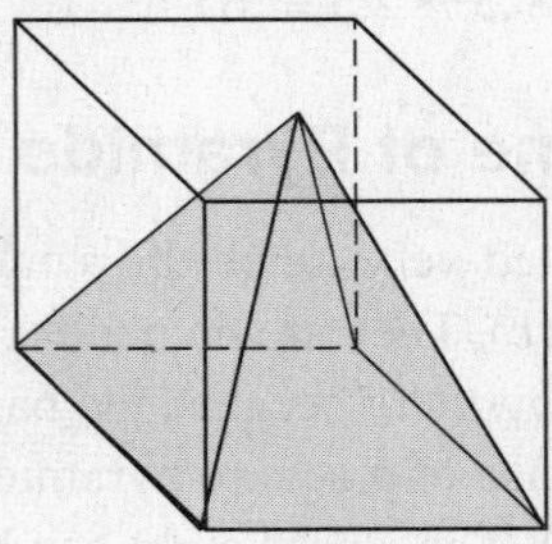

Example: Find the volume of the pyramid shown in the accompanying diagram.

$$V = \frac{1}{3}(lwh)$$
$$= \frac{1}{3}(3 \cdot 3 \cdot 4)$$
$$= 12 \text{ cm}^3$$

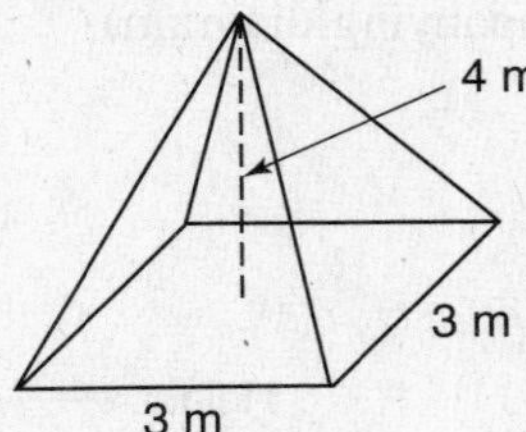

Sample Questions

1. Find the surface area of the pyramid shown here.

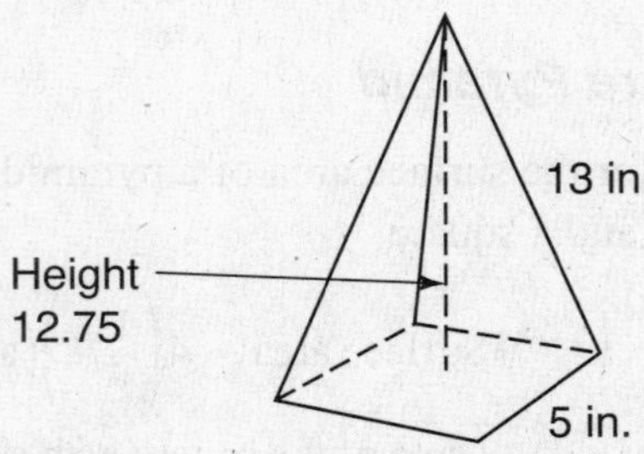

2. Find the volume of the pyramid shown in question 1.

Answers

1. $\text{SA} = 2l\ell + l^2$
   $= 2 \cdot 5 \cdot 13 + 5^2$
   $= 130 + 25 = 155 \text{ in.}^2$
2. $V = \frac{1}{3}(lwh)$
   $= \frac{1}{3} \cdot 5 \cdot 5 \cdot 12.75$
   $= 106.25 \text{ ft}^3$

Surface Area and Volume of Spheres

A sphere is the three-dimensional form of a circle. A baseball is an example of a sphere. The diameter of a sphere is measured as the distance from one side of the sphere to another, crossing the center. As in a circle, the radius is half the diameter. If you compare a sphere to any other solid with the same surface area, the sphere will have the largest volume.

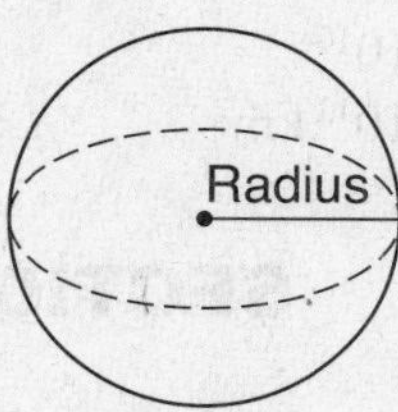

Surface Area of a Sphere

The surface area of a sphere is found using the formula surface area = $4\pi r^2$.

Example 1: Find the surface area of a sphere with a radius of 3.7 centimeters.

$SA = 4\pi r^2$.
$SA = 4 \cdot 3.14 \cdot 3.7^2$
$\approx 171.9 \text{ cm}^2$

Example 2: Find the surface area of a sphere with a diameter of 6 inches.

$SA = 4\pi r^2$
$= 4 \cdot 3.14 \cdot 3^2$
$\approx 113.04 \text{ in.}^2$

Volume of a Sphere

The formula for the volume of a sphere is $V = \frac{4}{3}\pi r^3$.

Example: Find the volume of a sphere with a radius of 3.7 centimeters.

$V = \frac{4}{3}\pi r^3$
$V = \frac{4}{3} \cdot 3.14 \cdot 3.7^3$
$\approx 212.1 \text{ cm}^3$

To multiply by $\frac{4}{3}$, first multiply by 4, then divide by 3.

Sample Questions

1. A circular asteroid made up of iron ore has a diameter of $\frac{1}{2}$ mile. To the nearest hundredth of a cubic mile, how much iron ore is that?

2. The sun has a radius of about 7×10^5 kilometers. What is its surface area? Express your answer in scientific notation.

Answers

1. $V = \frac{4}{3} \cdot 3.14 \cdot .25^3$
$\approx 0.065 \text{ mi}^3$
$\approx 0.07 \text{ mi}^3$

2. $SA = 4\pi r^2$
$= 4 \times 3.14 \times (7 \times 10^5) \times (7 \times 10^5)$
$\approx 12.56 \times 7 \times 7 \times 10^5 \times 10^5$
$\approx 12.56 \times 49 \times 10^{10}$
$\approx 615.44 \times 10^{10}$
$\approx 6.1544 \times 10^{12} \text{ km}^2$

SECTION 3: TRIANGLES

Triangles are three-sided polygons. They are generally classified in two ways, by the num-

Triangle Classification by Sides

Number of Congruent Sides	Name of Triangle	Diagram
Three	Equilateral triangle	
Two	Isosceles triangle	
None	Scalene triangle	

Triangle Classification by Angle

Type of Angle	Name of Triangle	Diagram
Three 60° angles	Equiangular	
One 90° angle	Right	
All angles <90°	Acute	
One angle >90°	Obtuse	

ber of congruent sides and by the measurement of their angles:

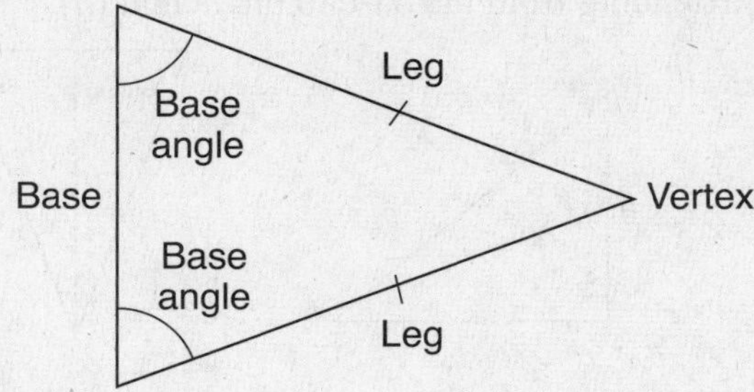

The triangle shown here is an isosceles triangle. The congruent sides are *legs*, and the remaining side is the *base*. In an isosceles triangle, the base angles, which are opposite the congruent sides, are always congruent. You can use this idea plus the fact that the angles add to 180° to find the measures of other angles in an isosceles triangle.

> ***Test Tip:***
> The sum of all angles in *any* triangle is 180°.

Example: If $m\angle R = 30°$, find $m\angle Q$.

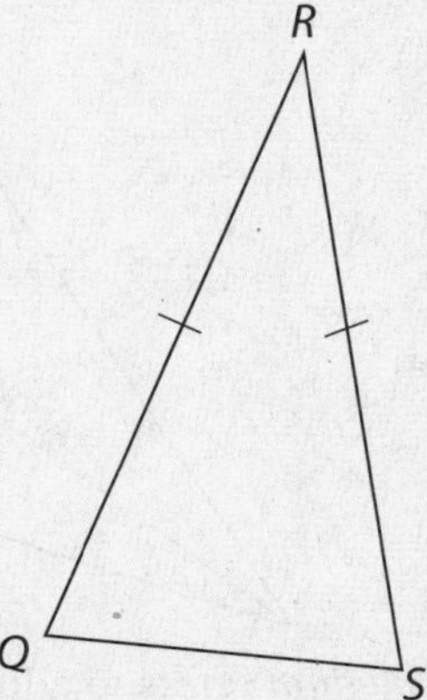

The sum of all angles in any triangle is 180°. Subtract: $180° - 30° = 150°$. The other two angles are congruent, so divide 150° by 2 to find each angle measurement. $m\angle Q = 75°$.

Example: If $m\angle D = 55°$, find $m\angle E$.

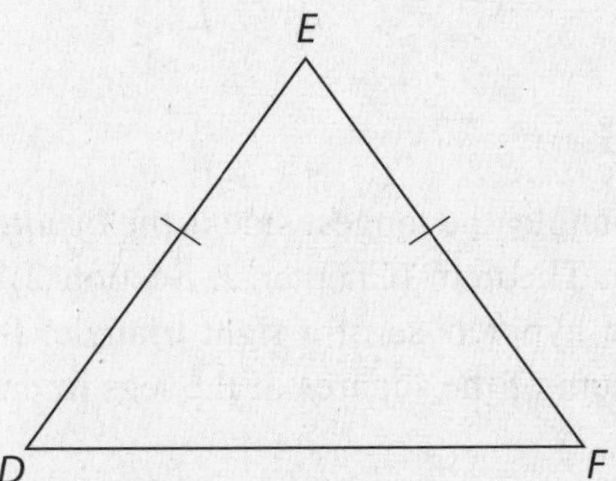

$m\angle D + m\angle E + m\angle F = 180°$. $\angle D \cong \angle F$, so substitute their measures: $55° + m\angle E + 55° = 180°$. $m\angle E = 70°$.

The height of a triangle is often called the *altitude*. It is a line segment perpendicular to the base extending from the base to the height (h).

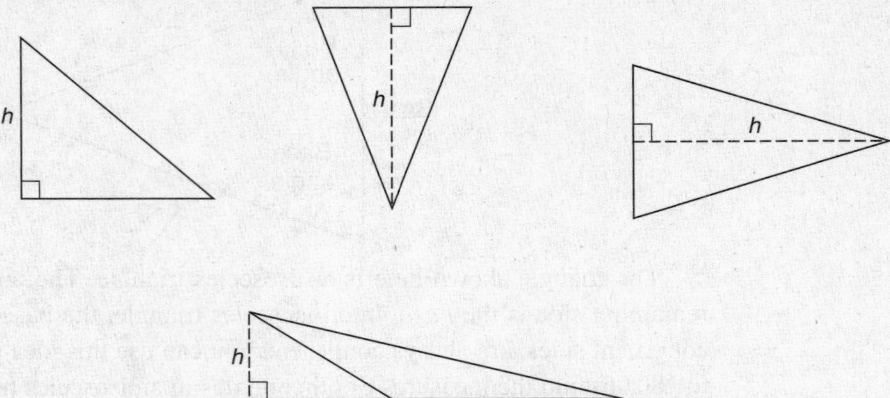

An *angle bisector* in a triangle is a line segment extending from a vertex of the triangle to the opposite side. The angle bisector cuts the angle at the vertex into two congruent angles but does *not* always cut the opposite side into two equal parts.

Bisect means to cut into two equal parts.

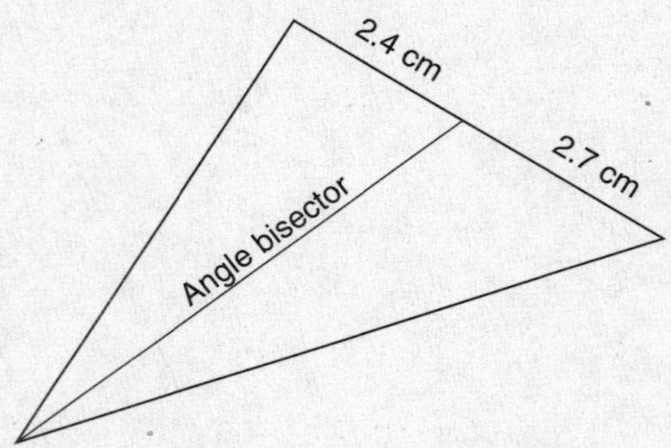

In a triangle, a *median* refers to a line segment drawn from the vertex of a triangle to the *midpoint* of the opposite side. In this diagram line segment $\overline{LM}$ is a median.

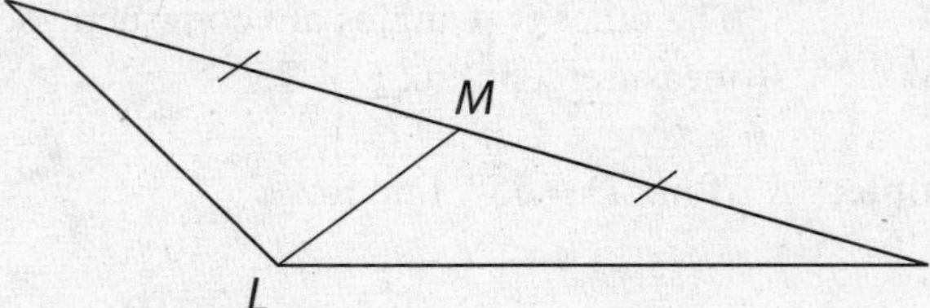

RIGHT TRIANGLES

In a right triangle, the longest side is the *hypotenuse*. The other two sides are called *legs*. The Pythagorean Theorem (Chapter 2, Section 2) describes a special relationship between the legs and the hypotenuse of a right triangle. It states that the square of the hypotenuse (c) equals the sum of the squares of the legs (a and b).

$c^2 = a^2 + b^2$

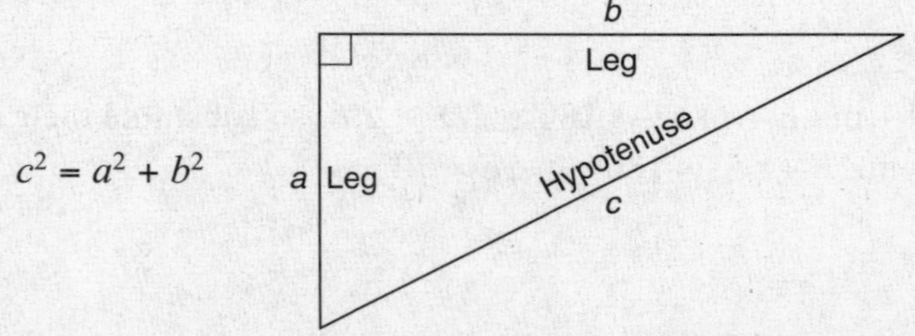

There are two special right triangles you should be familiar with: *45-45-90* and *30-60-90* triangles. The numbers 45-45-90 and 30-60-90 refer to the measures of the angles in these right triangles.

45-45-90 Right Triangles

A 45-45-90 triangle has two 45-degree angles and one 90-degree angle. Not only is it a right triangle but, because it has two congruent angles, it is also isosceles. We say that it is a *right isosceles triangle*.

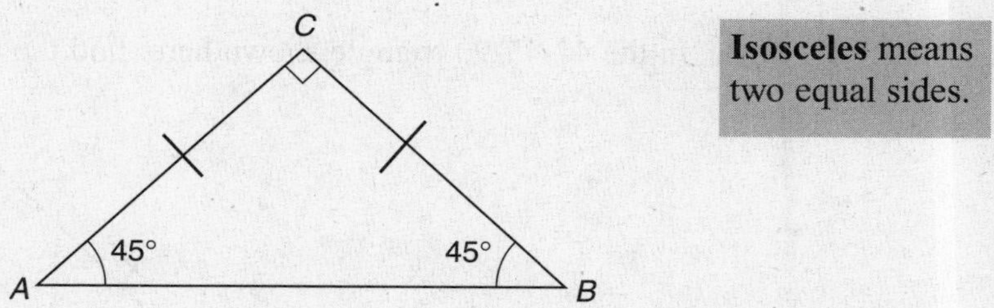

Isosceles means two equal sides.

In the 45-45-90 triangle shown here, the measure of the hypotenuse is equal to $\sqrt{2}$ times the measure of a leg: $c = a\sqrt{2}$ or $b\sqrt{2}$.

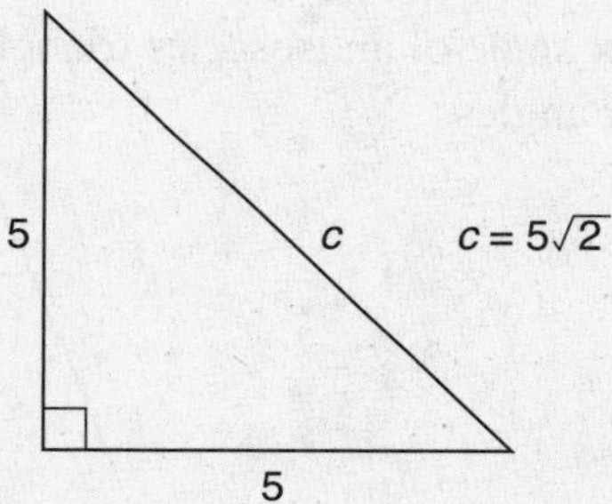

30-60-90 Right Triangles

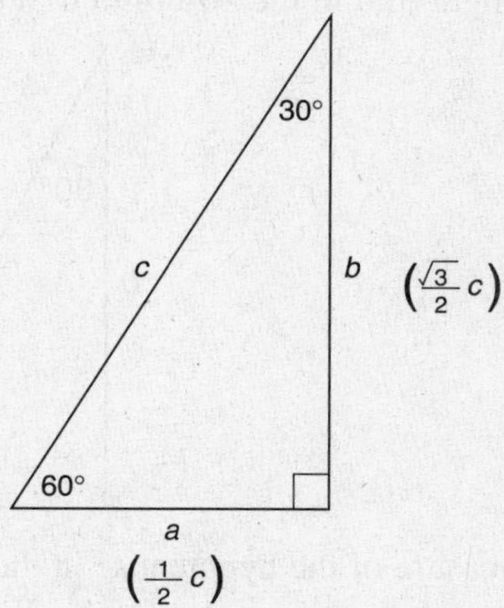

In the 30-60-90 triangle shown here, a, the measure of the side opposite the 30-degree angle is equal to $\frac{1}{2}$ times c, the measure of the hypotenuse. Also, b, the side opposite the 60-degree angle, is equal to $\sqrt{3}/2$ times the measure of c, the hypotenuse. (Multiply the hypotenuse by $\sqrt{3}$ and then divide by 2.) In the diagram, if $c = 12$, then $a = 6$ and $b = 12\sqrt{3}/2$, or $6\sqrt{3}$.

Sample Questions

1. In the diagram shown here, if $m\angle A = 50°$, find the measure of $\angle B$.

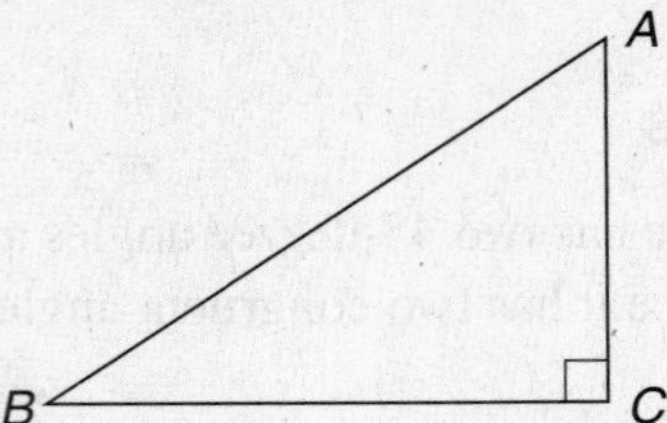

2. In the 45-45-90 triangle shown here, find the measure of x.

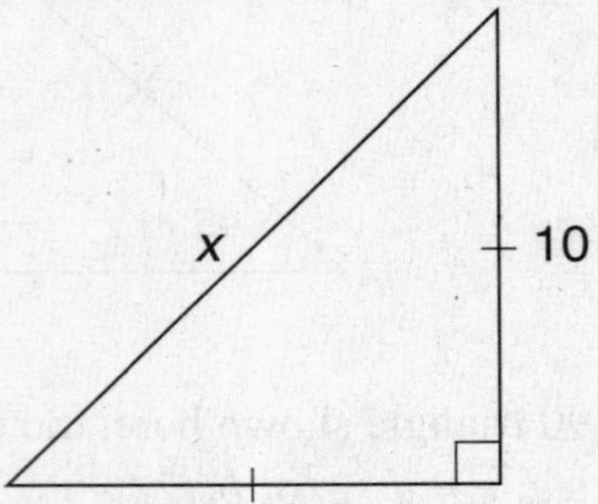

3. If the vertex angle of an isosceles triangle measures 40 degrees, find the measure of one of the other angles.

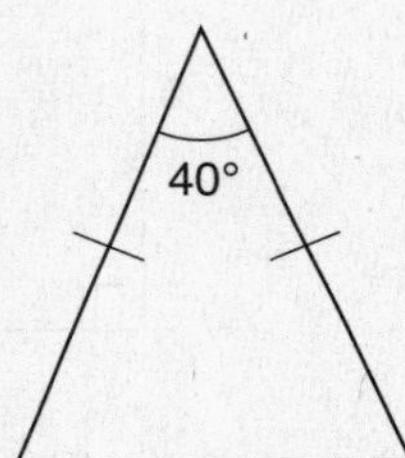

4. Find the length of b in the 30-60-90 triangle shown here.

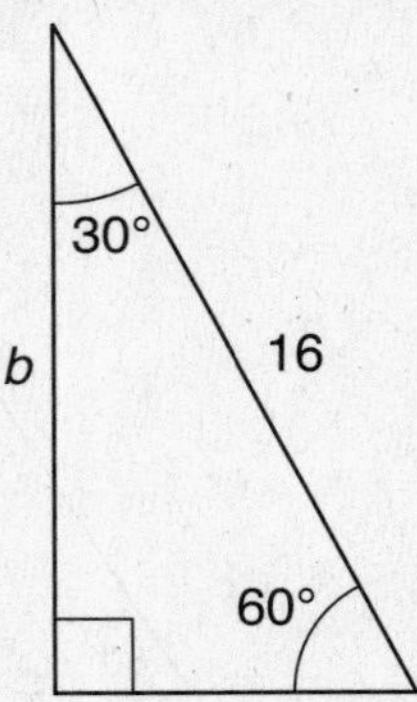

5. Find the measure of the hypotenuse in the 30-60-90 triangle shown here.

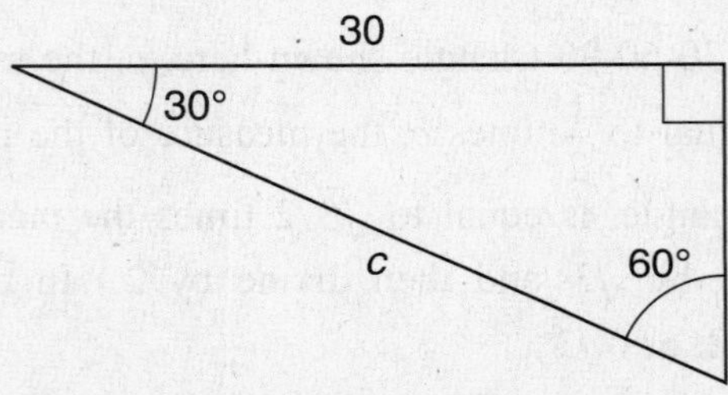

Answers

1. m$\angle B$ = 40°. Angles must add to 180 degrees.
2. $10\sqrt{2}$. Multiply the leg (10) by $\sqrt{2}$.
3. 70° 180 – 40 = 140. Divide 140 by 2.
4. $8\sqrt{3}$. Find $\frac{1}{2}$ of the hypotenuse: $\frac{1}{2} \times 16 = 8$. Multiply by $\sqrt{3}$.
5. $20\sqrt{3}$. $\frac{c\sqrt{3}}{2} = 30$; $c\sqrt{3} = 60$; $c = \frac{60}{\sqrt{3}}$; $c = \frac{60 \cdot \sqrt{3}}{\sqrt{3} \cdot \sqrt{3}} = \frac{60\sqrt{3}}{3} = 20\sqrt{3}$

ANGLE AND SIDE RELATIONSHIPS

In triangles, the longest side is opposite the largest angle.

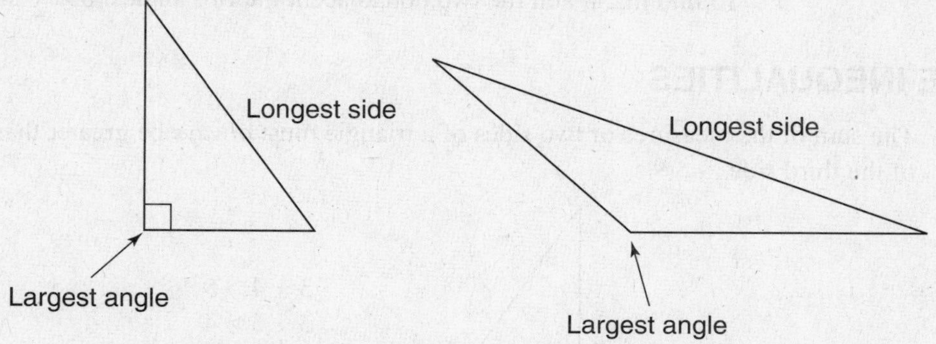

It is also true that the smallest angle is opposite the shortest side.

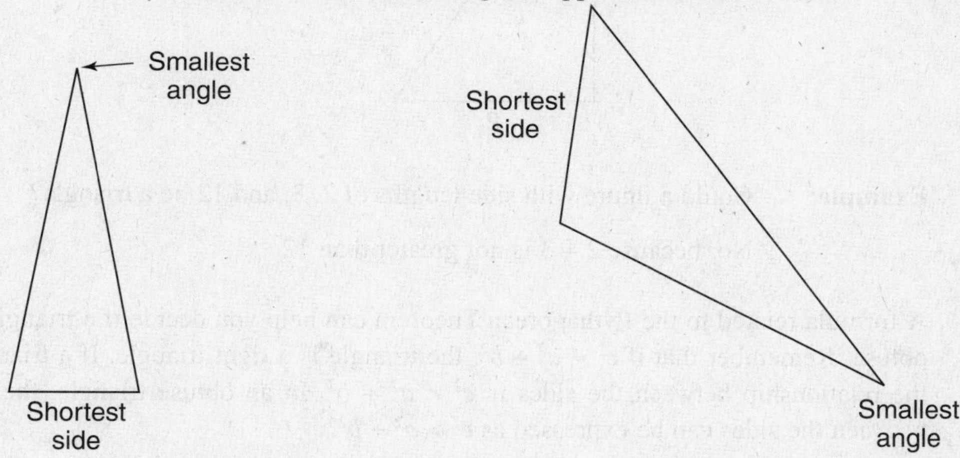

The exterior angle of a triangle can be found by adding the two nonadjacent interior angles. To find $\angle 4$ in the diagram shown, add $\angle 1$ and $\angle 2$ because $\angle 3$ is adjacent to $\angle 4$.

$$\angle 1 + \angle 2 = \angle 4$$

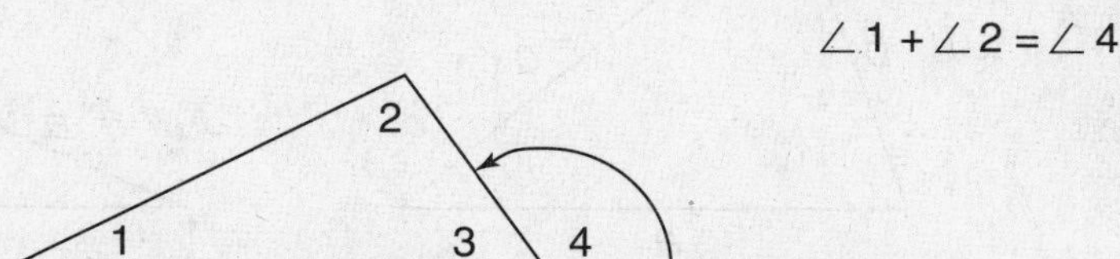

Example: Find m∠4 in the diagram.

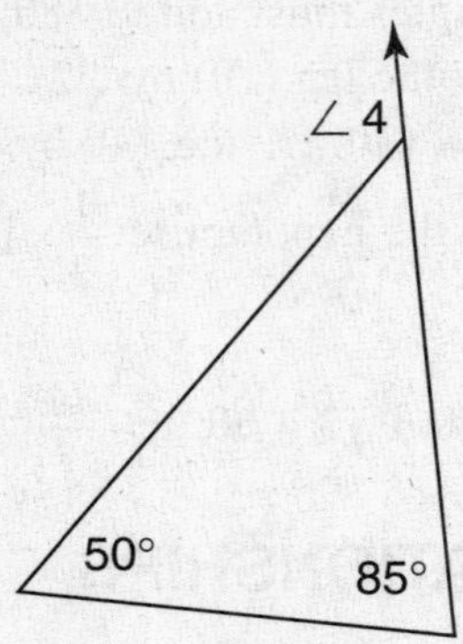

To find m∠4, add the two nonadjacent interior angles: 85° + 50° = 135°.

TRIANGLE INEQUALITIES

The sum of the measures of two sides of a triangle must *always* be greater than the measure of the third side.

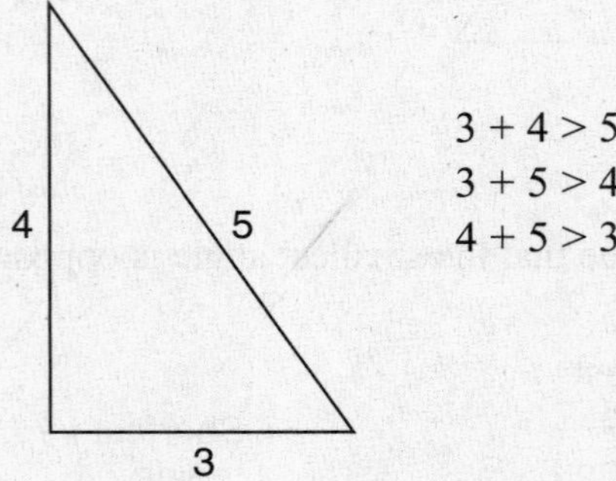

$3 + 4 > 5$
$3 + 5 > 4$
$4 + 5 > 3$

Example: Could a figure with side lengths of 2, 3, and 12 be a triangle?

No, because 2 + 3 is not greater than 12.

A formula related to the Pythagorean Theorem can help you decide if a triangle is acute or obtuse. Remember that if $c^2 = a^2 + b^2$, the triangle is a right triangle. If a triangle is acute, the relationship between the sides is $c^2 < a^2 + b^2$. In an obtuse triangle, the relationship between the sides can be expressed as $c^2 > a^2 + b^2$.

Obtuse means one angle is greater than 90°.

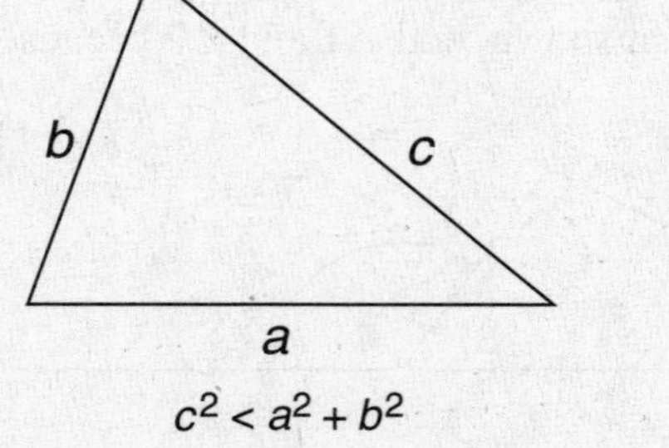

$c^2 < a^2 + b^2$

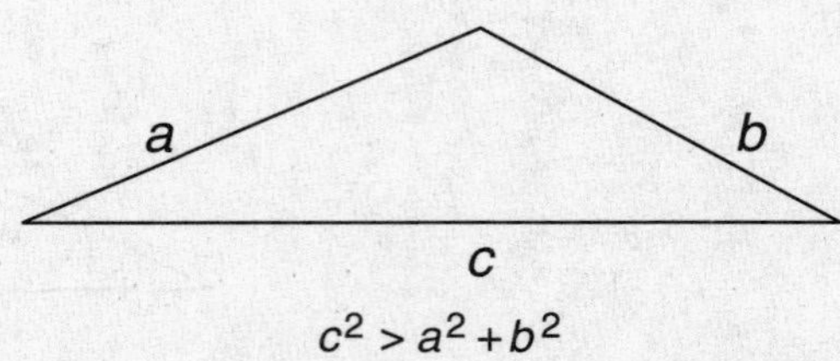

$c^2 > a^2 + b^2$

Sample Questions

1. Determine if a figure with side lengths of 6, 7, and 12 could be a triangle.
2. Is a triangle with side lengths of 9, 12, and 16 right, acute, or obtuse?
3. An obtuse triangle has two side lengths (a and b) of 7 and 12. What inequality statement about c describes the length of the third side?

Answers

1. Yes, because $6 + 7 > 12$, $6 + 12 > 7$, and $7 + 12 > 6$.
2. $9^2 + 12^2 < 16^2$. The triangle is obtuse.
3. $c > \sqrt{193}$. Because $c^2 > 193$.

SECTION 4: QUADRILATERALS

Quadrilaterals are four-sided polygons. Their names are based on the number of parallel sides, the number of equal sides, and their angles.

Parallelograms are quadrilaterals. Opposite sides are congruent (the same length) and parallel. In the accompanying diagram, you can see that the rhombus, the rectangle, and the square are all considered parallelograms.

Rectangles are quadrilaterals. Opposite sides are congruent and parallel, the angles are all 90°.

In a *rhombus*, all sides are congruent and opposite sides are parallel.

A *square* is both a rectangle and a rhombus. All sides are congruent, opposite sides are parallel, and the angles are all 90°.

Kites have two sets of congruent sides. None are parallel. Their diagonals are perpendicular.

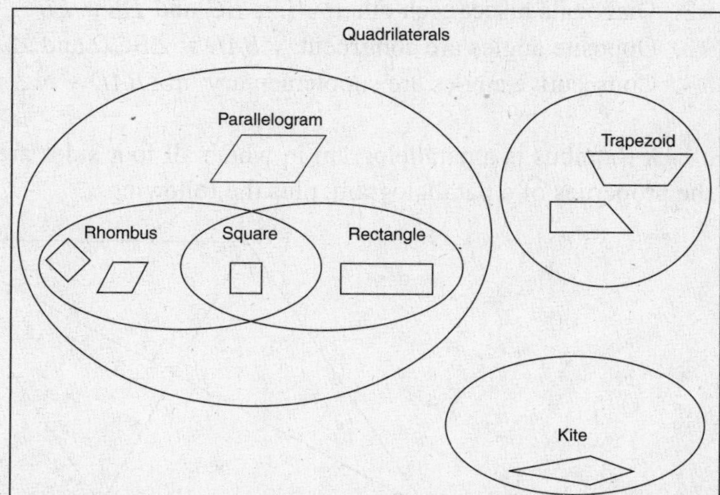

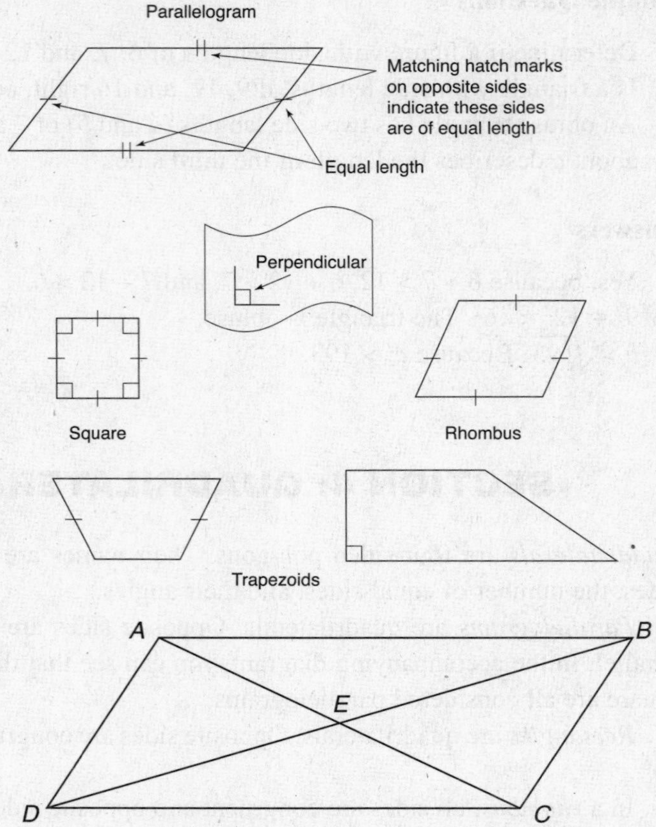

Parallelograms have specific properties:

1. Opposite sides are congruent: $\overline{AB} \cong \overline{DC}$ and $\overline{AD} \cong \overline{BC}$.
2. Diagonals bisect each other: $\overline{AE} \cong \overline{EC}$ and $\overline{DE} \cong \overline{EB}$.
3. Opposite angles are congruent: $\angle BAD \cong \angle BCD$ and $\angle ADC \cong \angle ABC$.
4. Consecutive angles are supplementary: $m\angle BAD + m\angle ADC = 180°$.

A rhombus is a parallelogram in which all four sides are congruent. A rhombus has all the properties of a parallelogram, plus the following.

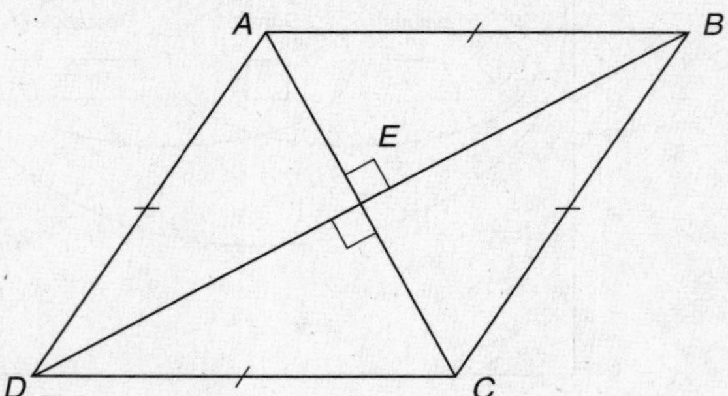

1. Diagonals bisect opposite angles; for example, $\angle DCE \cong \angle BCE$.
2. Diagonals are perpendicular to each other: $\overline{AC} \perp \overline{BD}$.

Rectangles are parallelograms. Rectangles have all the properties of a parallelogram, plus the following.

1. All angles are 90 degrees.
2. Diagonals are congruent.

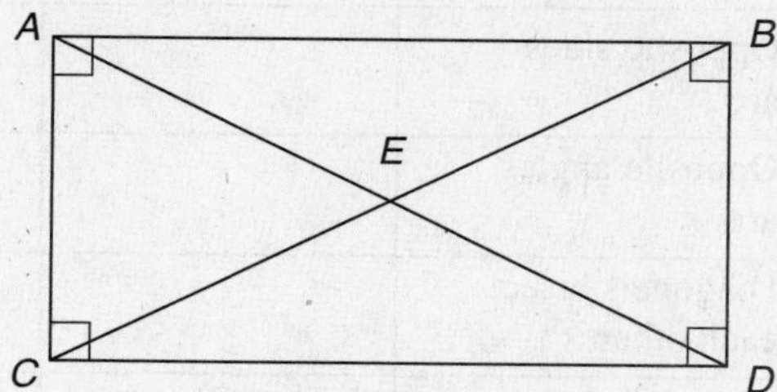

In addition to being a parallelogram, a square is both a rectangle and a rhombus. Therefore, squares have the properties of both a rectangle and a rhombus.

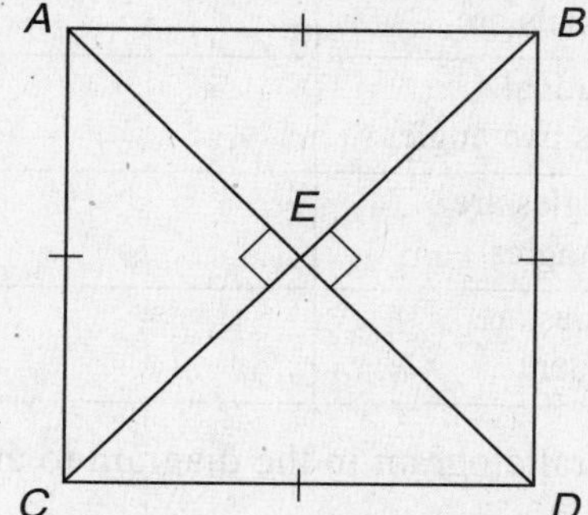

Trapezoids are quadrilaterals. Because they only have one set of parallel sides, they are in a special classification by themselves. The parallel sides are called the *bases*. The other sides are *legs*. Trapezoids with congruent legs are called *isosceles trapezoids*. In the case of isosceles trapezoids, the base angles are congruent. The *median* of the trapezoid is the line segment drawn between the midpoints of each of the legs. The median is parallel to the bases. The length of the median can be found by averaging the bases: $\frac{\text{base}_1 + \text{base}_2}{2}$.

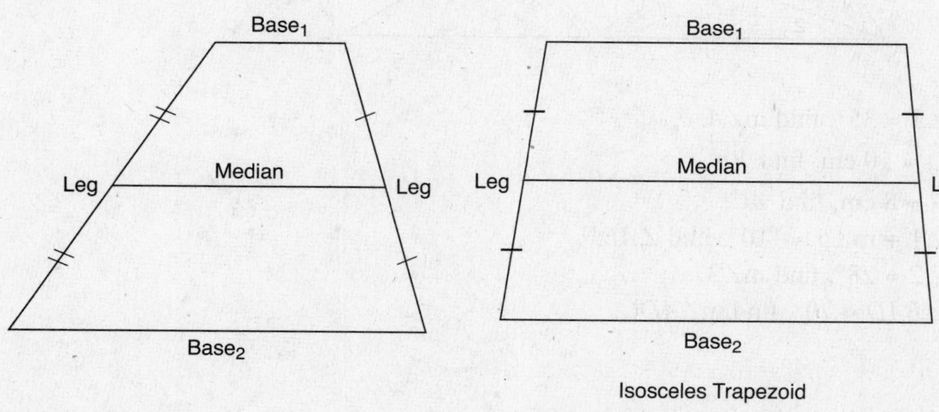

Isosceles Trapezoid

A trapezoid with two equal sides is called an isosceles trapezoid.

Test Tip: A shortcut for finding the area of a trapezoid is to multiply the median by the height.

Sample Questions

1. Place check marks in the appropriate spaces.

Property	Parallelogram	Rectangle	Rhombus	Square
Opposite sides are ∥				
Opposite sides are ≅				
Opposite angles are ≅				
Diagonals bisect each other				
A diagonal forms two ≅ triangles				
Diagonals are ≅				
Diagonals are ⊥				
A diagonal bisects two angles				
All angles are right angles				
All sides are congruent				

2. Use the parallelogram in the diagram to answer the following questions.

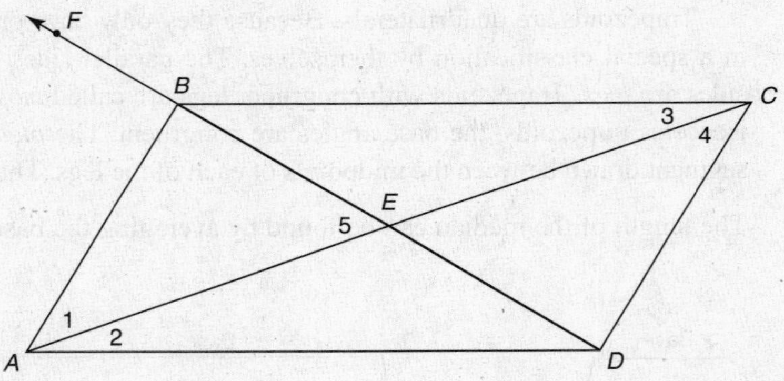

(*a*) If $m\angle 1 = 35°$, find $m\angle 4$.
(*b*) If $\overline{BD} = 10$ cm, find $\overline{BE}$.
(*c*) If $\overline{AE} = 8$ cm, find $\overline{AC}$.
(*d*) If $m\angle 1 + m\angle 5 = 110°$, find $\angle ABF$.
(*e*) If $m\angle 2 = 28°$, find $m\angle 3$.
(*f*) If $m\angle BAD = 70°$, find $m\angle ADC$.

3. Use the diagram of the trapezoid to answer the following questions.

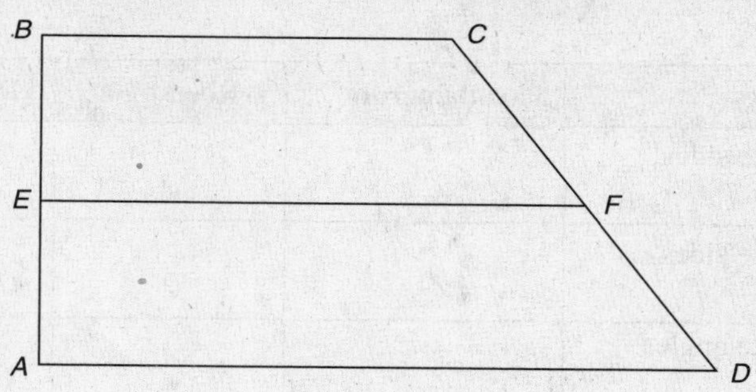

The length of the median = $\frac{\text{base}_1 + \text{base}_2}{2}$

(*a*) How long is $\overline{EF}$ if $\overline{BC} = 4$ and $\overline{AD} = 10$?

(*b*) If $\overline{EF} = 16$ and $\overline{AD} = 20$, find $\overline{BC}$.

(*c*) If $\overline{CD} = 22$, find $\overline{FD}$.

4. In the rhombus shown here, if m∠1 = 30°, find m∠3.

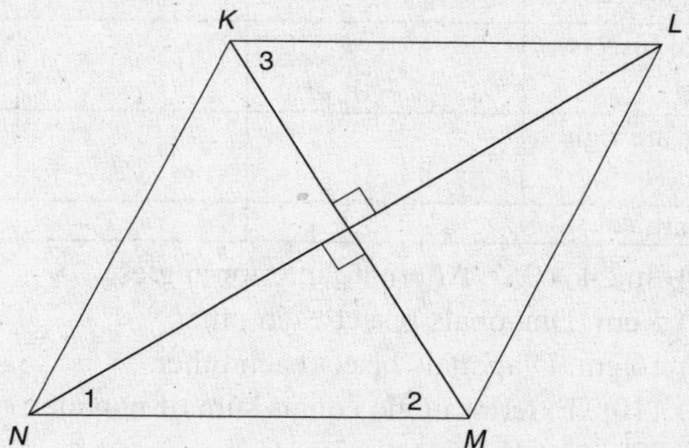

5. In the diagram below, $\overline{FC}$ is an altitude of △*ABC*. ∠1 and ∠2 are

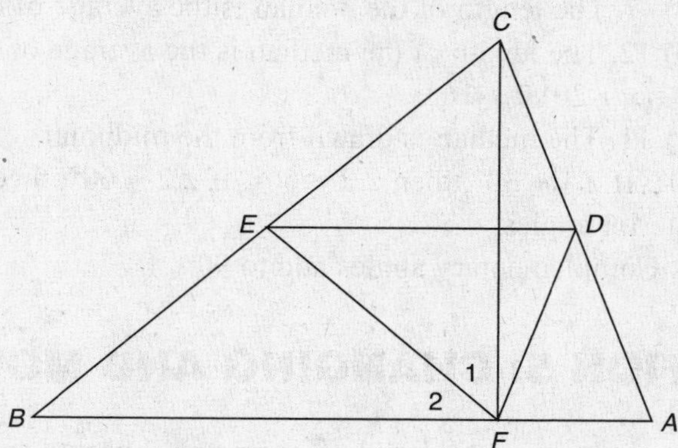

A. supplementary angles
B. right angles
C. congruent angles
D. complementary angles

Answers

1.

Property	Parallelogram	Rectangle	Rhombus	Square
Opposite sides are ∥	✓	✓	✓	✓
Opposite sides are ≅	✓	✓	✓	✓
Opposite angles are ≅	✓	✓	✓	✓
Diagonals bisect each other	✓	✓	✓	✓
A diagonal forms two ≅ triangles	✓	✓	✓	✓
Diagonals are ≅		✓		✓
Diagonals are ⊥			✓	✓
A diagonal bisects two angles			✓	✓
All angles are right angles		✓		✓
All sides are ≅			✓	✓

2. (*a*) m∠4 = 35°. Alternate interior angles.
   (*b*) 5 cm. Diagonals bisect each other.
   (*c*) 16 cm. Diagonals bisect each other.
   (*d*) 110°. Exterior angle equals sum of nonalternate interior angles in a triangle.
   (*e*) 28°. Alternate interior angles.
   (*f*) 110°. Sequential angles are supplementary.
3. (*a*) 7. The length of the median is the average of the bases.
   (*b*) 12. The length of the median is the average of the bases.
   $(x + 20)/2 = 16$
   (*c*) 11. The median is drawn from the midpoints of the legs, so $\overline{FD}$ is half of $\overline{CD}$.
4. 60°. If ∠1 = 30°, then ∠2 = 60°. If ∠2 = 60°, then ∠3 = 60° because they are alternate interior angles.
5. **D.** Complementary angles add to 90°.

SECTION 5: CHANGING AND MOVING SHAPES

In this section, we will build on geometry concepts reviewed in Section 1–4. We will look at various shapes in geometry and how they combine to form new shapes, how shapes can be broken into simpler forms, and how shapes can be changed. We will use terms and notation we have previously touched on such as *congruent* (≅), *perpendicular* (⊥), and *parallel* (||).

CONGRUENCE

Two figures are *congruent* when they have both the same size and the same shape. Each side on one figure corresponds to another side on the second figure. As mentioned in previous sections, we will use the geometry notation ≅ to show congruency. If corresponding parts (sides and angles) of two figures are congruent, then the two figures are congruent. Also, if two figures are congruent, then their corresponding parts are congruent. It works both ways.

Congruent Triangles

The triangles shown here are congruent ($\triangle ABC \cong \triangle FGH$). That is, they are the same size and they are the same shape. In addition, each angle and side on the first figure is congruent to an angle and side on the second figure.

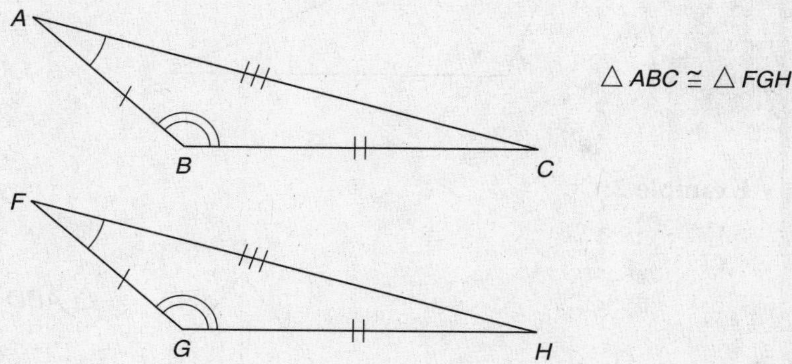

In the figure, $\angle A$ corresponds to $\angle F$; therefore $\angle A \cong \angle F$. Notice the markings on the sides and angles of each figure. When the markings match, the sides or angles are congruent. This is true in terms of both one figure and more than one figure. "Same markings" means "same measure."

For the FCAT, there are five methods you can use to prove two triangles are congruent: side-side-side (SSS), side-angle-side (SAS), angle-side-angle (ASA), angle-angle-side (AAS), and, for right triangles only, hypotenuse-leg (HL).

Side-Side-Side (SSS)

If all three sides of one triangle are congruent to three sides of a second triangle, the triangles are congruent.

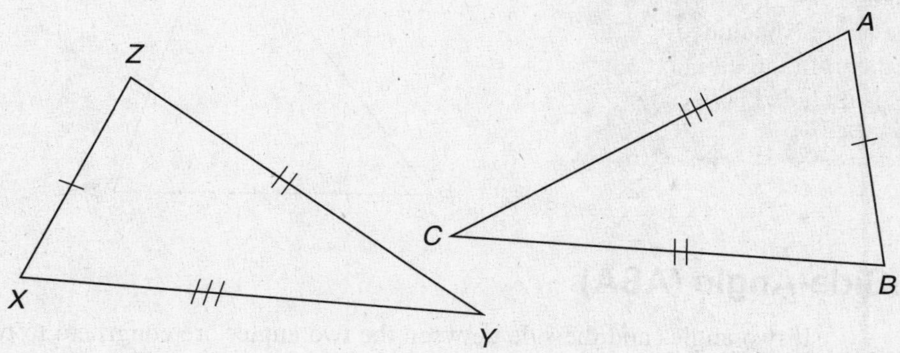

$\triangle XYZ \cong \triangle ACB$

Side-Angle-Side (SAS)

If two sides and the angle between the two sides (the *included* angle) are congruent to two sides and the included angle of another triangle, the triangles are congruent.

Example 1:

For SAS to work, the angles must be located at the intersection of the two corresponding sides.

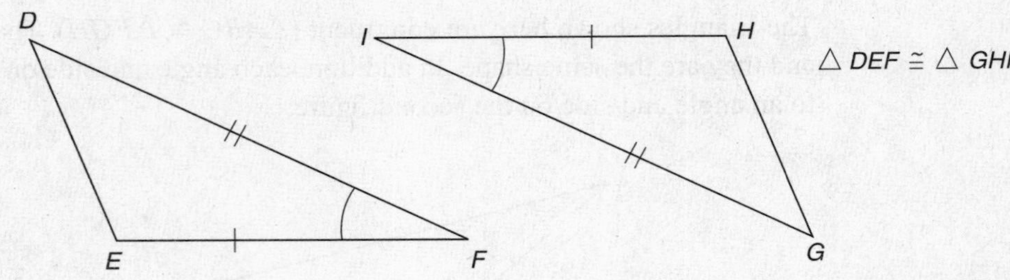

Example 2:

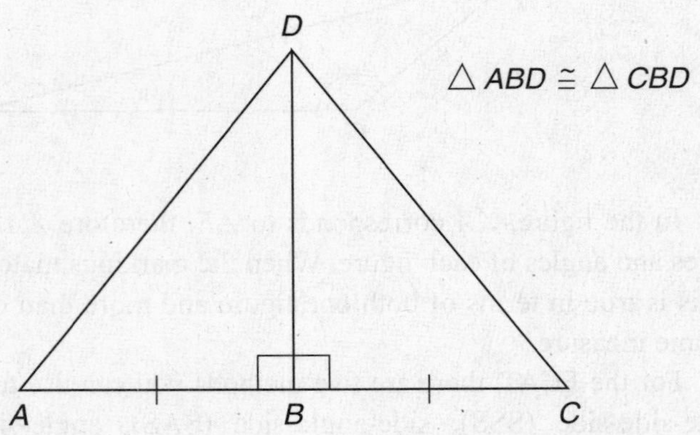

In Example 2, $\overline{BD}$ is "shared" and counts as a congruent side even though it is not marked as congruent.

Test Tip: You may still be having trouble deciding what *included* means. When two rays form an angle, the angle is included between them. In the diagram here, $\angle P$ is included between sides $\overline{PR}$ and $\overline{PQ}$.

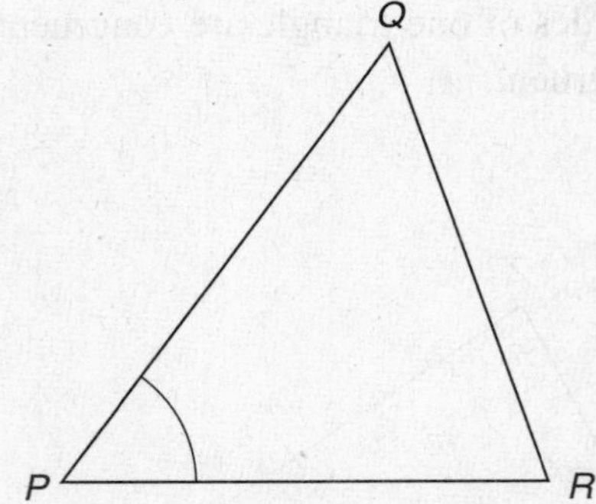

Angle-Side-Angle (ASA)

If two angles and the side between the two angles are congruent to two angles and the side between them, the triangles are congruent.

Example 1:

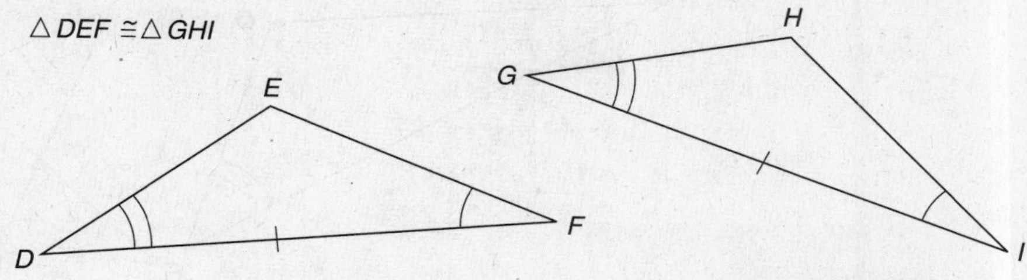

Example 2:

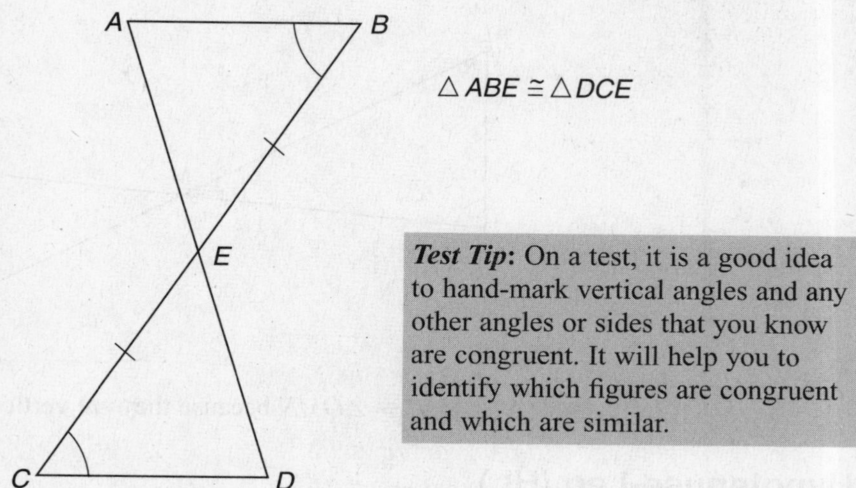

> ***Test Tip***: On a test, it is a good idea to hand-mark vertical angles and any other angles or sides that you know are congruent. It will help you to identify which figures are congruent and which are similar.

In Example 2, $\angle AEB \cong \angle DEC$ because they are vertical angles. On the FCAT, you will not generally find vertical angles marked as congruent because it is assumed that you know this (Section 1).

Angle-Angle-Side (AAS)

If two angles and a *nonincluded* side (a side not between the two angles) are congruent to two angles and a nonincluded side of another triangle, the two triangles are congruent.

Example 1:

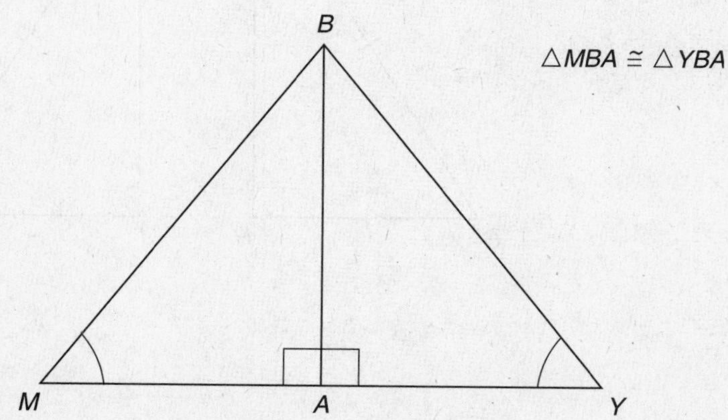

In Example 1, $\overline{BA}$ is shared and counts as a congruent side even though it is not marked.

Example 2:

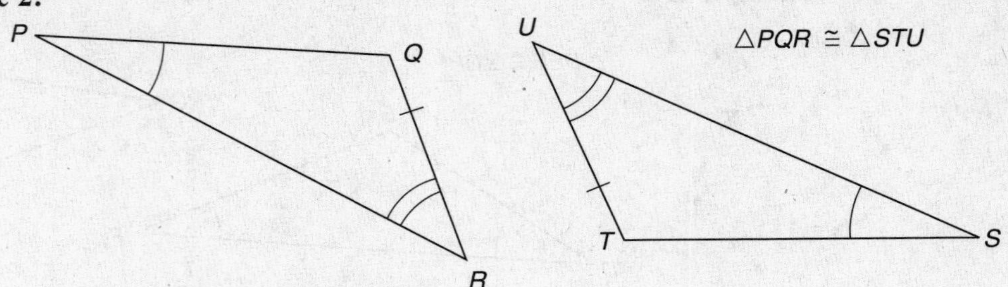

Example 3:

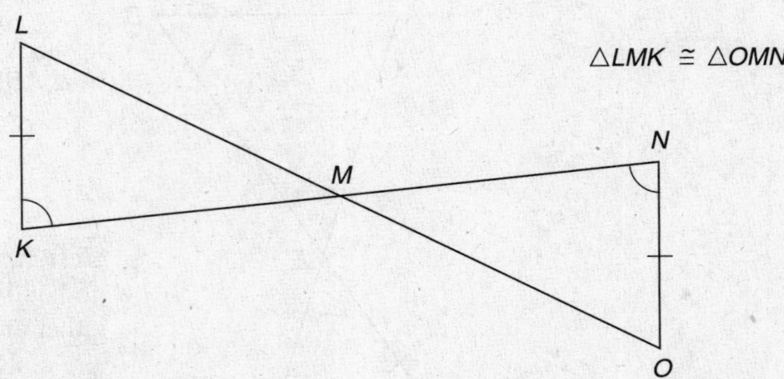

In Example 3, $\angle LMK \cong \angle OMN$ because they are vertical angles.

Hypotenuse-Leg (HL)

This congruence theorem works only for right triangles. If a hypotenuse and a leg of one right triangle are congruent with the hypotenuse and leg of a second triangle, the triangles are congruent.

Example 1:

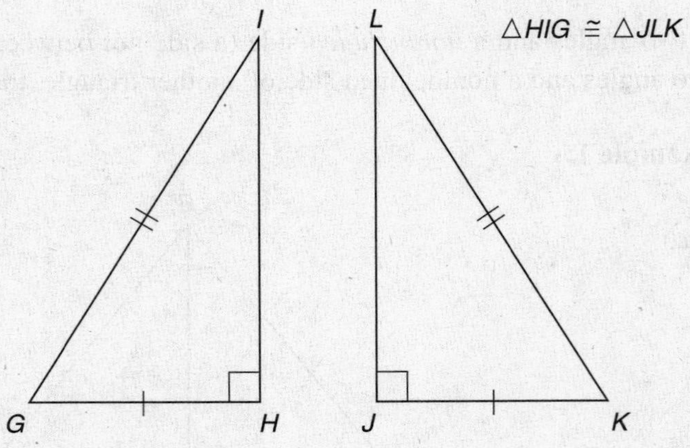

Example 2:

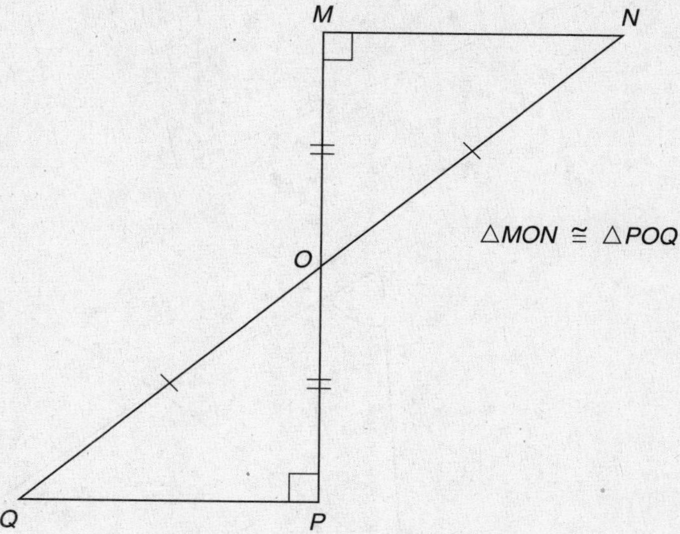

Congruent Polygons

The figures shown below are congruent. That is, they are the same size and shape. Each angle and side on the first figure corresponds to and is congruent to each angle and side on the second figure.

A **polygon** is a closed-in shape made up of straight lines.

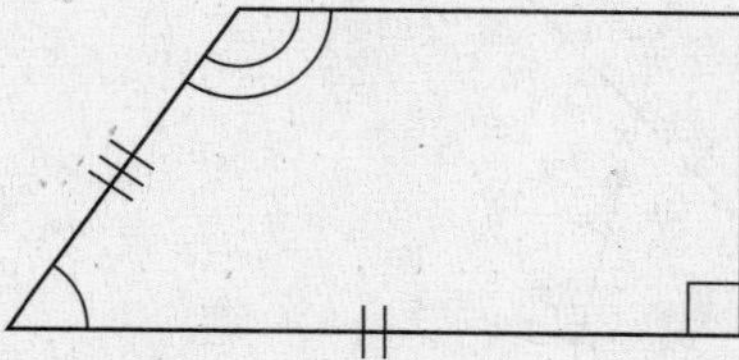

Sample Questions

Which of the following method proves the triangles shown are congruent—SSS, SAS, ASA, AAS, or HL?

1.

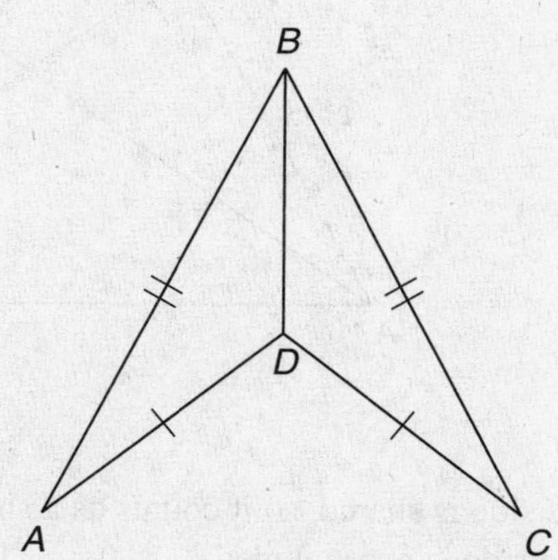

2.

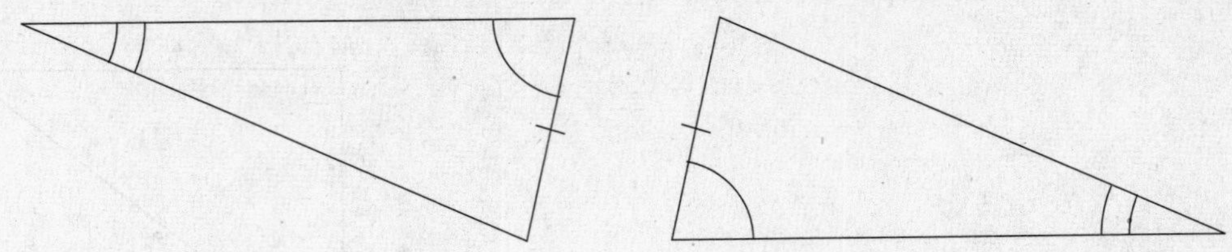

3.

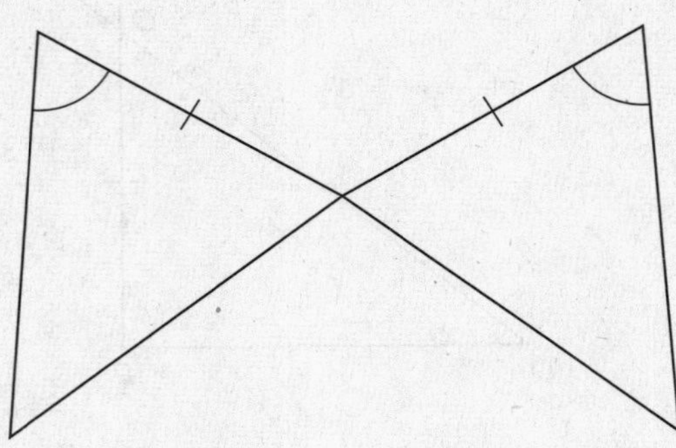

4.

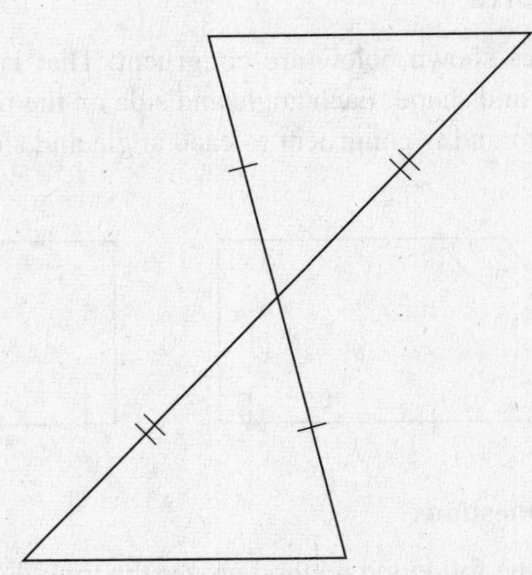

5.

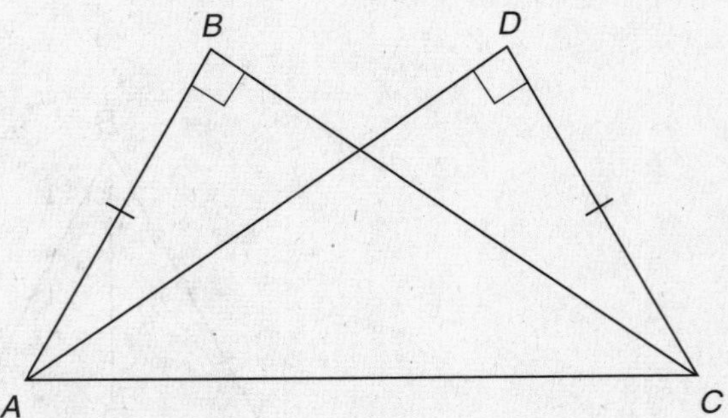

Answers

1. SSS. One side is shared, so it counts as a side for both triangles.
2. AAS. The side is not included.

3. ASA. The vertical angles are congruent.
4. SAS. The vertical angles are congruent.
5. HL. These are right triangles. They share a hypotenuse, so they have a leg and a hypotenuse that are congruent.

SIMILAR TRIANGLES

Two figures are *similar* if they have the same shape but not necessarily the same size. In the diagram shown, $\triangle ABC$ is similar to $\triangle DEF$. In geometry notation this is written as $\triangle ABC \simeq \triangle DEF$.

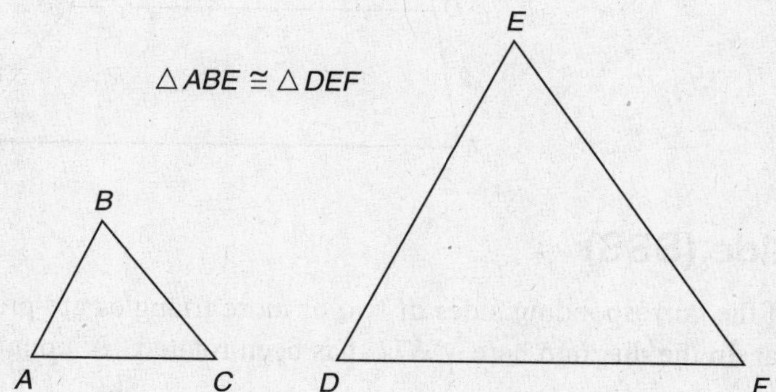

For the FCAT, you should be familiar with three methods of proving triangles are similar: angle-angle (AA), side-angle-side (SAS), and side-side-side (SSS).

Angle-Angle (AA)

If two angles in one triangle are congruent to two corresponding angles in a second triangle, the triangles are similar ($\simeq$).

Example

The geometry symbol for similar is $\simeq$.

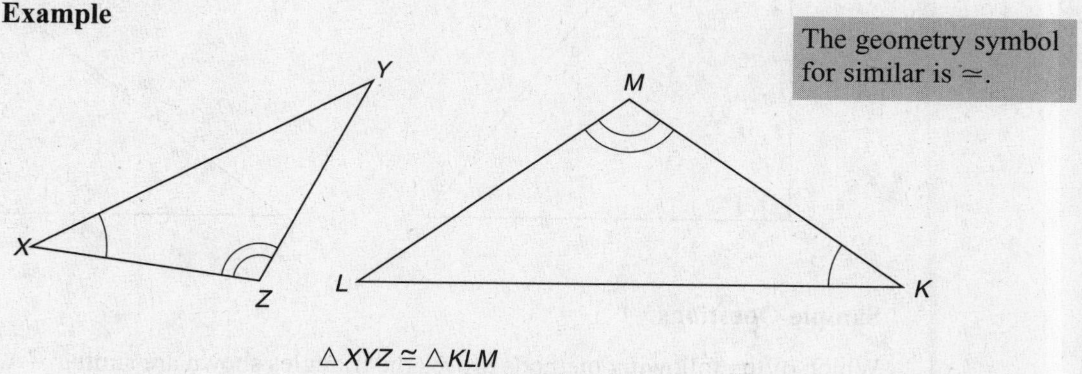

Side-Angle-Side (SAS)

If an angle of one triangle is congruent to an angle of a second triangle, and if the lengths of the sides on either side of the angles are proportional, the triangles are similar. In the diagram here, $\angle B$ is shared by both triangles and

$$\frac{\overline{BD}}{\overline{BA}} = \frac{\overline{BE}}{\overline{BC}} \quad \text{or} \quad \frac{4}{6} = \frac{6}{9}$$

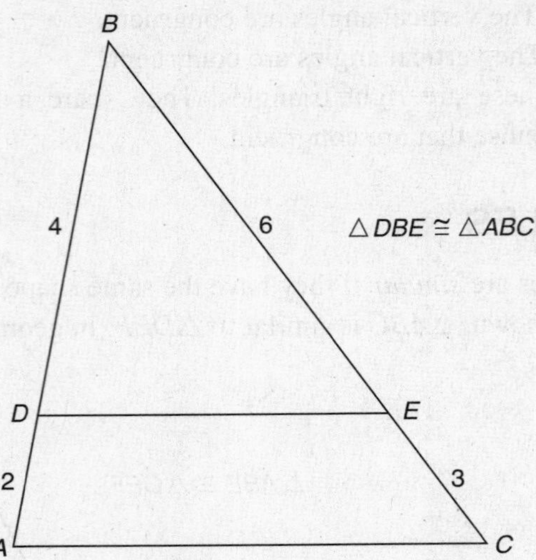

Side-Side-Side (SSS)

If the corresponding sides of two or more triangles are proportional, the triangles are similar. In the diagram here, $\triangle STU$ has been rotated 90° counterclockwise.

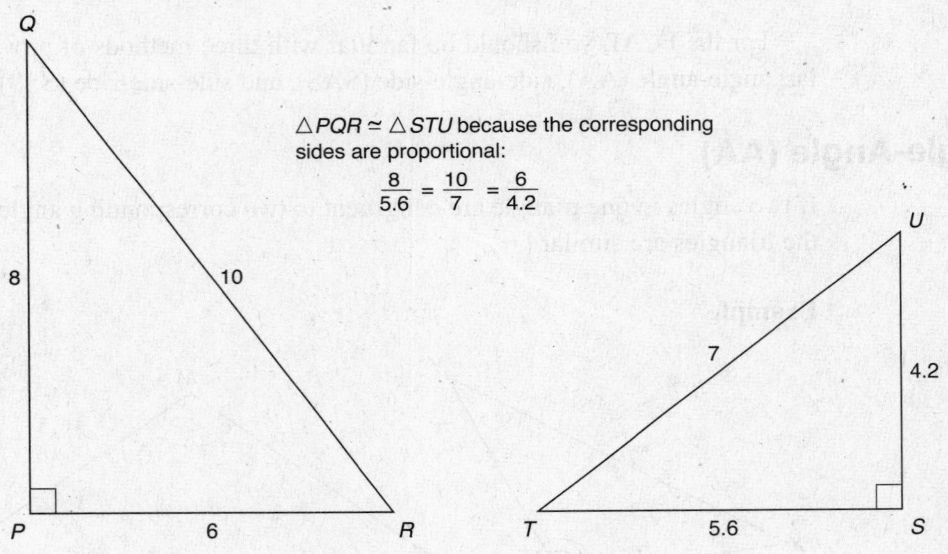

Sample Questions

Which of the following methods proves the triangles shown are similar—AA, SAS, or SSS?

1.

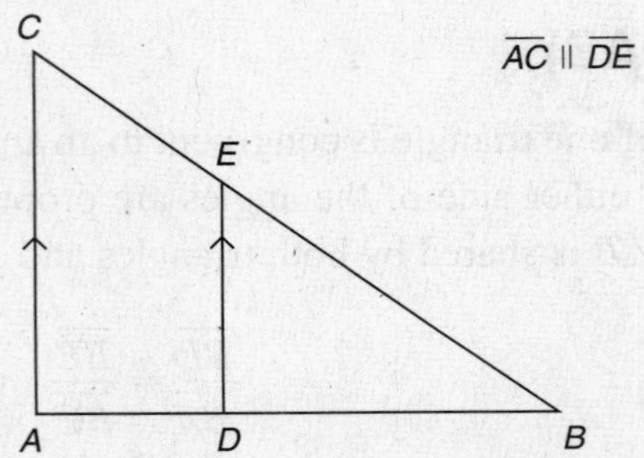

2.

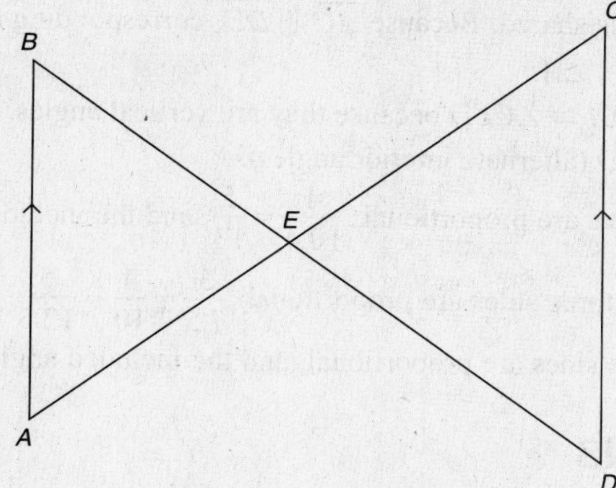

3.

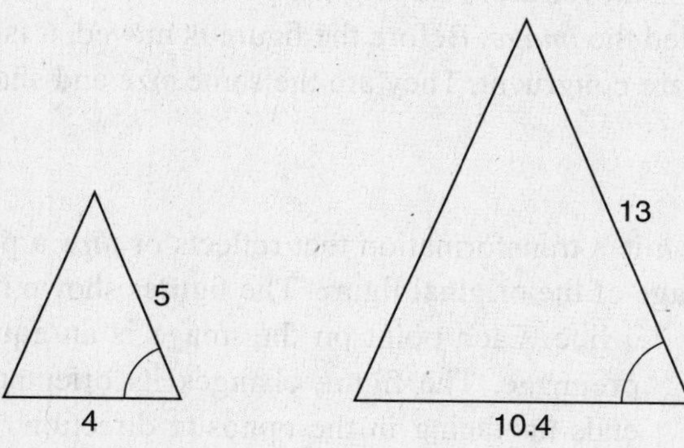

4.

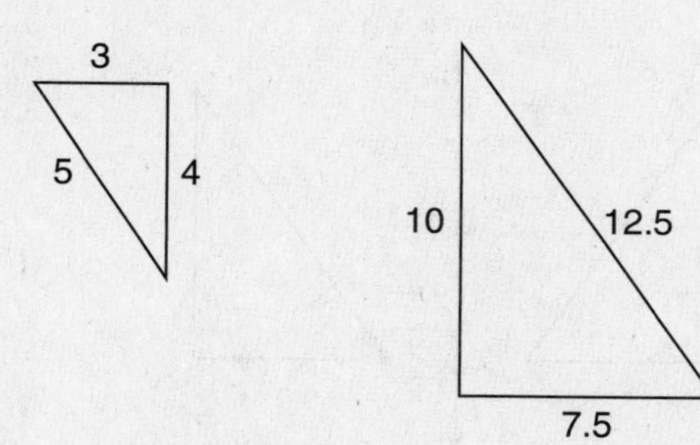

5.

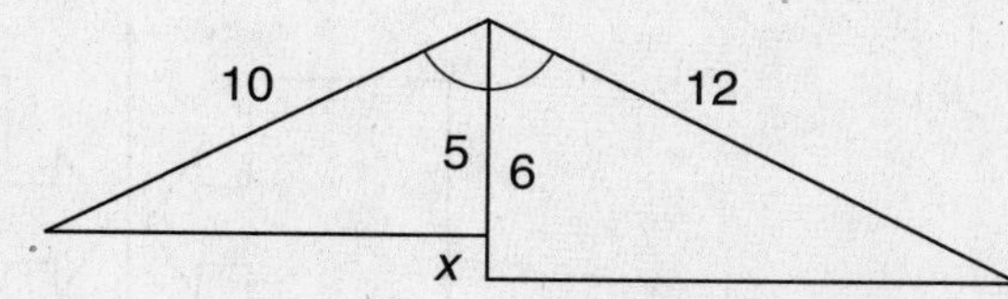

Answers

1. AA. $\angle B$ is shared. Because $\overline{AC} \parallel \overline{DE}$, corresponding angles are congruent. $\angle E \cong \angle C$ and $\angle D \cong \angle A$.
2. AA. $\angle BEA \cong \angle CED$ because they are vertical angles. $\overline{BA} \parallel \overline{CD}$, and so $\angle C \cong \angle A$ and $\angle B \cong \angle D$ (alternate interior angles).
3. SAS. Sides are proportional: $\frac{4}{10.4} = \frac{5}{13}$, and the included angles are congruent.
4. SSS. All three sides are proportional: $\frac{3}{7.5} = \frac{4}{10} = \frac{5}{12.5}$.
5. SAS. The sides are proportional, and the included angle is congruent.

TRANSFORMATIONS

Reflection, translation, and rotation are examples of *transformations*. When a figure is transformed, it is moved from its original position to a new position. The figure in the new position is called the *image*. Before the figure is moved, it is called the *preimage*. The image and preimage are congruent. They are the same size and shape.

Reflection

A *reflection* is a transformation that reflects or *flips* a point or figure over a line. It forms a mirror image of the original figure. The figures shown here are examples of reflections over a line. Each point on the image is an equal distance from each point on the preimage. The figure changes its orientation during this transformation and ends up facing in the opposite direction. The figures shown are examples of reflections.

A **translation** can also be called a "slide."

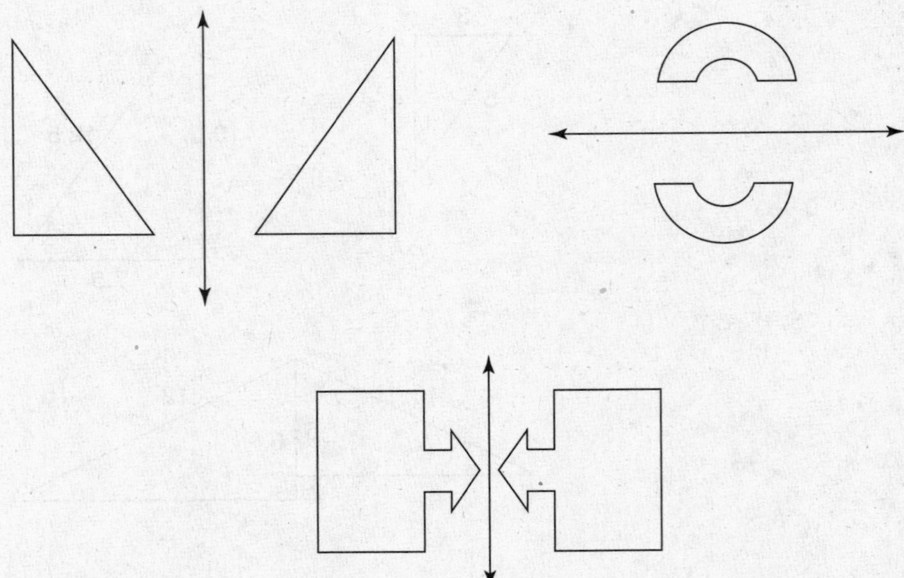

If you join a point on the image to a point on its preimage, the resulting line will be perpendicular to the reflecting line. In the accompanying diagram, $\triangle ABC$ is the preimage and $\triangle A'B'C'$ (read $\triangle A'B'C'$ as "triangle A-prime, B-prime, C-prime") is the image. Line segment $\overline{CC'}$ is perpendicular to the reflecting line.

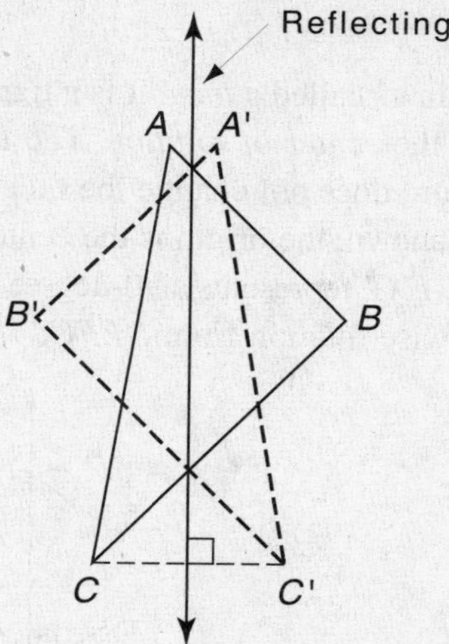

Figures can have *reflectional* or *line symmetry*. A figure has reflectional or line symmetry if you can draw a straight line through it in such a way that the two halves are mirror images of each other. The line is called the *line of symmetry* and can be vertical, horizontal, or diagonal. Many letters have line symmetry:

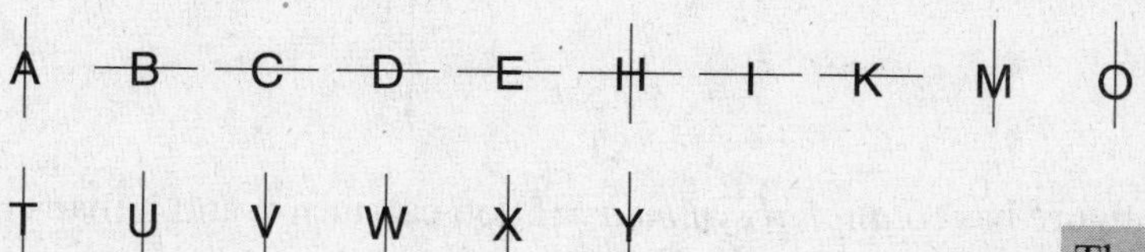

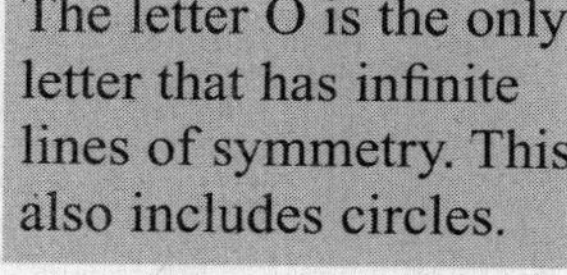
The letter O is the only letter that has infinite lines of symmetry. This also includes circles.

The following figures have reflectional symmetry.

Translations

A transformation that slides all points of a figure the same distance is called a *translation* or *slide*. It is a movement of the figure up, down, or diagonally. The figure does not change its orientation during this move, and its image is congruent to its preimage.

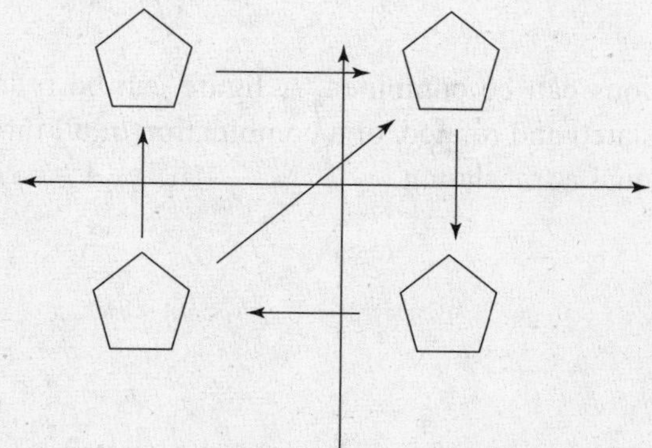

Moving a shape does not change its size.

Rotations

A *rotation* is sometimes called a *turn*. It is a transformation that turns a figure about a point. The point is called the *center of rotation*. The farther the turn, the larger the *angle of rotation*. Rotating a figure does not change the fact that the image and preimage are congruent.

In the diagram shown, the origin is the center of rotation for the trapezoid *DEFG*. Using prime notation, *D'E'F'G'* represents a 90-degree clockwise rotation and *D"E"F"G"* represents a 270-degree clockwise rotation from *DEFG*.

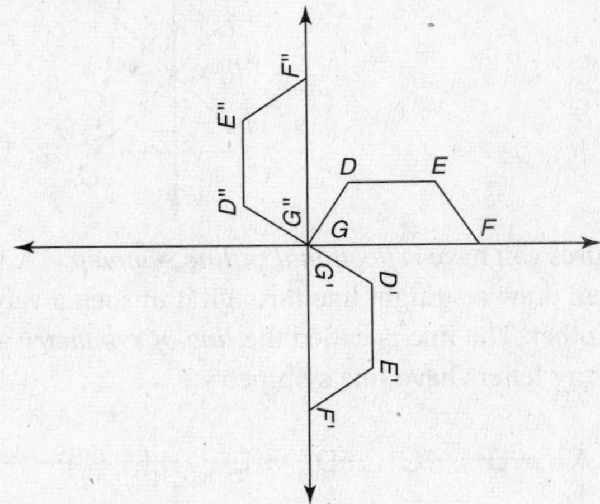

A figure has *rotational symmetry* if you can turn it and it maps back onto itself (matches up) before it makes a full turn. The figure shown here has rotational symmetry because after a 180-degree turn it maps back onto itself.

The next figure also has rotational symmetry. It maps back onto itself after 360 ÷ 8 or 45 degrees. The center of rotation is the center of the circle, and the angle of rotation is 45 degrees.

Transformations can be combined. A figure can be reflected and translated, reflected and rotated, translated and rotated, or a combination of all three transformations. Two sequential reflections equal a translation.

Sample Questions

1. Rotate this figure 90 degrees clockwise about the origin.

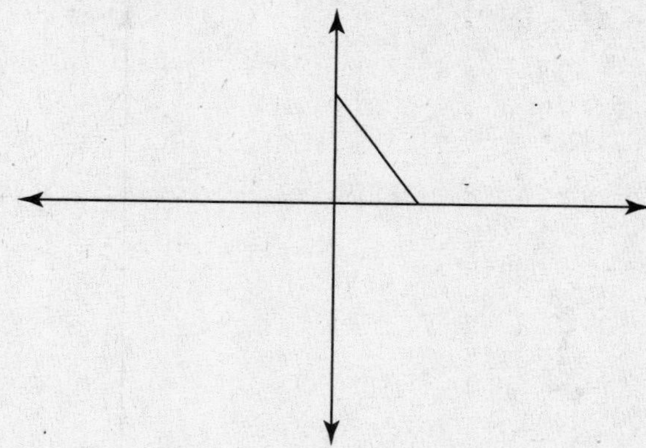

2. What transformation or transformations will turn this word into proper form?

3. What two transformations change A to A'?

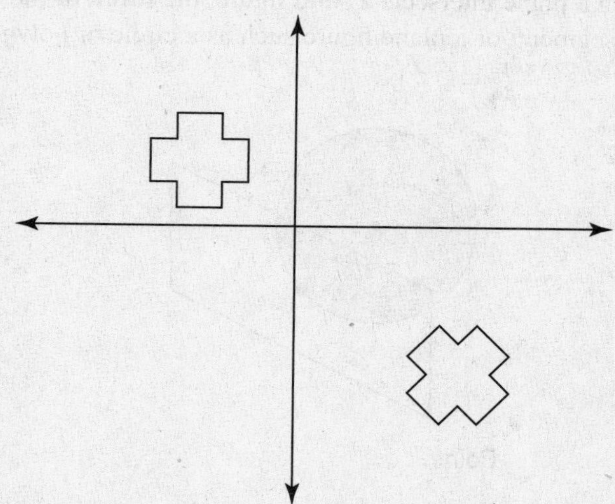

4. The following figure has undergone what transformation(s)?

A. reflection
B. reflection and translation
C. rotation and reflection
D. reflection, translation, and rotation

Answers

1.

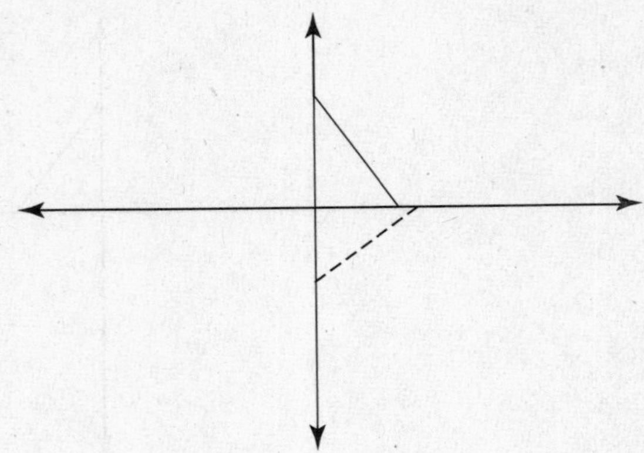

2. Reflection
3. Translation and rotation (45°)
4. **D.**

PLANAR CROSS SECTIONS

When a plane intersects a solid figure, the result of the intersection can be a point, a line or line segment, or a plane figure such as a circle or polygon.

Intersection is where one figure touches another.

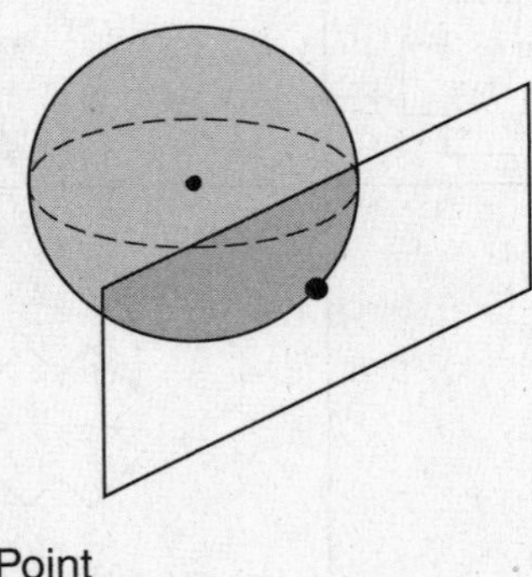

Point

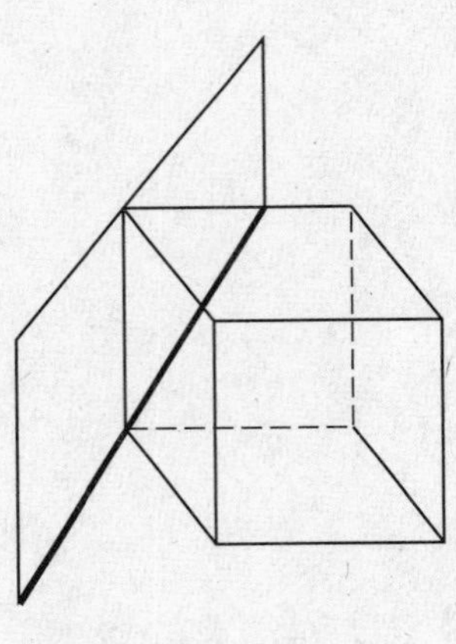

Line

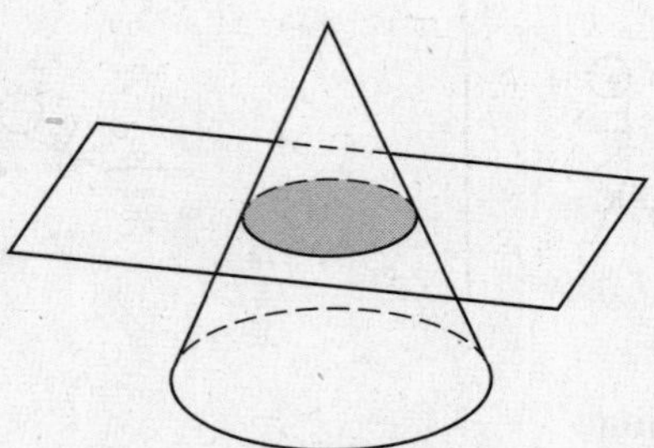

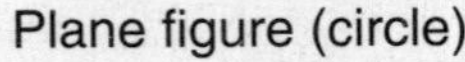

Plane figure (circle)

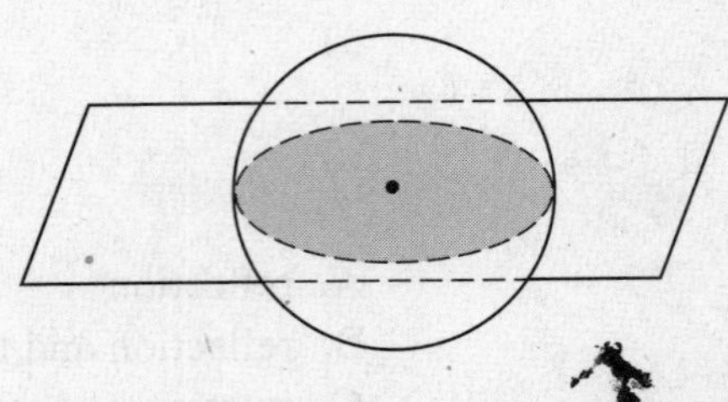

Plane figure (circle)

If the plane is parallel to the base of the solid, the plane figure formed will be similar or congruent to the base of the solid.

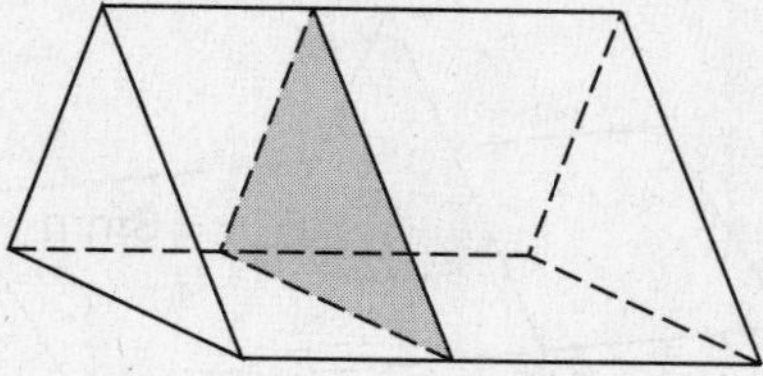

Plane figure (triangle)

For the FCAT, you should be able to analyze the shape and apply geometric properties to solve a problem involving a planar cross section.

Example: Find the area, in square meters, of the plane figure formed by the cross section of the cube in the diagram shown here.

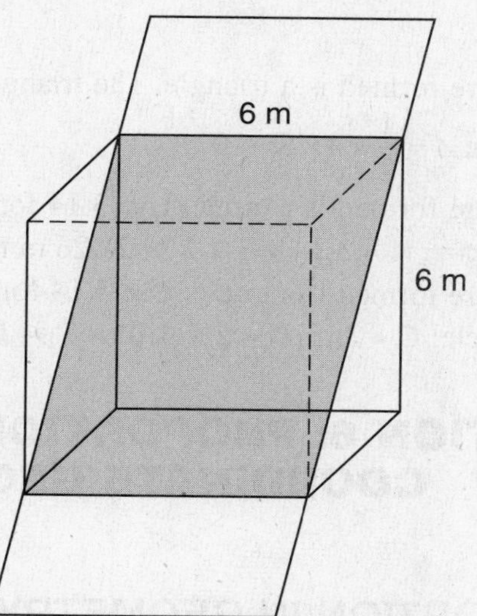

The Pythagorean theorem: $a^2 + b^2 = c^2$

The plane figure formed is a rectangle. One side of the rectangle matches the cube and is 6 meters long. The other side of the rectangle is found using the Pythagorean Theorem or using properties of 45-45-90 right triangles. It is $6\sqrt{2}$ meters long. To find the area, multiply: $6 \times 6\sqrt{2} = 36\sqrt{2} \approx 50.9$ m^2.

Sample Questions

1. A plane intersects a triangular prism parallel to the base. The prism has a base of 4 inches and a height of 5 inches. What is the area of the figure formed by the intersection?

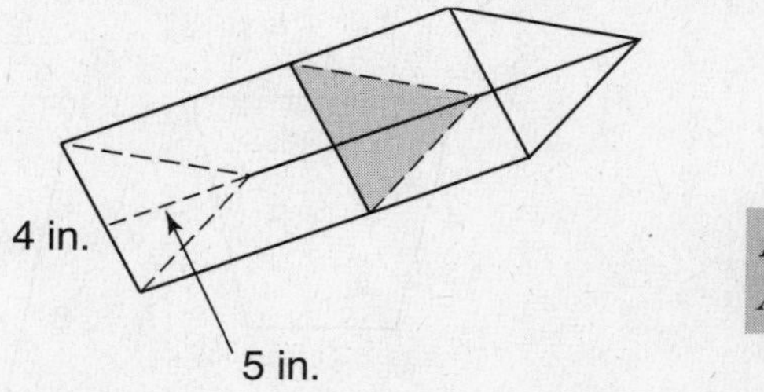

Area of a triangle: $A = .5\ bh$

2. Find the area of the figure formed by the intersection of a plane and a cone if the plane is parallel to the base of the cone.

Area of a circle:
$A = \pi r^2$

Circumference:
$C = 2\pi r$

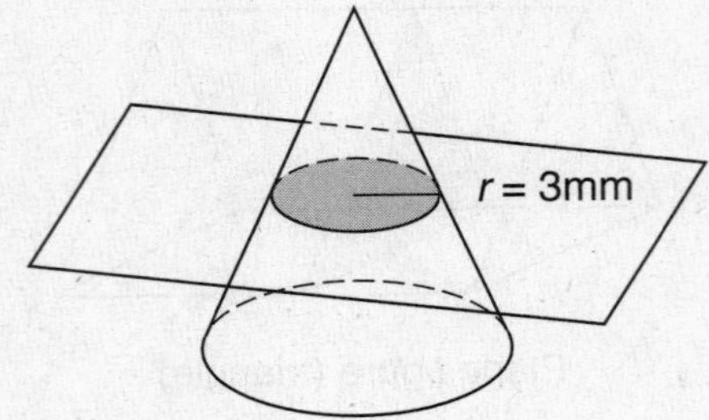

3. Find the circumference of the figure formed by the intersection of a plane and a cone as shown in question 2.

Answers

1. The figure formed is a triangle. The triangle has the same base and height as the plane. $A = \frac{1}{2}bh$; $A = \frac{1}{2} \times 4 \times 5 = 10$.
2. The figure formed is a circle. Use 3.14 for π. Substitute into the area formula for a circle: $A = \pi r^2$; $A = 3.14 \times 3 \times 3 \approx 28.26$ mm^2.
3. The figure formed is a circle. Use 3.14 for π. Substitute into the circumference formula for a circle: $C = 2\pi r$; $C = 2 \times 3.14 \times 3 \approx 18.84$ mm.

SECTION 6: PROPORTION, SCALE, AND COORDINATE GEOMETRY

RATIO AND PROPORTION IN GEOMETRY

Ratios are used to compare quantities to find out how many times greater one quantity is than another. They can be expressed in fraction form: $\frac{a}{b}$. For example, when comparing Bob's age (55) to Amy's age (33), we write $\frac{55}{33}$ or $\frac{5}{3}$. Bob is 1.67 times older than Amy. Two ratios that are equal are called *proportions.* If ratio $\frac{a}{b}$ is equal to ratio $\frac{c}{d}$, we write $\frac{a}{b} = \frac{c}{d}$. We say that geometric figures are *proportional* when the ratios of their corresponding sides are equal.

Example:

4.5
9
4
8

Proportion is one of the most useful ideas in mathematics.

In the diagram, $\frac{4}{4.5} = \frac{8}{9}$. This is read as "four is to four and five-tenths as eight is to nine." The corresponding parts of this trapezoid are proportional. You can check the proportion by cross-multiplying: $4 \times 9 = 4.5 \times 8$. Since both sides of the equation equal 36, the proportion is true.

The parts of a proportion have special names. To best see this, we can write the proportion in ratio notation. The numbers on the ends are called the *extremes*, and the numbers in the middle are called the *means*.

Extremes

$$4:4.5 = 8:9$$

Means

The Geometric Mean

The *geometric mean* is a special proportion in which the middle terms (means) of the proportion are equal. For example, $\frac{2}{4} = \frac{4}{8}$. To check, you can multiply the means (4×4) and compare to the extremes (2×8). Both equal 16.

To solve a proportion in which the geometric mean is unknown, use the same technique: $1/x = x/25$. To find the geometric mean of 1 and 25, cross-multiply: $1 \bullet 25 = x^2$. If $25 = x^2$, to solve, take the square root of both sides: $\sqrt{25} = \sqrt{x^2}$. The geometric mean between 1 and 25 is 5.

Example: Find the length of side x in the diagram if x represents the geometric mean of 10 and 6.

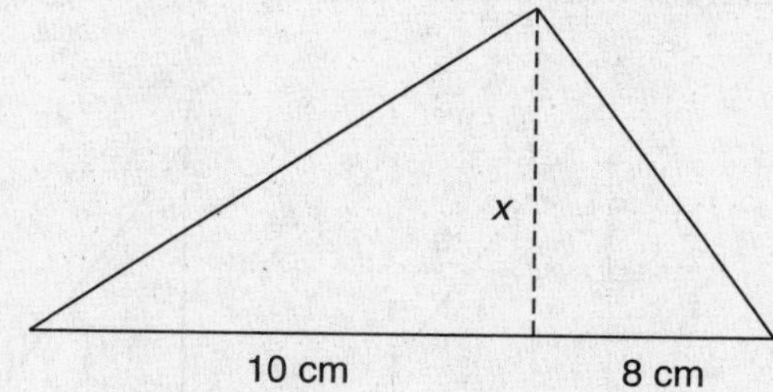

Set up a proportion: $10/x = x/8$. Solve using geometric means: $10 \bullet 8 = x^2$; $\sqrt{80} = \sqrt{x^2}$; $x \approx 8.9$. The side length of the triangle is approximately 8.9 cm.

Sample Questions

Find the geometric mean of the following pairs of numbers.

1. 2 and 32
2. 5 and 20
3. 4 and 100
4. 7 and 12

Answers

1. $\frac{2}{x} = \frac{x}{32}$; $64 = x^2$; $\sqrt{64} = \sqrt{x^2}$; $8 = x$
2. $\frac{5}{x} = \frac{x}{20}$; $100 = x^2$; $\sqrt{100} = \sqrt{x^2}$; $10 = x$
3. $\frac{4}{x} = \frac{x}{100}$; $400 = x^2$; $\sqrt{400} = \sqrt{x^2}$; $20 = x$
4. $\frac{7}{x} = \frac{x}{12}$; $84 = x^2$; $\sqrt{84} = \sqrt{x^2}$; $9.2 \approx x$

Ratio and Proportion in Right Triangles

There are three theorems that relate proportions to right triangles. For the purposes of these three theorems, we will use $\triangle ABC$ and draw an altitude ($\overline{BD}$).

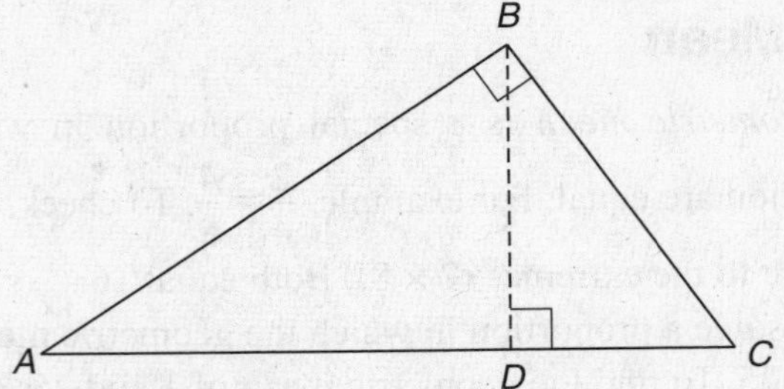

Theorem 1: If an altitude ($\overline{BD}$) is drawn to the hypotenuse of a right triangle, the two new triangles formed will be similar to the original triangle and to each other.

To help you more easily visualize this theorem and the two following theorems, examine the next diagram which shows the three resulting triangles arranged side by side.

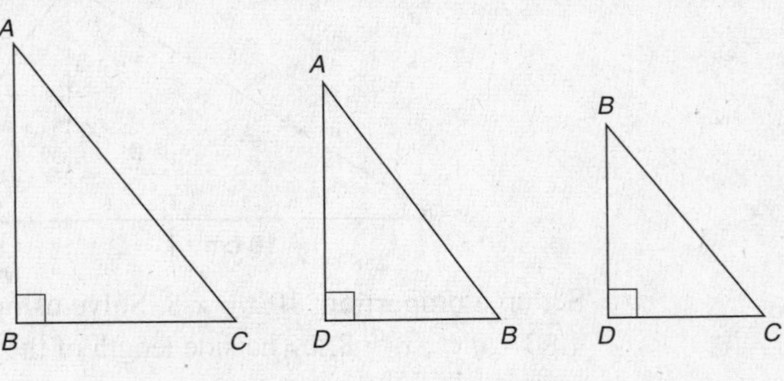

$\triangle ABC \simeq \triangle ADB \simeq \triangle BDC$

Theorem 2: In a right triangle, the altitude divides the hypotenuse into two segments ($\overline{AD}$ and $\overline{CD}$). The altitude ($\overline{BD}$) is the geometric mean of the two segments.

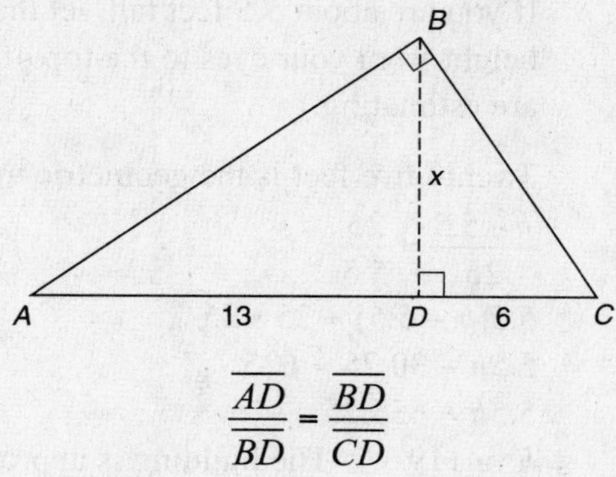

$$\frac{\overline{AD}}{\overline{BD}} = \frac{\overline{BD}}{\overline{CD}}$$

Example 1: Find the value of x in the diagram shown above.

Use the geometric mean:

$$\frac{13}{x} = \frac{x}{6}$$

$$78 = x^2$$

$$\sqrt{78} = \sqrt{x^2}$$

$8.8 \approx x$ The value of x is approximately 8.8.

Example 2: To estimate the height of a building, you can use a stiff sheet of paper or cardboard. Hold one of the right angles of the paper up to your eye and back up until you can sight along the top edge of the paper to the top edge of the building. The bottom edge of the paper should be sighted to the bottom edge of the building at ground level. Use the distance between yourself and the building as the "altitude." Set up and solve the proportion using a geometric mean.

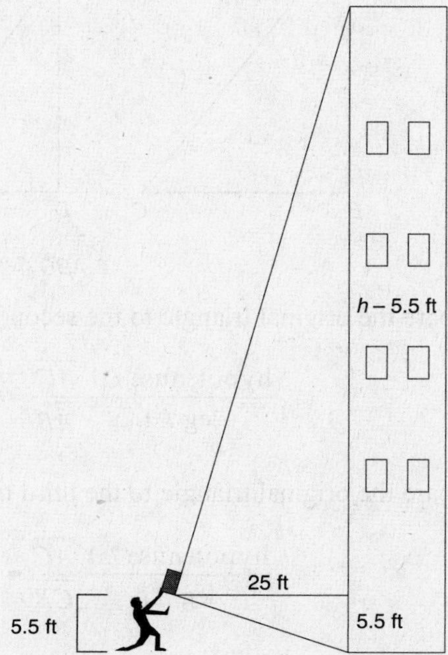

If you are about 5.5 feet tall, let the height of the building be h and let the height from your eyes to the top of the building be $h - 5.5$. (Remember, you are estimating.)

Twenty five feet is the geometric mean between your height and $h - 5.5$:

$$\frac{h-5.5}{25} = \frac{25}{5.5}$$

$$5.5(h - 5.5) = 25 \cdot 25$$

$$5.5h - 30.25 = 625$$

$$5.5h = 655.25$$

$h \approx 119$ The building is approximately 119 feet tall.

Theorem 3: In a right triangle, the altitude ($\overline{BD}$) divides the hypotenuse into two segments ($\overline{AD}$ and $\overline{CD}$). Each leg of the original triangle ($\overline{AB}$ and $\overline{CB}$) is the geometric mean of the hypotenuse and the segment of the hypotenuse adjacent to that leg.

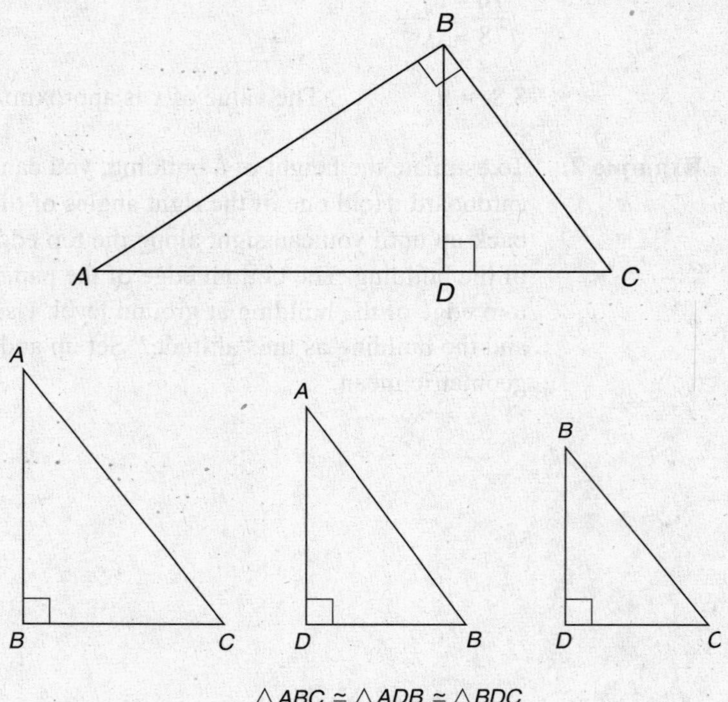

$\triangle ABC \simeq \triangle ADB \simeq \triangle BDC$

Compare the original triangle to the second triangle in the diagram above.

$$\frac{\text{hypotenuse } \triangle 1}{\text{leg } \triangle 1} \frac{\overline{AC}}{\overline{AB}} = \frac{\overline{AB}}{\overline{AD}} \frac{\text{hypotenuse } \triangle 2}{\text{leg } \triangle 2}$$

Compare the original triangle to the third triangle in the diagram above.

$$\frac{\text{hypotenuse } \triangle 1}{\text{leg } \triangle 1} \frac{\overline{AC}}{\overline{CB}} = \frac{\overline{CB}}{\overline{CD}} \frac{\text{hypotenuse } \triangle 3}{\text{leg } \triangle 3}$$

Example 1: In the right triangle shown here, an altitude is drawn to the hypotenuse. Find a, b, and c.

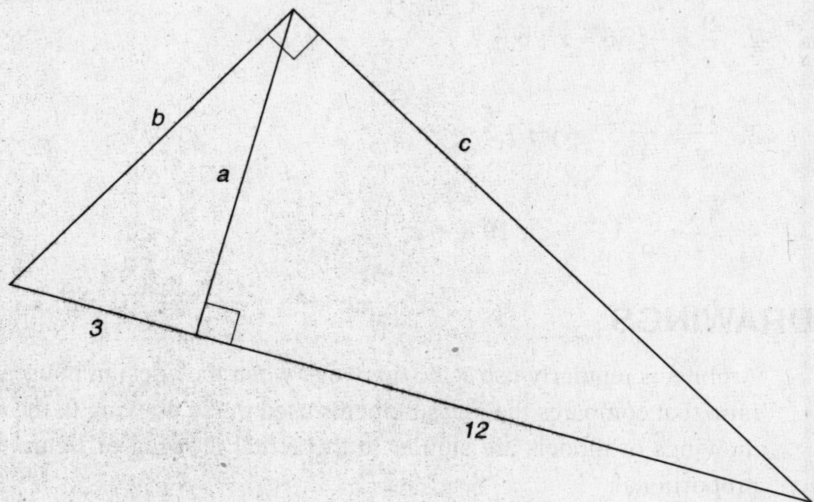

First, find a: $\frac{3}{a} = \frac{a}{12}$

$36 = a^2$

$6 = a$

Next, find b: $\frac{15}{b} = \frac{b}{3}$ (The hypotenuse of the original triangle = 3 + 12.)

$45 = b^2$

$6.7 \approx b$

Last, find c: $\frac{15}{c} = \frac{c}{12}$

$180 = c^2$

$13.4 \approx c$

Sample Questions

Use the diagram shown to answer the following questions.

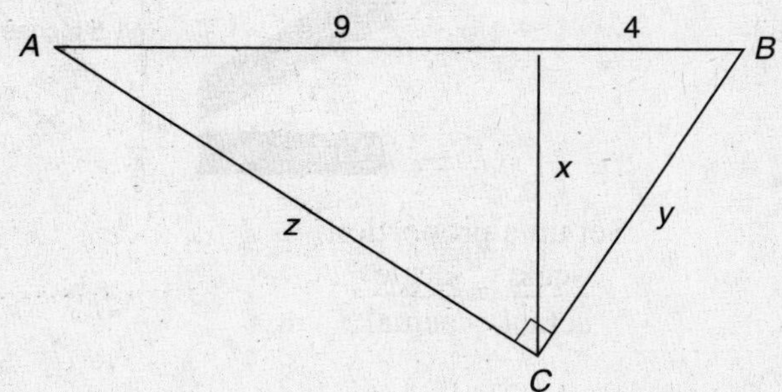

1. Find the length of $\overline{AB}$.
2. Find the value of x.
3. Find the value of y.
4. Find the value of z.

Answers

1. $9 + 4 = 13$
2. $\frac{9}{x} = \frac{x}{4}$; $36 = x^2$; $6 = x$
3. $\frac{13}{y} = \frac{y}{4}$; $52 = y^2$; $7.2 \approx y$
4. $\frac{13}{z} = \frac{z}{9}$; $117 = z^2$; $10.8 \approx z$

SCALE DRAWINGS

Architects regularly use scale drawings when they design houses and buildings. A *scale* is a ratio that compares the measurements used in the drawing to the actual measurements. Scale drawings or models are similar to the actual drawing or figure, and therefore the sides are proportional.

The ratio you should use when working with scale drawings or models is

When setting up a proportion, words should match: $\frac{\text{in.}}{\text{ft.}} = \frac{\text{in.}}{\text{ft.}}$

$$\frac{\text{scale measurement}}{\text{actual measurement}}$$

This is called a *scale ratio*. For example, if a model car measures 2 cm for every 1 ft of the actual car, the scale ratio is 2 cm/1 ft. Knowing this ratio allows you to set up a proportion to figure out how long the actualcar would be if the scale model was 36 cm: $\frac{2\text{ cm}}{1\text{ ft}} = \frac{36\text{ cm}}{x\text{ ft}}$. By solving the proportion, you can determine that the actual car length is 18 ft.

Example: The height of the symbol in the accompanying drawing is 1.5 inches. The actual symbol will be 12 inches tall once it is placed on a sign. If the width of this symbol is 1 inch, how wide will the actual symbol be?

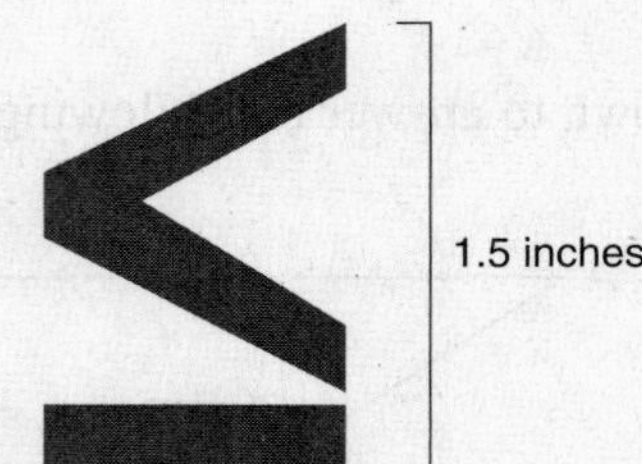

Set up a proportion:

$$\frac{\text{scale}}{\text{actual}} = \frac{\text{scale}}{\text{actual}}$$

$$\frac{1.5}{12} = \frac{1}{x}$$

$$1 \bullet 12 = 1.5x$$

$$12 \div 1.5 = x$$

$$8 = x$$

The actual symbol will be 8 inches wide.

Maps are drawn with the scale included so that you can figure actual distances. If a map scale shows that 1 inch = 50 miles, you would use the scale 1 in./50 mi. With this scale, you could calculate the actual distance represented by $4\frac{3}{4}$ inches by setting up a proportion:

$$\frac{1\text{ in.}}{50\text{ mi}} = \frac{4\frac{3}{4}\text{ in.}}{x\text{ mi}}$$

$$50 \bullet 4\frac{3}{4} = 1x = 237.5\text{ mi}$$

When solving a proportion, use cross-multiplication.

Sample Questions

1. A photograph measures 4 inches wide and 5 inches long. If you have the photograph enlarged to fit a frame 36 inches long, what is the widest the photograph can be?
   A. 36 inches
   B. 32 inches
   C. 30 inches
   D. 28 inches
2. A Florida map has a scale of 1 inch = approximately 22.8 miles. If the distance on the map between Vero Beach and Boynton Beach is 3.5 inches, what is the actual distance?

Use “scale” as one of your two ratios in the proportion.

3. The smallest mammal, Kitti’s hog-nosed bat, has a head-body length of $1\frac{7}{50}$ inches and a wingspan of about $5\frac{1}{10}$ inches. A scale drawing of this little bat is made showing it in full flight. The wingspan on the drawing is 15 inches. What should the length of the bat be in the drawing?
   A. $3\frac{1}{3}$ inches
   B. $5\frac{1}{2}$ inches
   C. 4 inches
   D. 67 inches
4. The Sears Tower is 1450 feet tall. Jeff wanted to make a scale model for his class project, but his mother’s car can hold only a 36-in.-tall model. Which of the following scales is closest to the scale Jeff needs?
   A. $\frac{1\text{ in.}}{25\text{ ft}}$
   B. $\frac{1\text{ in.}}{40\text{ ft}}$
   C. $\frac{1\text{ in.}}{45\text{ ft}}$
   D. $\frac{2\text{ in.}}{45\text{ ft}}$

5. Use the scale 2 in. = 15 ft to find the measurement closest to the actual length of the white shark shown here.

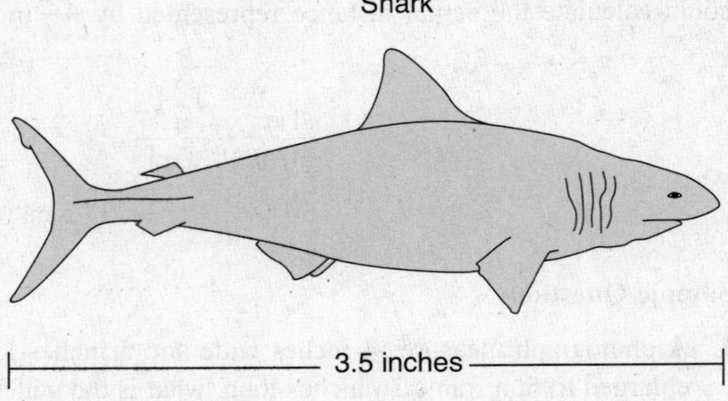

A. 24 feet
B. 25 feet
C. 26 feet
D. 27 feet

Answers

1. **D.** For the picture to be proportional, $\frac{\text{width}}{\text{length}} = \frac{\text{width}}{\text{length}}$; $\frac{4}{5} = \frac{x}{36}$. Cross-multiply: $4 \cdot 36 = 5x$; $x = 28.8$. The picture cannot be wider than 28.8 inches.

2. Set up the proportion $\frac{1\text{ in.}}{22.8\text{ mi}} = \frac{3.5\text{ in.}}{x\text{ mi}}$. Cross-multiply: $3.5 \cdot 22.8 = 79.8$ miles.

3. **A.** Use $\frac{\text{length}}{\text{wingspan}} = \frac{\text{length}}{\text{wingspan}}$. Change $1\frac{7}{50}$ inches to 1.14 inches, and $5\frac{1}{10}$ inches to 5.1. Use the proportion $\frac{1.14}{5.1} = \frac{x}{15}$. Cross-multiply: $1.14 \cdot 15 \div 5.1 = 3.35$ inches.

4. **A.** $1 \div 40 \approx .025$, which is closest to $36 \div 1450 \approx .0248$.

5. **C.** The scale drawing is 3.5 inches long. Set up a proportion: $\frac{2\text{ in.}}{15\text{ ft}} = \frac{3.5\text{ in.}}{x\text{ ft}}$. Cross-multiply: $3.5 \cdot 15 = 2x$; $52.5 = 2x$. The shark is 26.25 ft long.

INDIRECT MEASUREMENT USING SIMILAR FIGURES

It is possible to measure figures indirectly using the concepts covered under similarity. Recall that similar figures are alike in shape, are proportional, and have congruent corresponding angles. You can measure the height of tall or distant objects by using these concepts.

Example: To measure the height of a tree you can use a person's shadow and the tree's shadow in a proportion.

Compare the actual height to the shadow for both the tree and the person:

$$\frac{\text{actual height of tree}}{\text{shadow length of tree}} = \frac{\text{actual height of person}}{\text{shadow length of person}}$$

$$\frac{\text{actual height }(h)}{315\text{ inches}} = \frac{71\text{ inches}}{60\text{ inches}}$$

Cross-multiply: $315 \bullet 71 = 60h$; $22{,}365 = 60h$; $372.75 = h$. The height of the tree is about 373 inches, or 31 feet.

When using similar figures to perform indirect measurement, be certain you are comparing corresponding parts.

Example: Surveyors took measurements as shown here to find the width of a pond. Use these measurements to determine the width of the pond to the nearest foot.

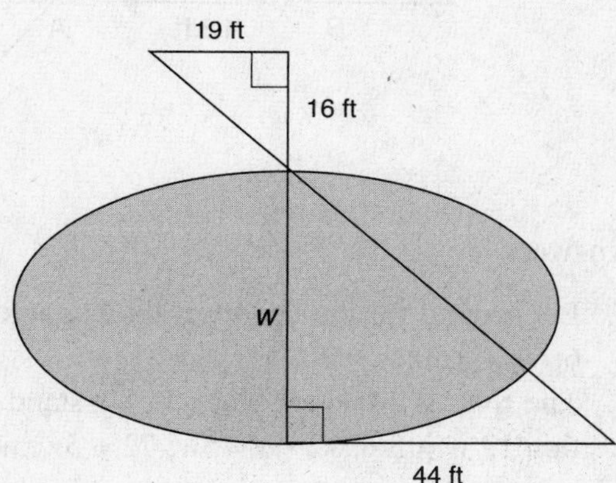

Compare the two triangles.

Let w = the width of the pond. Then $\frac{16}{19} = \frac{w}{44}$; $16 \bullet 44 = 19w$; $704 = 19w$; $w \approx 37.05$. The pond is approximately 37 feet wide.

Sample Questions

1. A flagpole stands in front of a school. A 6-foot student standing next to the flagpole casts a 9-foot shadow. The shadow of the flagpole is 30 feet long. How tall is the flagpole?

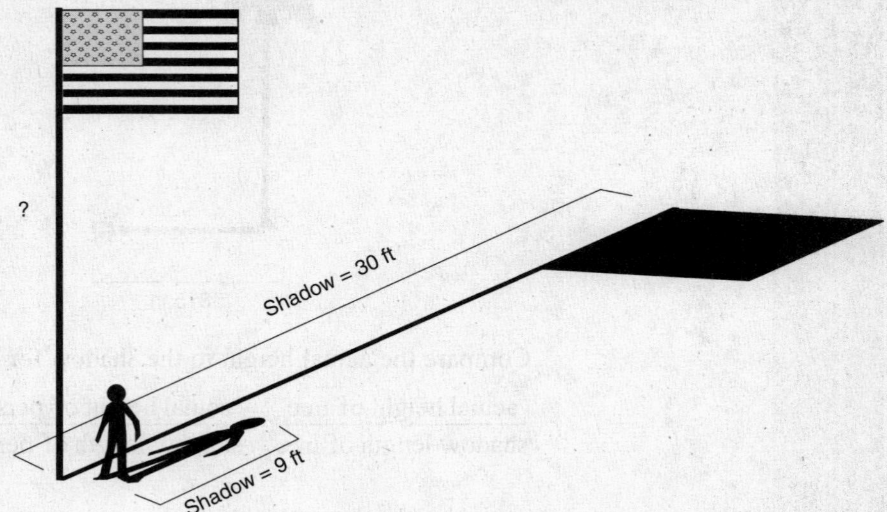

2. Forrest wants to estimate the width of a stream. He measures three distances and notes that $\overline{AB} = 12$ ft, $\overline{AD} = 5$ ft, and $\overline{DE} = 6$ ft. How wide is the river?

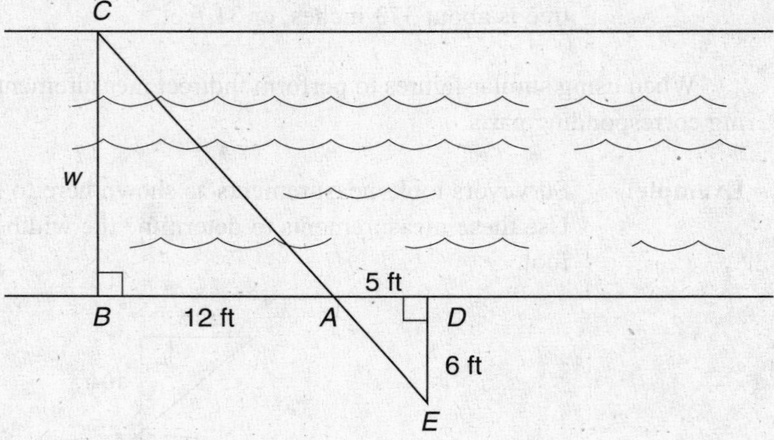

Answers

1. Let h stand for the height of the flagpole. Then use the proportion $h/30 = 6/9$. $9h = 30 \bullet 6$; $9h = 180$; $h = 20$.
2. The river is 14.4 feet wide. Let w stand for the width of the river. Then use the proportion $12/w = 5/6$. $12 \bullet 6 = 5w$; $72 = 5w$; $w = 14.4$.

COORDINATE GEOMETRY

In coordinate geometry we look at graphing points on a graph and review the relationships between points and lines once they are graphed.

Cartesian Plane

The system we will use to graph is called the *Cartesian plane* or the *coordinate plane* (sometimes called a *grid*). This plane or grid has two axes, an x-axis and a y-axis. The axes divide the plane into four *quadrants* which are numbered counterclockwise beginning at the upper right. Typically, quadrants are numbered using roman numerals. The intersection of these axes is called the *origin* and is located at the point (0, 0).

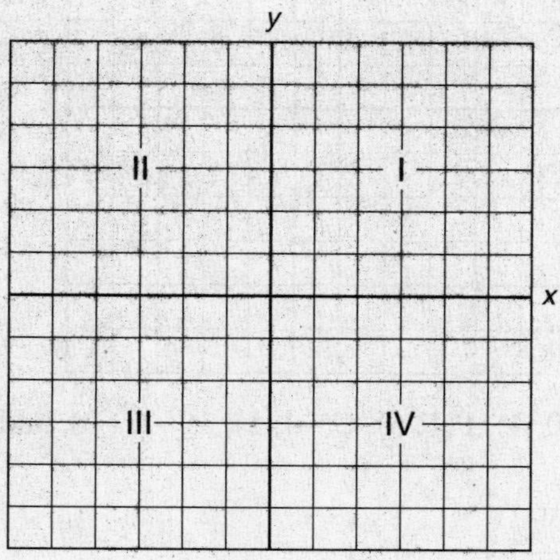

The origin is located at the center of this grid and is the starting place for plotting points.

Ordered Pairs

Points graphed on a plane require two coordinates to place them (x, y). These two coordinates are called *ordered pairs*. The order in which you graph the coordinates is extremely important to the position of the point on the graph.

On the coordinate plane shown, the ordered pairs $A(0, 1)$, $B(-2, 4)$, $C(3, 2)$, $D(-4, -5)$, and $E(3, -3)$ are graphed. The *x-coordinate* is the first number in an ordered pair. This number tells you how far to move horizontally away from the origin. Positive x-coordinates are to the right, and negative coordinates are to the left. The *y-coordinate* is the second number in an ordered pair. This number tells you how far to move vertically away from the origin. Positive y-coordinates are in an upward direction, and negative coordinates are in a downward direction.

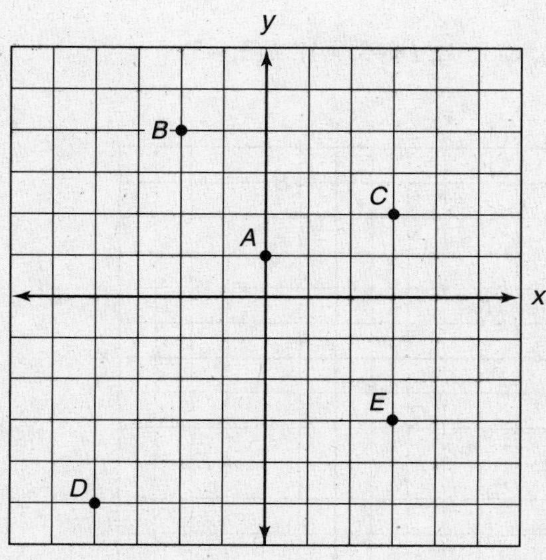

The ordered pair (2, −3) is an instruction to move right two spaces and down three spaces before placing the dot that represents your point.

Sample Questions

1. Give the coordinates of the points labeled on the diagram shown.

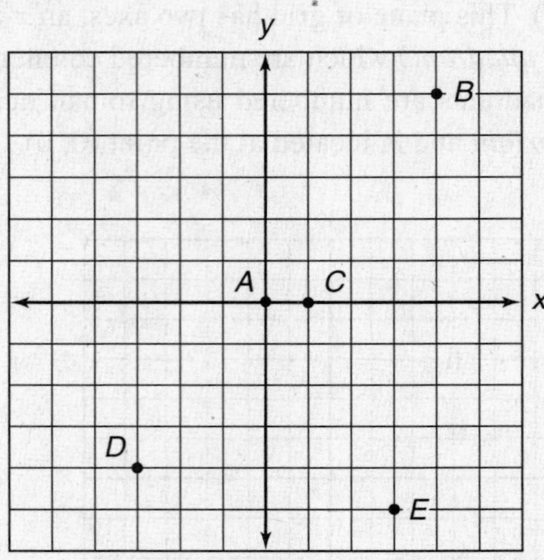

2. Plot the points *A*(0, 4), *B*(1, 5), *C*(−4, 1), *D*(−2, −2), and *E*(3, −3) using this graph.

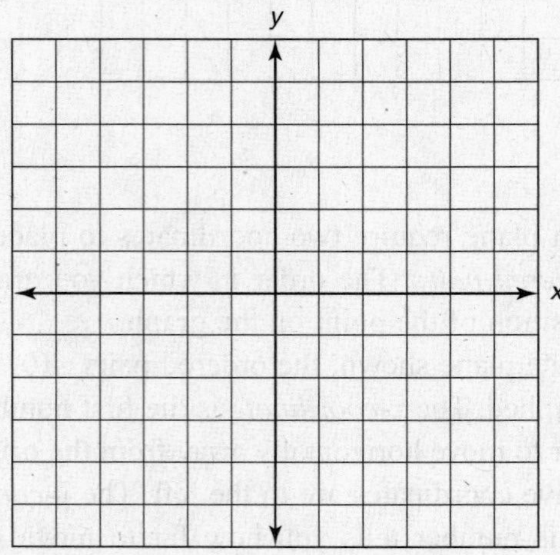

Answers

1. *A*(0, 0), *B*(4, 5), *C*(1, 0), *D*(−3, −4), *E*(3, −5)
2.

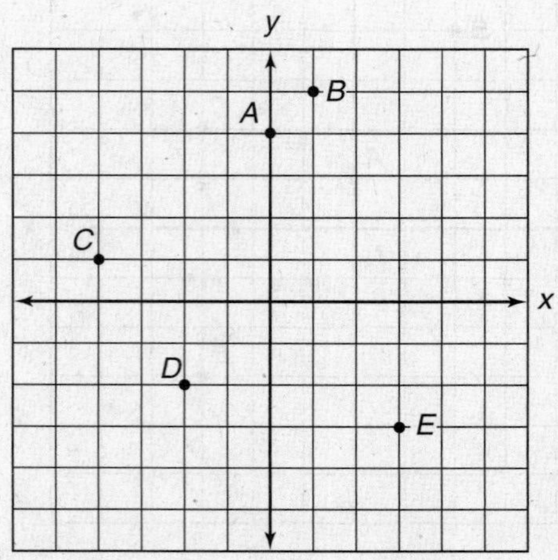

Midpoint of a Line

As mentioned previously, the midpoint of a line segment is the point exactly halfway between the endpoints. Because finding an ordered pair requires movement in two directions, to calculate the midpoint between two ordered pairs, you need to go halfway between each of the x-coordinates and the y-coordinates.

Example: The midpoint of (2, 8) and (0, 4) is (1, 6). Add the two x-coordinates together and divide by 2. Add both y-coordinates and divide by 2.

As with many things in mathematics, there is a formula for finding the midpoint: $\frac{x_1 + x_2}{2}, \frac{y_1 + y_2}{2}$. Here x_1 is the x-coordinate from the first ordered pair and x_2 is the x-coordinate from the second ordered pair. Similarly, y_1 and y_2 are the first and second y-coordinates.

Example 1: Find the midpoint of $\overline{AB}$, the line segment joining $A(-5, 2)$ and $B(2, -4)$.

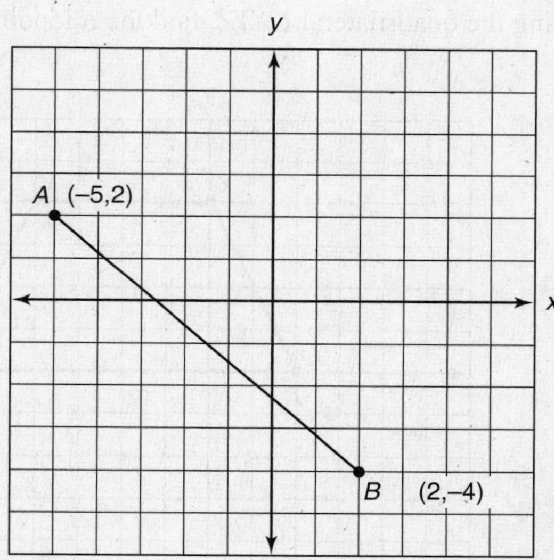

To find the midpoint: Add the x-coordinates and divide by 2. Add the y-coordinates and divide by 2.

$$\frac{x_1 + x_2}{2}, \frac{y_1 + y_2}{2}$$

$$\frac{-5+2}{2}, \frac{2+(-4)}{2}$$ Substitute the given coordinates.

$$\frac{-3}{2}, \frac{-2}{2} = (-1.5, -1)$$

The midpoint M of the ordered pair is (−1.5, −1).

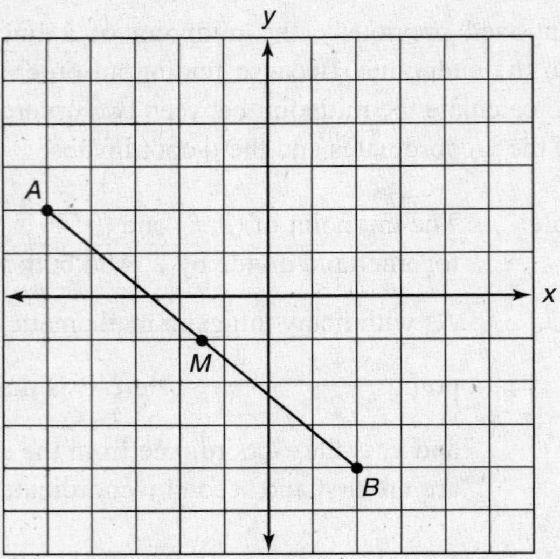

Example 2: Using the quadrilateral $OXYZ$, find the midpoint of $\overline{XZ}$.

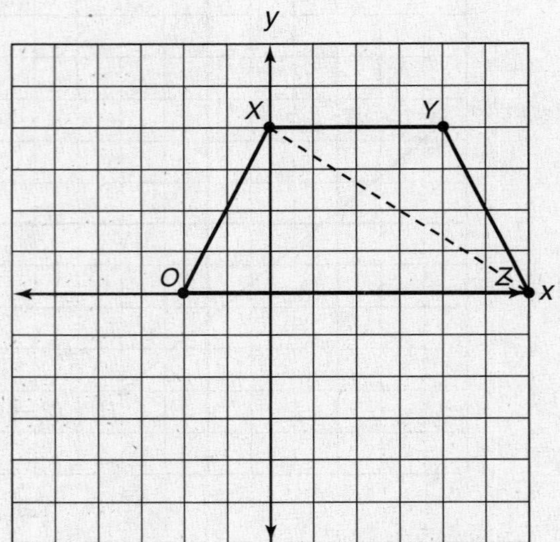

The coordinates of X are (0, 4) and the coordinates of Z are (6, 0). Use the midpoint formula and average the x-coordinates and y-coordinates:

$$\frac{x_1 + x_2}{2}, \frac{y_1 + y_2}{2}$$

$$\frac{0+6}{2}, \frac{4+0}{2}$$

$$\frac{6}{2}, \frac{4}{2} = (3, 2)$$

The coordinates of the midpoint $\overline{XZ}$ are (3, 2).

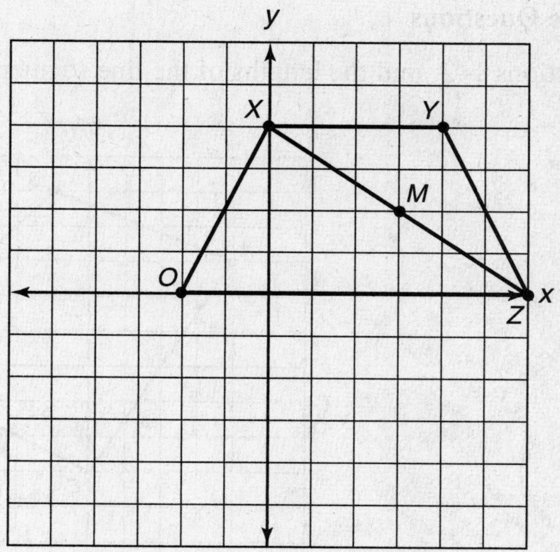

The Distance Formula

Finding distances on a coordinate plane is actually an extension of the Pythagorean Theorem. To find distances on a coordinate plane, the Pythagorean Theorem is rewritten as the Distance Formula: $\sqrt{(x_1 - x_2)^2 + (y_1 - y_2)^2}$. Use the Distance Formula to find the length of a line segment.

Example: In the accompanying diagram, find the distance between point A and point B. In other words, find the length of $\overline{AB}$.

Test Tip: It is a good idea to verify visually that the distance between the two points is about 5 by counting or estimating.

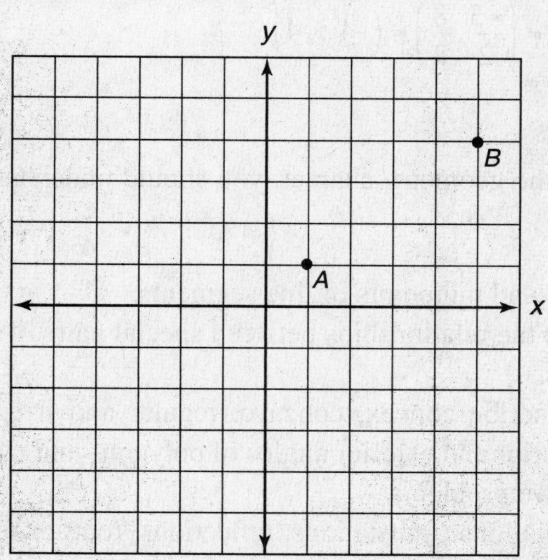

$$\sqrt{(x_2 - x_1)^2 + (y_2 - y_1)^2}$$
$$\sqrt{(5-1)^2 + (4-1)^2} = \sqrt{16+9} = \sqrt{25} = 5$$

Sample Questions

In questions 1–2, find the lengths of the line segments in the accompanying diagram.

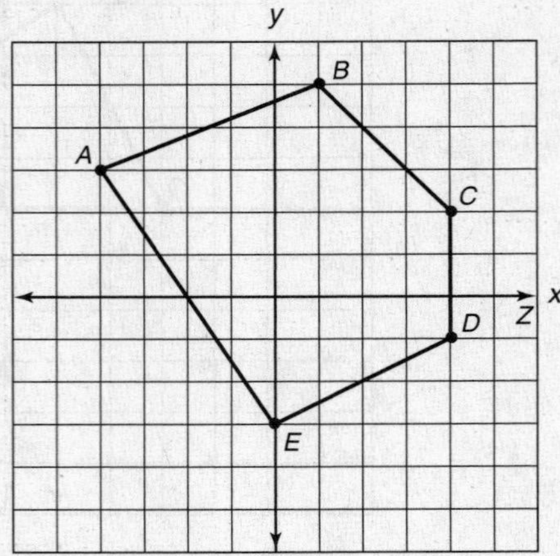

1. $\overline{AB}$
2. $\overline{CD}$
3. Find the coordinates of the midpoint of $\overline{AB}$ in the diagram above.

Answers

1. $\sqrt{(-4-1)^2+(3-5)^2} = \sqrt{(-5)^2+(-2)^2} = \sqrt{25+4} = \sqrt{29} \approx 5.39$
2. The distance formula is not needed. Just count the spaces. The answer is 3.
3. $\left(\frac{-4+1}{2}, \frac{3+5}{2}\right) = \left(\frac{-3}{2}, \frac{8}{2}\right) = (-1.5, 4)$

SUMMARY

As you complete the geometry chapter, you should understand or be able to do most of the following:

- Find the lengths and midpoints of line segments.
- Identify and use the relationships between special pairs of angles formed by parallel lines and transversals.
- Identify and describe convex, concave, regular and irregular polygons. Determine the measures of interior and exterior angles of polygons and use properties of congruency and similarity to solve problems.
- Apply transformations (translations, reflections, rotations, dilations, and scale factors) to polygons to determine congruence, similarity and symmetry.
- Find perimeter, and area of polygons. Find surface and volume of polyhedrons. Determine how changes in dimensions affect perimeter, area, surface area, and volume of common geometric figures.
- Describe, classify, and compare relationships among quadrilaterals including the square, rectangle, rhombus, parallelogram, trapezoid, and kite on the basis of their properties.

- Classify and describe triangles that are right, acute, obtuse, scalene, isosceles, equilateral, and equiangular. Identify altitudes, medians, angle bisectors, and perpendicular bisectors.
- Use properties of congruent and similar polygons to solve problems involving lengths and areas.
- Identify: circumference, radius, diameter, arc, arc length. Find circumference and area of a circle and solve real-world problems using measures of circumference and areas of circles and sections of a circle.

PRACTICE PROBLEMS

1. Which of the following can be classified as an obtuse scalene triangle?

A.

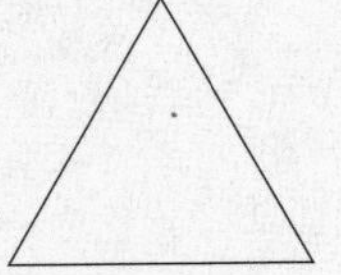

B.

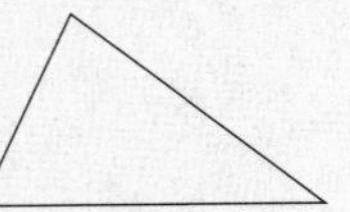

C.

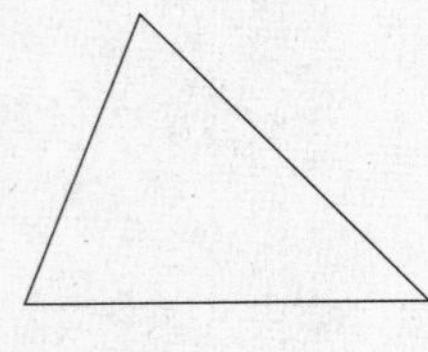

D.

1. ________

2. Find the measure of $\angle PQS$ in the parallelogram shown.

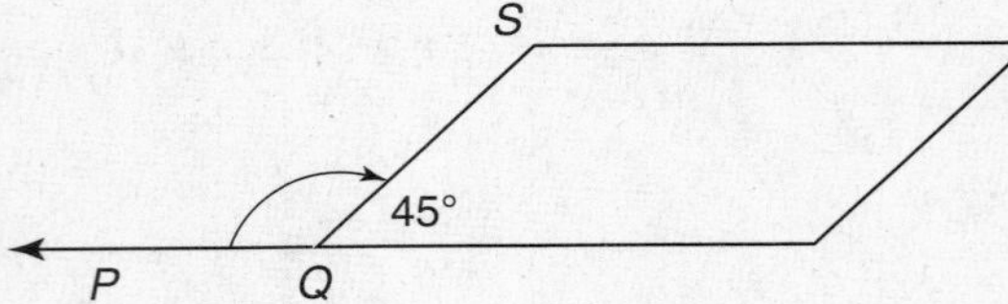

2.

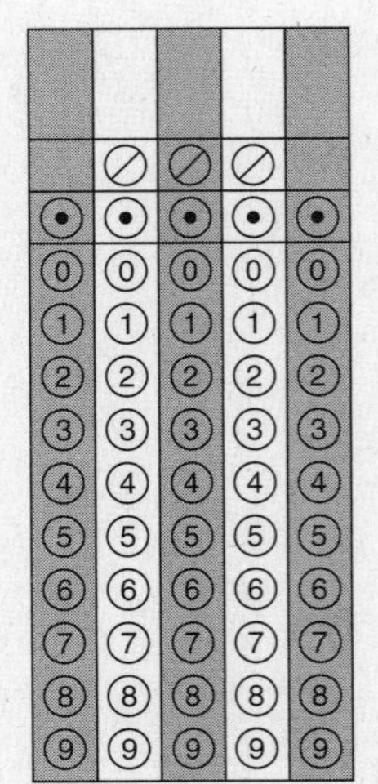

3. Find the sum of the measures of $\angle ABD$ and $\angle CBE$.

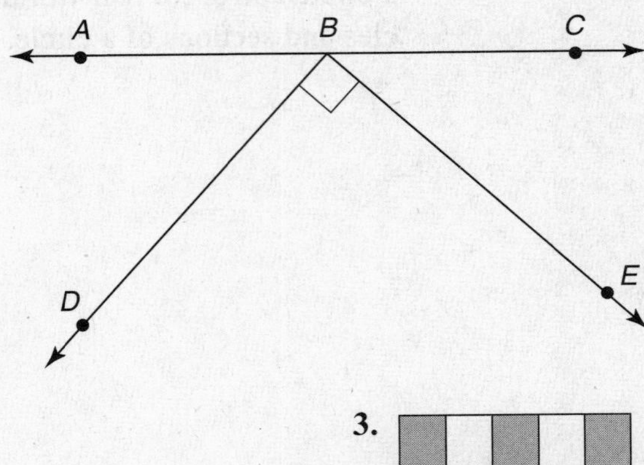

3.

4. $\angle X$ and $\angle Y$ are supplementary. m$\angle X$ is 20 more than 3 times m$\angle Y$. Find m$\angle Y$ and show or explain how you got your answer.

THINK

SOLVE

EXPLAIN

4. ________

5. Find the measure of $\angle ABC$ in the isosceles triangle shown here.

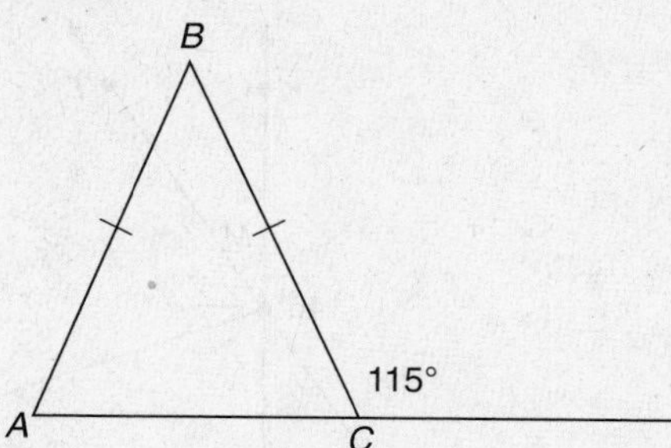

5.

	⊘	⊘	⊘	
⊙	⊙	⊙	⊙	⊙
0	0	0	0	0
1	1	1	1	1
2	2	2	2	2
3	3	3	3	3
4	4	4	4	4
5	5	5	5	5
6	6	6	6	6
7	7	7	7	7
8	8	8	8	8
9	9	9	9	9

6. In the figure shown, $\angle L \cong \angle N$ and $\overline{LM} \cong \overline{MN}$.

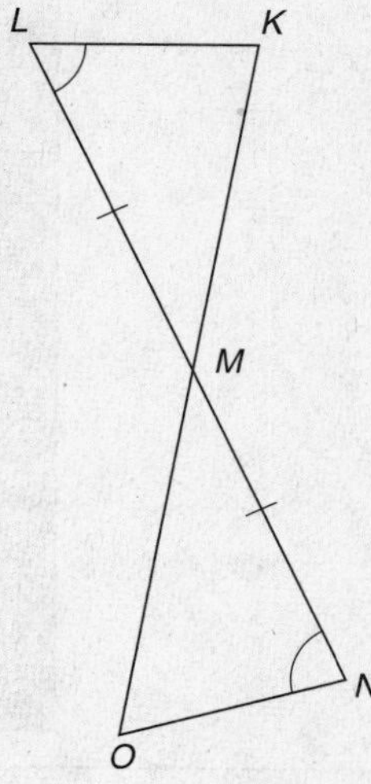

(*a*) Is $\overline{KM} \cong \overline{MO}$?

(*b*) Show formally or informally how you arrived at your conclusion.

THINK

SOLVE

EXPLAIN

6. (*a*) ________

(*b*) ________

7. Which of the following represents an obtuse angle?

F. $m\angle A = 15°$

G. $m\angle B = 180°$

H. $m\angle C = 90°$

I. $m\angle D = 121°$

7. ________

8. What is the measure of the angle complementary to an 81-degree angle?

8.

9. Find the measure of $\angle a$ in the accompanying figure.

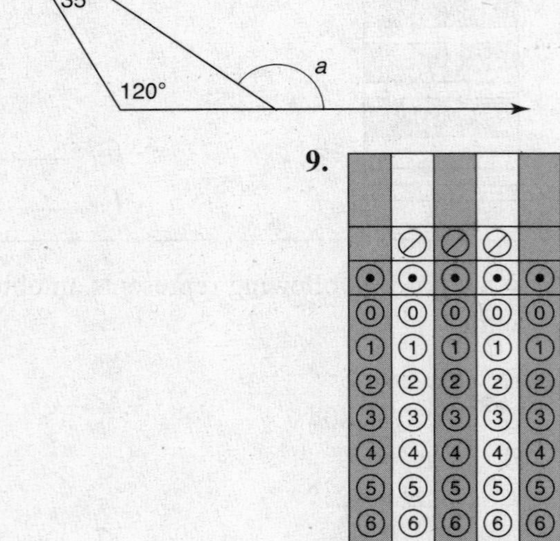

9.

10. Find the measure of $\angle 2$ in the figure shown if $\overline{MN}$ bisects $\angle LMI$.

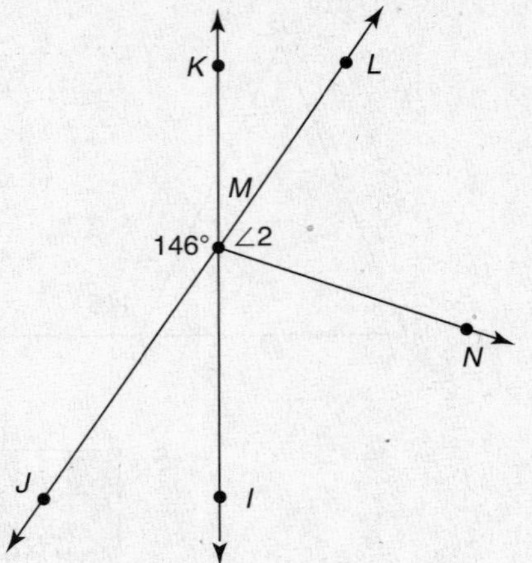

10.

11. Which of the following is a 90-degree counterclockwise rotation about point P in the given figure?

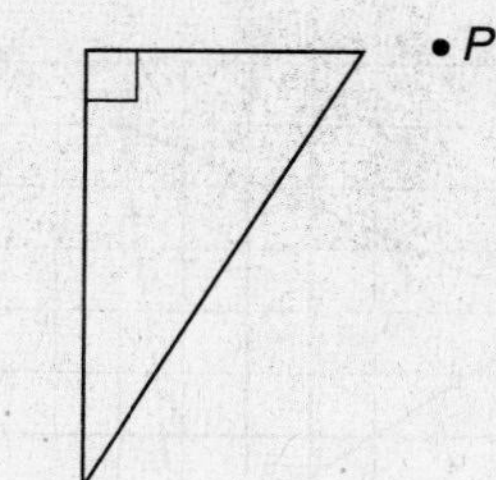

A.

B.

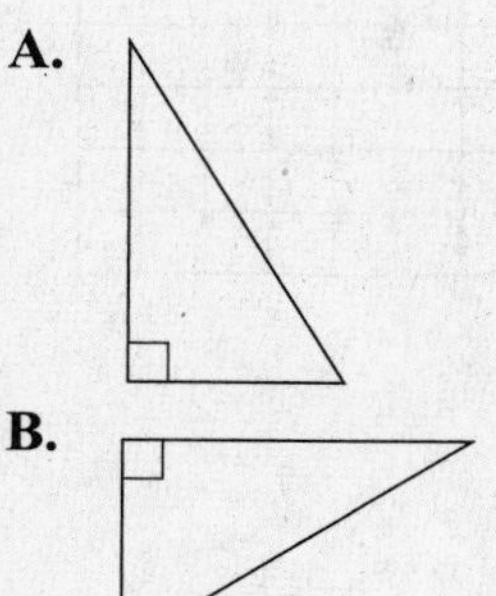

C.

D.

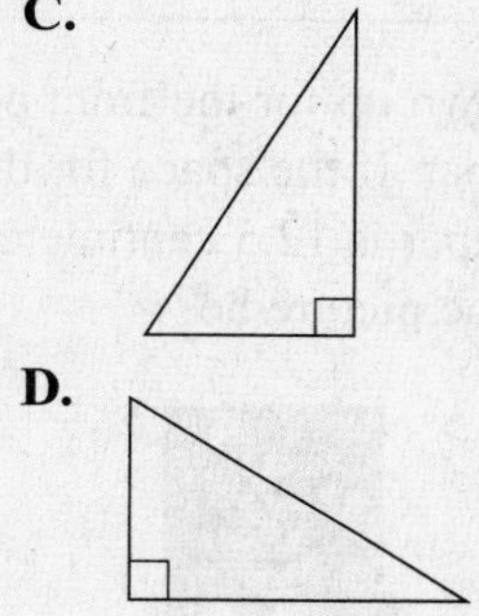

11. ________

12. Sandi is using the accompanying drawing as a design for a wall stencil. If the stencil is reflected over a vertical line of symmetry, which design will result from the reflection?

F.

G.

H.

I.

12. ________

13. Find the area in square feet of the planar cross section of the rectangular solid shown.

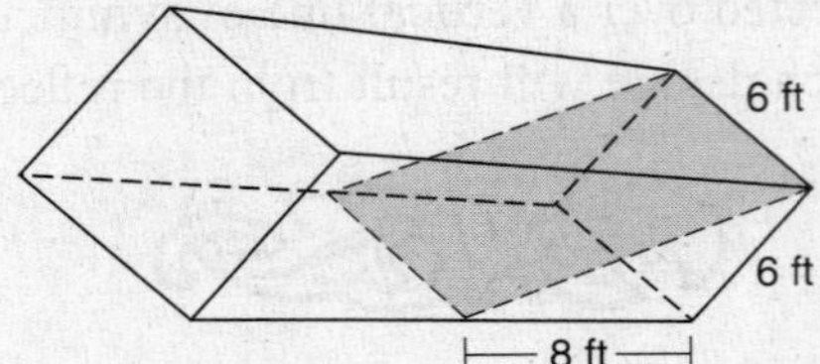

13.

14. The car shown here is a scale model. If the length of the actual car is 17.5 ft, how wide is the actual car?

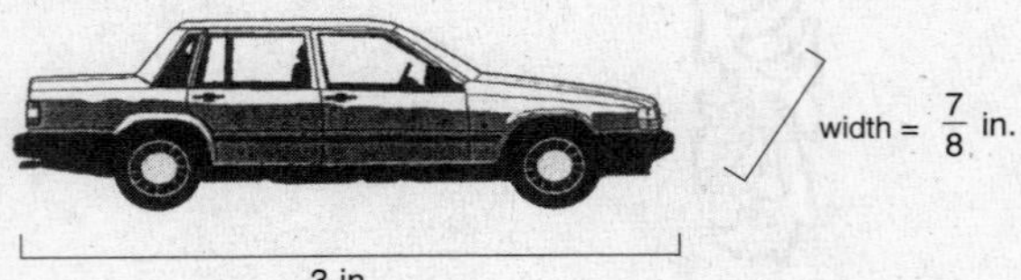

A. 5 ft 1 in.
B. 5 ft 2 in.
C. 5 ft 10 in.
D. 5 ft

14. ________

15. If the figure shown here is translated so that point A is at the origin, what will be the coordinates of point B?

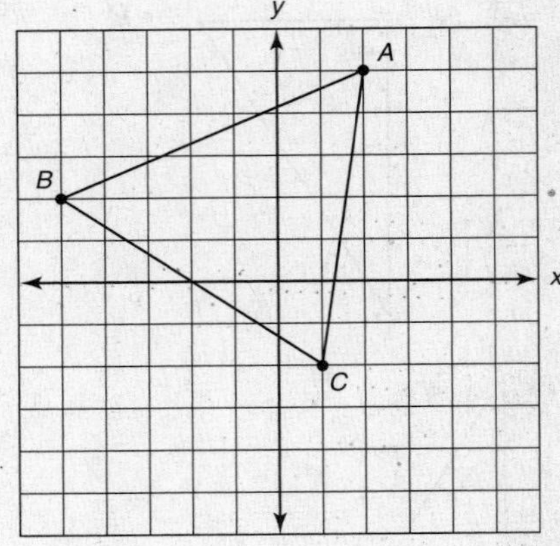

F. (0, 0)
G. (−5, 2)
H. (−7, −3)
I. (−5, −2)

15. ________

16. A cartoon is to be blown up for the front page of the school newspaper. If the space for the picture on the newspaper is 12.5 centimeters high, how wide can the picture be?

16.

17. A lighthouse that is 80 feet tall casts a shadow 50 feet long. Find the height of a tree that casts a shadow 22.5 feet long at the same time of day.

17.

18. Find the measure of x in the diagram shown.

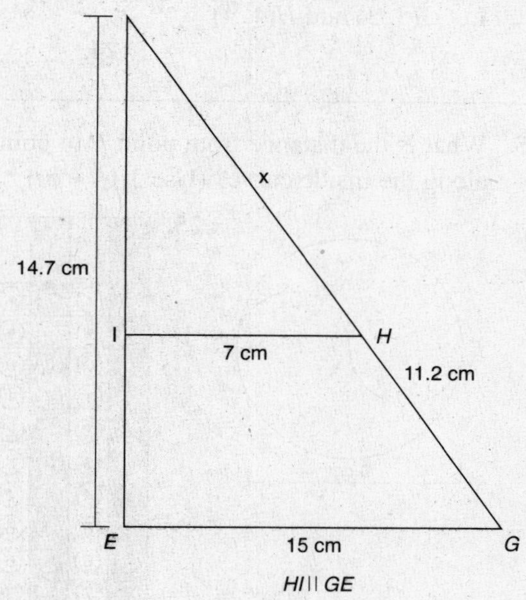

A. 13 cm
B. 14.7 cm
C. 21 cm
D. 9.8 cm

18. ________

19. Find the coordinates of the midpoint of $\overline{AB}$ when A is located at (–7, 4) and B is at (11, –2).

F. (2, 2)
G. (2, 1)
H. (–2, 2)
I. (9, 3)

19. ________

20. Indirect measurement can be used to find distances that cannot be measured directly. Use indirect measurement to find the distance across the lake ($\overline{AD}$).

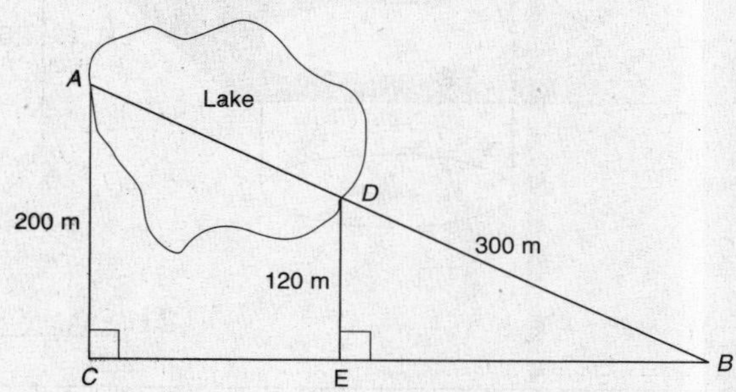

(*a*) Explain formally or informally why $\triangle DBE$ is similar to $\triangle ABC$.
(*b*) Write a proportion that can be used to find the distance across the lake.
(*c*) Solve the proportion and show your work.

THINK
SOLVE
EXPLAIN

20. (*a*) ________
(*b*) ________
(*c*) ________

21. In the figure shown, the shaded plane is parallel to the base of the cylinder. The height of the cylinder is 20 cm, and its volume is 2260 cm^3. Which of the following represents the area of the cross section of the cylinder if its radius is 8 cm?

A. 354.95 cm^2
B. 18.84 cm^2
C. 36 cm^2
D. 200.96 cm^2

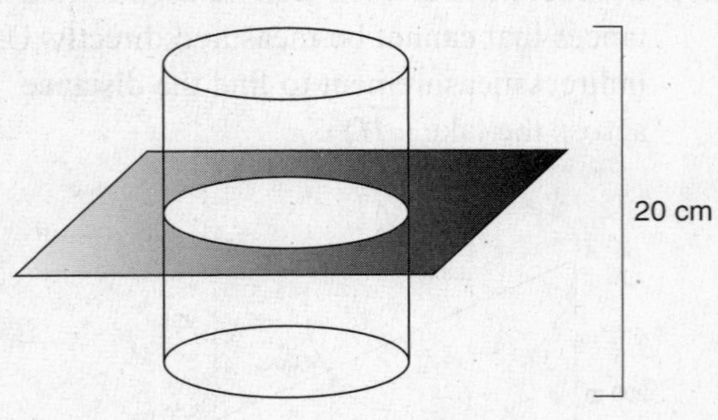

21. ________

22. Quadrilateral *DEFG* is a square with coordinates at $D(-2, -2)$, $E(-2, 4)$, $F(4, 4)$, $G(4, -2)$. How far is it from point *D* to point *F*?

F. 6
G. 72
H. $6\sqrt{2}$
I. $8\sqrt{2}$

22. ________

23. Which pair of equations represent lines perpendicular to each other?

A. $y = 2x + 4$
$y = 2x - 7$
B. $y = 3x - 6$
$y = \frac{1}{3}x + 6$
C. $y = 4x - 2$
$y = -4x + 3$
D. $y = \frac{2}{5}x + 1$
$y = -2.5x - 8$

23. ________

24. Which of the following sets of coordinates lie on the same vertical line?

F. $A(-2, 4)$ and $B(-2, -6)$
G. $C(2, 1)$ and $D(-3, 1)$
H. $E(2, 5)$ and $F(-2, -5)$
I. $G(3, 3)$ and $H(4, 4)$

24. ________

25. What is the distance from point *P* to point *Q* along the inside curve? (Use $3.14 \approx \pi$.)

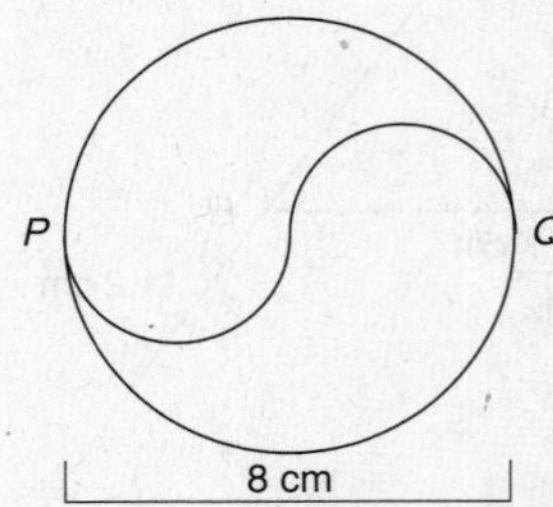

25.

26. Angus drills two holes in a piece of wood for dowels. Find the area of the shaded region remaining in the figure shown. (Use $3.14 \approx \pi$.)

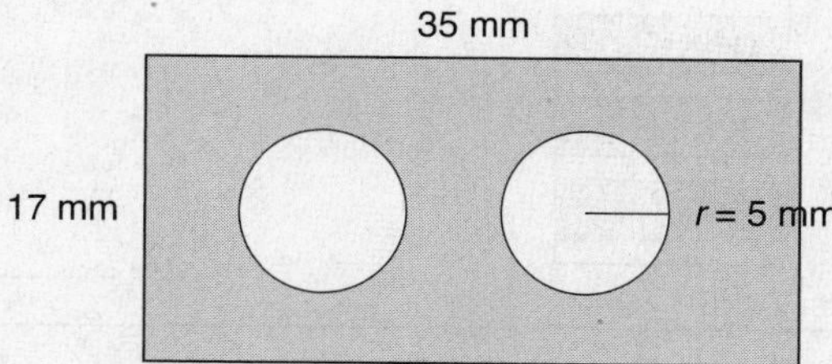

26.

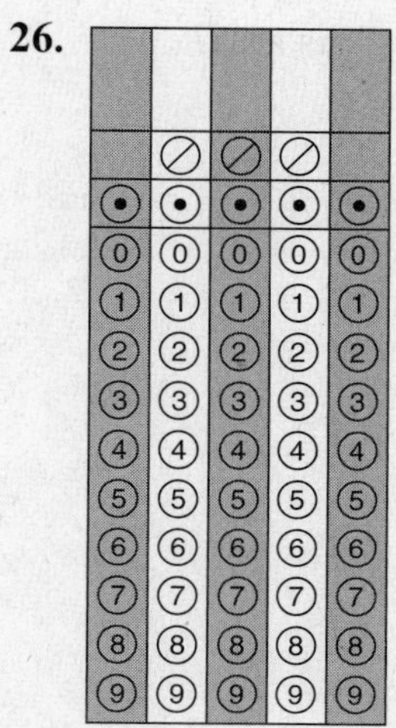

27. Tara is buying flooring for her house as indicated in the diagram. How many feet of flooring will she need to purchase?

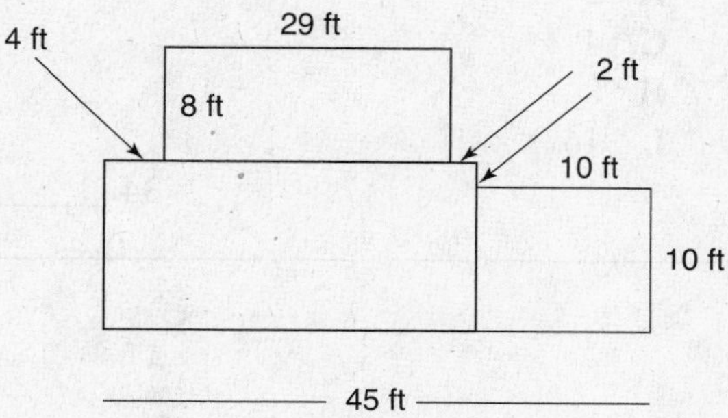

27.

28. The lake shown here has a circumference of approximately 3760 feet. If Raven runs around the lake three times each morning, to the nearest mile, how many miles will she have run at the end of a week?

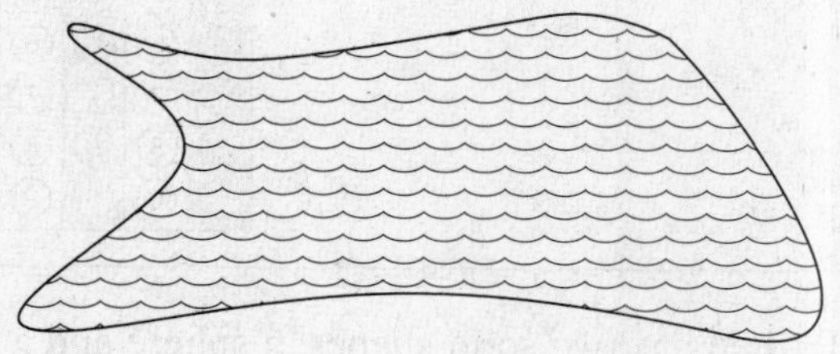

28.

29. What is the surface area of a cylinder that has a radius of 9 cm and a height of 22 cm?

A. 876.06 cm^2
B. 310.86 cm^2
C. 5849.82 cm^2
D. 1752.12 cm^2

29. ________

30. If the circumference of a circle is 95.456 meters, what is the radius in meters?

30.

31. Lee has two solid shapes, a sphere and a cylinder. If he pours all the liquid from the cylinder into a sphere of the same radius, will the sphere overflow? Explain how you would determine your answer.

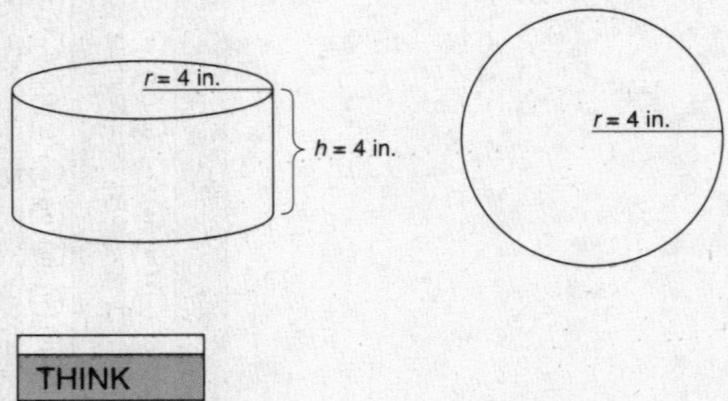

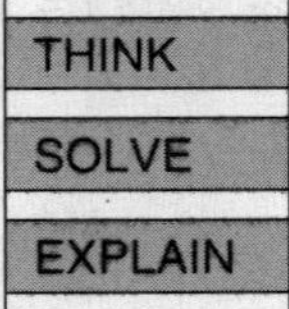

31. ________

32. Alicia needs to design a label for a can that has a radius of 1.5 in. and a height of 4 in. What should the dimensions of the label be? Show how you arrived at your answer.

THINK
SOLVE
EXPLAIN

32. ________

33. The perimeter of a square is 200 mm. What is its area?

33.

34. If the radius of a right cylinder is doubled, its volume increases by a multiple of

F. 2
G. 4
H. 8
I. 16

34. ________

35. The space shuttle travels at approximately 17,500 miles per hour. Which of the following represents the distance d that the shuttle has traveled after t hours?

A. $t = \dfrac{d}{17{,}500}$

B. $t = \dfrac{17{,}500}{d}$

C. $d = 17{,}500t$

D. $d = \dfrac{t}{17{,}500}$

35. ________

36. Find the area of the rectangle shown.

5x

3x

F. $15x$
G. $15x^2$
H. $16x$
I. $16x^2$

36. ________

37. A map is drawn to a scale of 2 cm = 60 m. If two cities are 8.4 cm apart on the map, what is the actual distance between them?

37.

38. Kevin can nail up to six fence rails in 20 minutes. At this rate, how many hours will it take him to put up 360 fence rails?

38.

39. Find the height of the trapezoid in the diagram if the area is 66.5 cm^2.

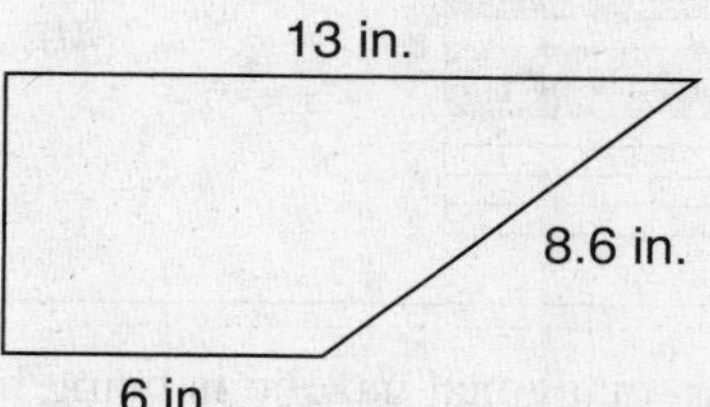

39.

40. Mrs. Modlin needs to add a dechlorinator to her outdoor fish pond. The dechlorinator is added at a rate of 5 drops per gallon of water. The circular pond has a radius of 28 inches and an average depth of 14 inches. If 1728 cu. in. = 7.5 gallons, about how many drops of the dechlorinator does she need to add?

(*a*) What formula should you use to calculate the amount of water in the pond?
(*b*) Calculate the number of cubic inches in the pond. Show all your steps and calculations.
(*c*) Use the answer from part (b) to calculate the number of drops needed. Show your steps and calculations.

THINK
SOLVE
EXPLAIN

40. (*a*) ________
(*b*) ________
(*c*) ________

41. An octagonal gazebo is being built for Leigh Ann's backyard. What is the measure, in degrees, of an interior angle of the gazebo?

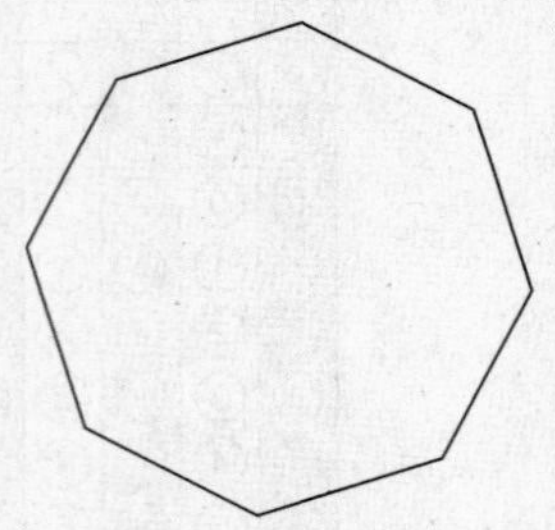

41.

42. Calculate the length of the arc shown in the diagram. (Use $3.14 \approx \pi$.)

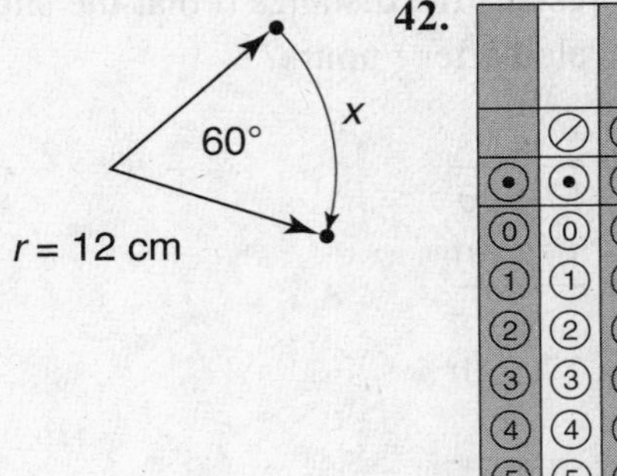

42.

43. A leaky faucet loses 4 ounces of water in 30 minutes. At this rate, how many gallons are lost in a day if there are 128 ounces per gallon?

43.

44. The area of a room is 650 square feet. If the width of the room is 25 feet, what is the length of the room?

44.

	⊘	⊘	⊘	
⊙	⊙	⊙	⊙	⊙
0	0	0	0	0
1	1	1	1	1
2	2	2	2	2
3	3	3	3	3
4	4	4	4	4
5	5	5	5	5
6	6	6	6	6
7	7	7	7	7
8	8	8	8	8
9	9	9	9	9

45. Pat's aquarium fish are multiplying at an alarming rate. His aquarium is 20 inches high, 14 inches wide, and 24 inches long, and he decides to build a new one. If the new aquarium is twice as long and twice as wide as the old aquarium, how will the volume of the new aquarium compare to that of the old?

(*a*) Sketch the new aquarium and label all dimensions.

(*b*) Explain how the volume of the new aquarium compares to the volume of the old aquarium. Justify your answers by using the volume of both.

THINK

SOLVE

EXPLAIN

45. (*a*) ________

(*b*) ________

Answers Explained

1. **D.** An obtuse scalene triangle has one obtuse angle, and all sides are different lengths.

2.

1	3	5		

Interior + exterior angles = 180 degrees.

3.

9	0			

$\angle DBE$ is a right angle. Therefore, the other two angles must add up to 90 degrees.

4. $\angle X + \angle Y = 180°$. $\angle X = 20 + 3Y$. Substitute $20 + 3Y$ for X in the first equation: $20 + 3Y + Y = 180$; $4Y = 160$; $m\angle Y = 40$.

5.

5	0			

$\angle C = 65$ degrees (interior + exterior angles = 180 degrees). $m\angle A = m\angle C$ because this is an isosceles triangle. Subtract: $180 - 65 - 65 = 50$. $\angle B = 50$ degrees.

6. (*a*) The two line segments are congruent.
   (*b*) The triangles are congruent, by ASA. The vertical angle plus the given information gives enough information to use ASA. If the two triangles are congruent, their corresponding sides are congruent. $\overline{KM}$ is congruent to $\overline{MO}$?

7. **1.** Obtuse angles are greater than 90 degrees but less than 180 degrees.

8.

9				

Complementary angles add to 90 degrees.

9.

1 5 5

The measure of an exterior angle is equal to the sum of the two nonadjacent interior angles. Add 120 + 35.

10.

7 3

m∠2 = 73 degrees. m∠*LMI* = 146 degrees because it is a vertical angle to ∠*KMJ*. $\overline{MN}$ bisects ∠*LMI*, and so ∠*LMN* = half of 146 degrees.

11. D. You can simulate a 90-degree counterclockwise rotation by rotating your paper 90 degrees to the left.

12. F. A vertical line of symmetry reflects the figure sideways as if it were reflected in a mirror.

13.

6 0

The length of the cross section is found using the Pythagorean Theorem $c^2 = 6^2 + 8^2$, where $c = 10$. Multiply the length by the width, or $10 \times 6 = 60$, to find the area. The area is 60 square feet.

14. A. Use the proportion: $\frac{3 \text{ in.}}{17.5 \text{ ft}} = \frac{.875 \text{ in.}}{x \text{ ft}}$; $x = 5.10$. 5 feet 1 inch is closest.

15. H. Point *A* must shift left two spaces and down five spaces. Point *B* should move exactly the same number of spaces (left 2, down 5).

16.

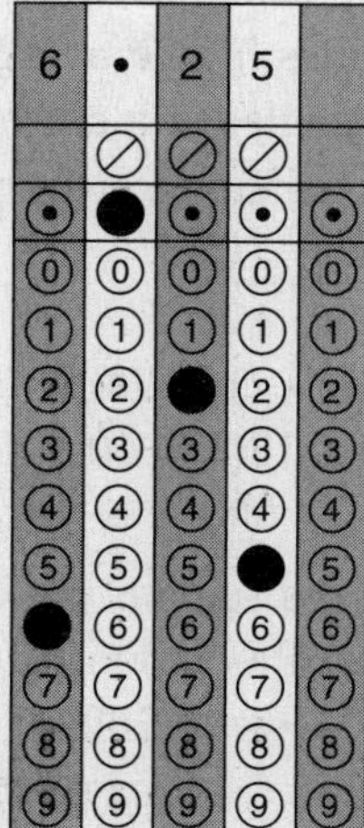

The figures are similar. Use the proportion: $\frac{5}{12.5} = \frac{2.5}{x}$; $2.5 \cdot 12.5 = 5x$; $31.25 = 5x$; $x = 6.25$.

17.

3	6			

Use the proportion: $\frac{80}{50} = \frac{x}{22.5}$;
$80 \bullet 22.5 = 50x$; $1800 = 50x$; $36 = x$.

18. D. Use the Pythagorean Theorem to solve the problem. The hypotenuse of the larger triangle is $c^2 = 14.7^2 + 15^2$; $c = \sqrt{14.7^2 + 15^2} = 21$. To find the missing side, subtract: $21 - 11.2 = 9.8$.

19. G. The midpoint is found by

$$\left(\frac{-7+11}{2}, \frac{4+(-2)}{2}\right) = \left(\frac{4}{2}, \frac{2}{2}\right) = (2, 1).$$

20. (*a*) The two triangles are similar by AA (if two angles in one triangle are congruent to two angles in a corresponding triangle, the two triangles are similar).

(*b*) $\frac{300 + x}{200} = \frac{300}{120}$

(*c*) $120(300 + x) = 200 \bullet 300$;
$36{,}000 + 120x = 60{,}000$;
$120x = 24{,}000$; $x = 200$, $\overline{AD} = 200$ meters

21. D. The cross section is a circle. The formula for the area of a circle is $A = \pi r^2$. This means that $A = 3.14 \bullet 8^2 = 200.96$. You need go no further to find the area of the circle.

22. H. Use the Distance Formula. The distance between these two points can be expressed as:
$\sqrt{(-2-4)^2 + (-2-4)^2} = \sqrt{6^2 + 6^2} = \sqrt{36 \times 2} = 6\sqrt{2}$ or (8.48).

23. D. Perpendicular lines have slopes that are opposite reciprocals of each other. For lines of the form $y = mx + b$, the slope is represented by m. In this case, $\frac{2}{5}$ is the opposite reciprocal of -2.5 (or $-\frac{5}{2}$).

24. F. Points that lie on the same vertical line have the same x-coordinate.

25.

1	2	•	5	6

The two parts of the curve together will form a circle. Use the circumference. The diameter of the curve is 4 cm. $C = \pi d = 3.14 \bullet 4$ cm $=$ 12.56 cm

26. Find the area of the rectangle minus the areas of the two circles.
$(35 \bullet 17) - 2\,(3.14 \bullet 5^2) = 595 - 2(78.5) =$ 438

27.

7	5	2		

Add the areas of each individual rectangle after working with the measurements you are given to find the missing lengths and widths.

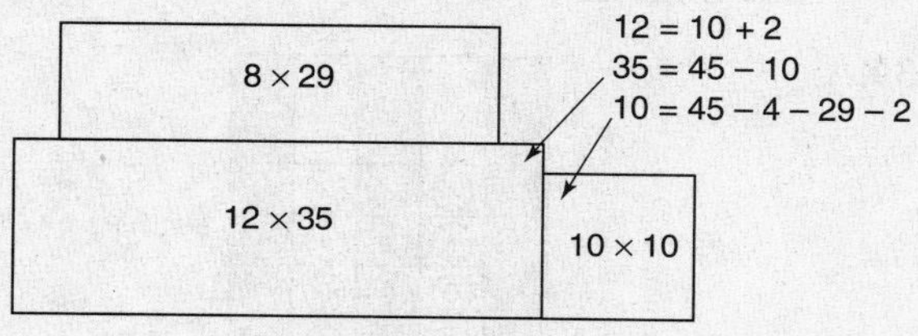

8 × 29 = 232
12 × 35 = 420
10 × 10 = 100
752

28.

1	5			

Multiply the circumference by 3 to get feet per day, and then by 7 to get feet per week. Then divide by 5280 to convert feet to miles.
3760 • 3 • 7 ÷ 5280 = 15 miles.

29. D. The formula for the surface area of a cylinder is used.

$$\begin{aligned} SA &= 2\pi rh + 2\pi r^2 \\ &= 2 \cdot 3.14 \cdot 9 \cdot 22 + 2 \cdot 3.14 \cdot 9 \cdot 9 \\ &= 1243.44 + 508.68 = 1752.12 \end{aligned}$$

30.

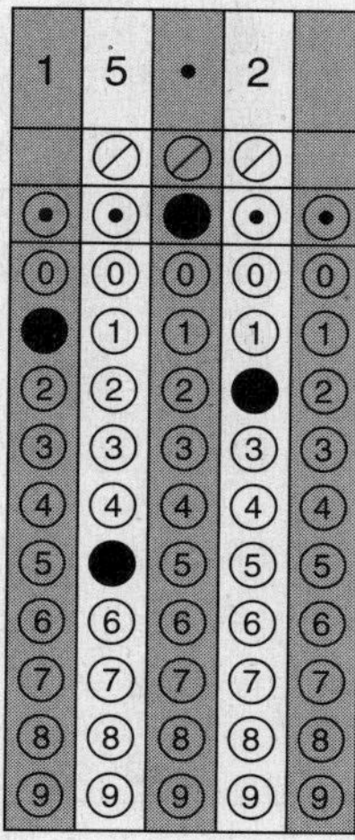

$C = 2\pi r$
$= 2 \cdot 3.14 \cdot r$ Substitute 95.456 for C.
$95.456 = 6.28r$ Divide both sides by 6.28.
$15.2 = r$

31. No, the cylinder is smaller in volume than the sphere.

Volume of sphere	Volume of cylinder
$V = \frac{4}{3}\pi r^3$	$V = \pi r^2 h$
$= \frac{4}{3} \cdot \pi \cdot 4^3$	$= \pi \cdot 4^2 \cdot 4$
$= 85.3\pi$ in.^{3}	$= 64\pi$ in.^{3}
$= 267.94$ in.^{3}	$= 200.96$ in.^{3}

32. The area of the label is represented by circumference × height.
$2\pi rh = 2 \cdot 3.14 \cdot 1.5 \cdot 4 = 37.68$ in.^{2}.

33.

2	5	0	0	

The square has 4 equal sides.
200 divided by 4 = 50. Since each side has a length of 50, the area = 50 • 50 or 2500.

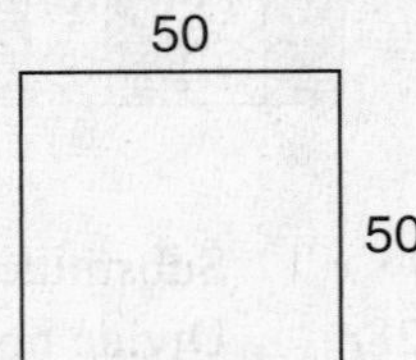

34. G. The radius r must be squared to find the volume. If the radius is doubled ($2r$), then $(2r)^2 = 4r^2$. The effect of this change will be to make the volume 4 times larger.

35. C. Use $d = rt$ and substitute 17,500 for r.

36. G. $(3x)(5x) = 15x^2$

37.

2	5	2		

Set up a proportion: $\frac{2 \text{ cm}}{60 \text{ mi}} = \frac{8.4 \text{ cm}}{x \text{ mi}}$.
$60 \times 8.4 \div 2 = 252$.

38.

2	6			

Set up a proportion: $\frac{6 \text{ rails}}{20 \text{ minutes}} = \frac{360 \text{ rails}}{x \text{ mi}}$.
$20 \times 360 \div 6 = 1200$ minutes.
1200 minutes $\div 60 = 20$ hours.

39.

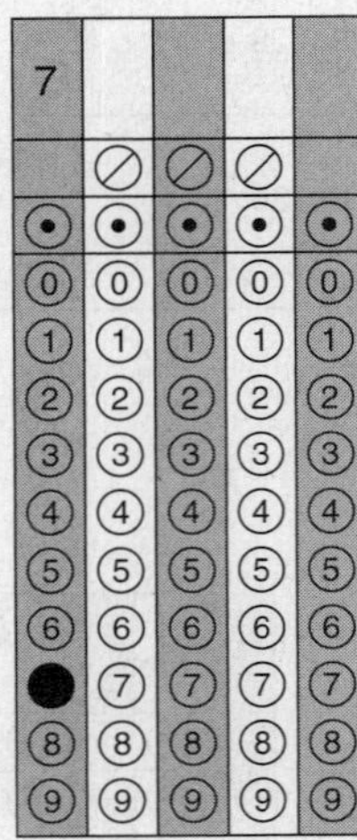

Use the formula for the area of a trapezoid and substitute the given values.

$$A = \frac{1}{2}(b_1 + b_2)h$$

$$66.5 = \frac{1}{2}(13+6)h \quad \text{Simplify.}$$

$$66.5 = \frac{1}{2} \bullet 19 \bullet h$$

$$66.5 = 9.5h \quad \text{Divide by 9.5.}$$

$$7 = h$$

40. (*a*) Use the formula for the volume of a cylinder: $V = \pi r^2 h$.

(*b*) $V = \pi r^2 h$. $V = 3.14 \cdot 28 \cdot 28 \cdot 14 =$ 34,464.64 cubic inches.

(*c*) $34{,}464.74 \div 1728 \cdot 7.5 = 149.55$ gallons. (There are 7.5 gallons in 1728 cubic inches.) Use 5 drops per gallon. Round 149.55 to 150 and multiply by 5. About 750 drops are needed.

41.

1 3 5

Use the formula for the sum of the angles of a polygon: $180(n - 2)$. Substitute the number of sides for n: $180\,(8 - 2) = 180 \cdot 6 = 1080°$. The octagon has a total of 1080°. To find the number of degrees in just one angle, divide by 8: $1080 \div 8 = 135$.

42.

1 2 • 5 6

First find the circumference of the whole circle represented by radius $r = 12$: $C = 2 \cdot 3.14 \cdot 12 = 75.36$. Find what portion or fraction of the entire circle the arc represents: $\frac{60}{360}$ or $\frac{1}{6}$. Multiply by the circumference:

$\frac{1}{6} \cdot 75.36 = 12.56$.

43.

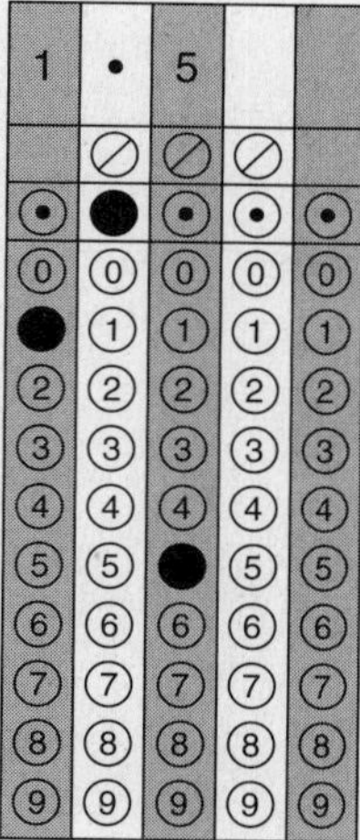

Set up a proportion:

$$\frac{4 \text{ oz}}{30 \text{ min}} = \frac{x \text{ oz}}{1440 \text{ min}}$$

(There are 1440 minutes in a day.)

Solve by multiplying $4 \cdot 1440$ and dividing by 30. There are 192 ounces. Convert 192 ounces to gallons by dividing by 128 (128 ounces per gallon): 1.5 gallons per day are lost.

44.

2	6			

$A = lw$	Substitute given values.
$650 = 25l$	Divide both sides by 25.
$26 = l$	The length is 26 feet.

45. (*a*) The new aquarium should be 20 inches high, 28 inches wide, and 48 inches long. The sketch should show these new dimensions.

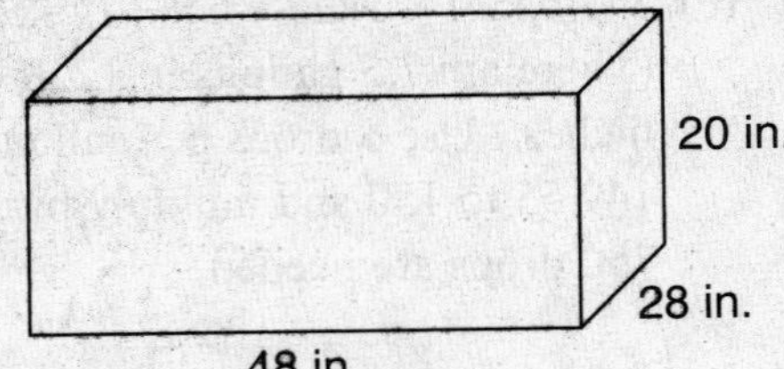

(*b*) The new volume of the aquarium will be four times the volume of the old aquarium.

Old volume: $20 \times 14 \times 24 = 6720$.
New volume: $20 \times 28 \times 24 = 26{,}880$. Two dimensions were doubled so the new volume is four times that of the old volume.

Chapter 3

DATA ANALYSIS, PROBABILITY, AND DISCRETE MATH

KEY TERMS/CONCEPTS

- Data analysis and measures of central tendency
- Probability and odds
- Numerical patterns
- Venn diagrams

SECTION 1: DATA ANALYSIS AND MEASURES OF CENTRAL TENDENCY

COLLECTING AND ANALYZING DATA

When you gather information about a group, you are collecting *data*. The US census, taken every ten years, attempts to gather data on the entire US population. A *population* is the entire group you are gathering information (data) about or studying. Populations do not have to be as large as the US population; they can be as small as a classroom. In fact, scientists study populations of animals, so your population doesn't even have to be human!

Many times, the population is so large that you cannot collect information from every individual. Even if you are collecting data from a classroom, someone may be absent on the day you need to gather that information. When you collect information from a portion of a population, you are taking a *sample*. It is important that the size of the sample be large enough so that it allows you to make accurate assumptions about the entire population.

For example, if you want to see how much time your classmates spend on homework, you would want to survey on a day when not many students were absent. If you asked on a day when the entire school band was out at a concert, and many of your classmates were in the band, your data might not represent the entire class very well.

Random Samples

A *random sample* is one in which everyone has an equal chance to participate. The sample should closely represent the entire population or group you are studying. If you want to know what percentage of the students in your school liked math, you would not ask the members of the Math Club. This would certainly give you a distorted view of the entire school's fondness for math. Instead, you might wait outside the main entrance and ask students about whether they liked math as they were coming to or leaving school.

Collecting Data

As you collect data, you need to find some organized way to keep track of it. One way is a frequency table. In a frequency table, like the one shown here, you use tallies and then count them up at the end (called the *frequency*).

Using intervals makes graphing large amounts of data easier.

Bird Spotted	Tallies	Frequency
Cardinal	𝍸 𝍸 ǁ	12
Mockingbird	𝍸 ǁ	7
Dove	𝍸 𝍸 𝍸 𝍸 ||||	24
Blue heron	|	1
Ibis	|||	3

The data collected above is about spotting birds. Some data, however, is strictly numerical. If there is a lot of numerical data, it can be helpful to organize it into *intervals*. Intervals are used to group data into spaces of a particular length. For example, the following data—0, 5, 8, 19, 2, 23, 8, 5, 7, 11, 24, 3, 1, 6, 20, 18, 5, 14, 28—can be grouped into intervals of 5 in the following way.

Interval	Tallies	Frequency
0–4	||||	4
5–9	𝍸 ǁ	7
10–14	ǁ	2
15–19	ǁ	2
20–24	|||	3
25–29	|	1

Sample Questions

1. According to the data in the table shown here, which is the most popular registered dog in the United States?

Breed	No. of Dogs Registered (thousands)
Golden retriever	63
German shepherd	57
Dachshund	50
Beagle	49
Poodle	46
Chihuahua	42
Rottweiler	42
Yorkshire terrier	41
Boxer	35
Shih tzu	35
Pomeranian	34
Cocker spaniel	30

2. This data represents the number of dogs (in thousands) registered with the American Kennel Club in 2007. Organize the data into a frequency chart using intervals of 10,000. Show the number of registrations of each breed in each interval.

Use your frequency table to answer questions 3–5.

3. Which interval contains the most registrations?
4. How many breeds of dogs have more than 49,000 dogs registered?
5. How many breeds of dogs have less than 40,000 dogs registered?
6. Ann's company employs 1500 people. Ann surveyed 46 employees in the finance department to see if a local restaurant would be the best place to hold the annual Christmas party. In the space provided below, explain why the results of Ann's survey are or are not representative of the entire company.

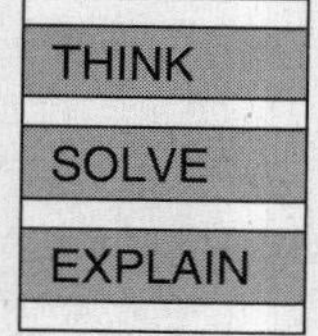

Answers

1. Golden retriever
2.

Number of Registrations	Tallies	Types of Dogs
30–39	||||	4
40–49	|||||	5
50–59	||	2
60–69	|	1

3. 40,000–49,000
4. Three
5. Four
6. The results are not representative. The finance department is not large enough to give a representative sample of the entire company, nor should the sample be limited to only one department. For the sample to be truly random, everybody in the company should have an equal chance to participate.

VENN DIAGRAMS

A Venn diagram is an additional way to organize and display data. Venn diagrams organize *sets* of numbers or things. The sets can then be compared to see what they have in common.

In the accompanying Venn diagram, Mr. Jones has recorded the number of students wearing shorts and the number of students wearing tennis shoes in his class. He has 35 students in the class.

35 Students in Class

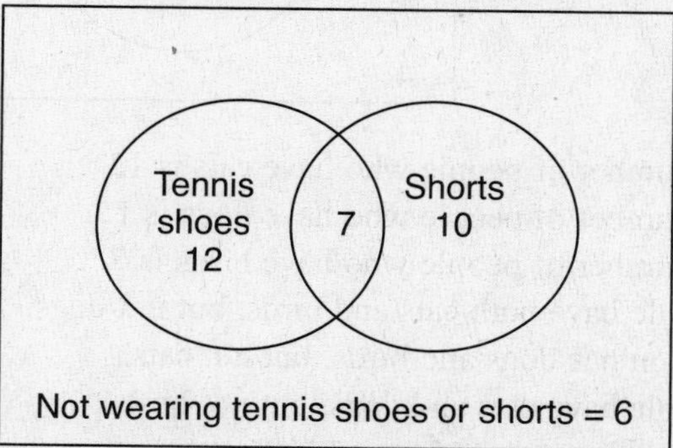

- The number of students wearing tennis shoes is 19. Count *all* the students in the circle containing the word *Tennis*.
- The number of students wearing shorts is 17. Count *all* the students in the circle containing the word *Shorts*.
- The number of students wearing *both* tennis shoes and shorts is 7. Notice the circles *share* these seven students. This is the place where the two sets *intersect.*
- The number of students wearing tennis shoes but not shorts is 12.
- The number of students wearing shorts but not tennis shoes is 10.
- The number of students wearing neither tennis shoes nor shorts is 6. (Subtract the number of students inside both circles from 35.)

It is possible for Venn diagrams to have one circle completely inside another. When this happens, one set is entirely a part of another set. In the Venn diagram shown here, the featured sets are A = {all the letters in the English alphabet} and B = {all the vowels in the English alphabet}. Since all the vowels are also part of the English alphabet, the circle for B is inside the circle for A.

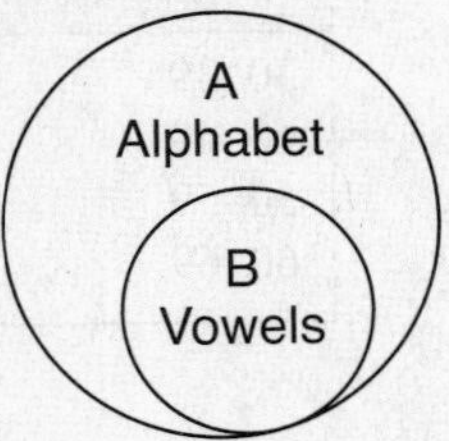

Venn diagrams can contain three circles. Their intersection is located at the center where the three circles overlap.

In the next example, Mark asked his neighbors what their favorite pet was: dogs, cats, or birds. He surveyed 40 neighbors. The results are shown here.

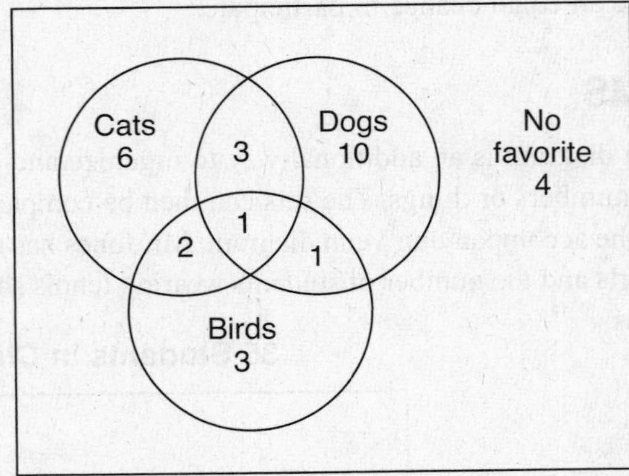

- The number of people who have cats is 12.
- The number of people who have dogs is 15.
- The number of people who have birds is 7.
- 2 people have both cats and birds, but not dogs.
- 1 person has dogs and birds, but not cats.
- 3 people have cats and dogs, but not birds.
- 1 person has cats, dogs, and birds.
- 4 people have no favorite (40 minus everybody else).

Sample Questions

1. Medford School for the Arts has 250 students. The accompanying Venn diagram shows the number of students enrolled in art classes at Medford. How many students are enrolled in Painting?

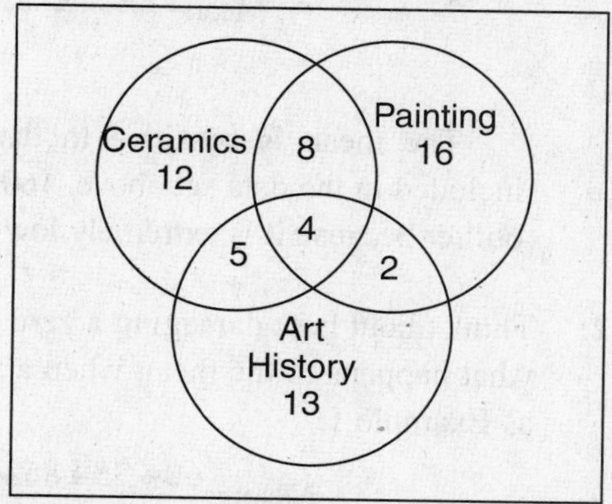

2. How many students are enrolled in both Ceramics and Painting?
3. How many students are enrolled in Art History but not Painting or Ceramics?
4. How many students are enrolled in all three classes?
5. How many students had none of these classes?

Answers

1. 30 students (count *all* the students in the Painting circle)
2. 12 students are enrolled in both Painting and Ceramics. Note that 4 of these students are also enrolled in Art History.
3. 13 students are enrolled in just Art History.
4. 4 students are enrolled in all three classes.
5. 190 students. Add all the students shown in the Venn diagram and subtract from 250.

MEASURES OF CENTRAL TENDENCY

Suppose you were keeping track of the temperatures for a week and you collected the following data: {81, 82, 86, 87, 89, 90, 92}. If you were asked to pick just one temperature that represented the week's temperatures, you might average the numbers. Another way you might select this temperature is to pick a number near the center after the data had been arranged in order (87).

Measures of central tendency allow us to use a single number to describe a set of data. Three measures of central tendency we will discuss here are called the *mean*, *median*, and *mode*. Because sets of data can vary greatly, each of these measures of central tendency can be useful in different situations.

Mean

Most students understand the word *average*. The *mean* is a measure of central tendency calculated by finding what most students think of as an average. In other words, add all the data points and divide by the number of data points available.

If the data set does not have extremely high or extremely low numbers, the mean should best describe the set.

Example 1: The mean is appropriate for the following test scores because none are extremely high or low: {60, 75, 85, 70, 85, 90, 65}.

$$\text{Mean}=\frac{60+75+85+70+85+90+65}{7}\approx 75.7$$

An **outlier** is a number that is either much higher or much lower than the other numbers in the set.

The mean is sensitive to data sets containing outliers. If zero were included in the data set above, {60, 75, 85, 70, 85, **0**, 90, 65}, it would be an outlier because it is extremely low when compared to the other test scores.

Example 2: Think about how damaging a zero on a test is to your grade and look at what happens to the mean when a zero test score is included in the same set as Example 1.

If the data set has one or more outliers, the median would be a better measure of central tendency than the mean.

$$\text{Mean}=\frac{60+75+85+0+70+85+90+65}{8}=66.25$$

Notice that the mean drops by more than 11 points!

Median

The median is another measure of central tendency. If you put a data set in order from least to greatest (or greatest to least), the *median* is the number that falls exactly in the middle. If the data set contains an odd number of data points, the median will be the middle number. If there is an even number of data points in the set, the median will be the mean (average) of the two middle numbers.

Example 1: Find the median of the following data set: {8, 6, 4, 7, 5, 6, 10, 4, 5}.

Step 1: Put the numbers in order from least to greatest: {4, 4, 5, 5, 6, 6, 7, 8, 10}.

Step 2: Locate the number in the middle: {4, 4, 5, 5, **6**, 6, 7, 8, 10}. That number is the median.
The median is 6.

Example 2: Find the median of the following data set: {5, 7, 9, 8, 7, 4, 2, 3, 8, 6, 7, 6}.

Step 1: Put the numbers in order from least to greatest: {2, 3, 4, 5, 6, 6, 7, 7, 7, 8, 8, 9}.

Step 2: Locate the numbers in the middle: {2, 3, 4, 5, 6, **6**, **7**, 7, 7, 8, 8, 9}.

Step 3: Find the mean of the two numbers in the middle: $\frac{6+7}{2}=6.5$
The median is 6.5.

Example 3: The number of people who bought bait at a local bait shop during a holiday week is shown in this table.

Date	July 1	July 2	July 3	July 4	July 5	July 6	July 7
Buyers	21	23	18	65	16	22	19

For this data set, the best measure of central tendency is the median. The mean best represents the number of people who *usually* buy bait during a *typical* week. More people bought bait on July 4th (a holiday) than on the other days of the week. The holiday caused the results to be different than normal so the median is a better measure to use in this case.

16, 18, 19, **21**, 22, 23, 65

The median is 21.

Mode

The third measure of central tendency is called the mode. The *mode* describes the value in a data set that occurs most often. For example, in the data set {5, 6, 3, 4, 2, 4, 7, 4, 8, 4}, the 4 occurs most often. The mode is a good choice for a measure of central tendency if a data set contains a lot of numbers that are identical.

Data sets can have *no mode*, for example, {4, 6, 3, 7, 2, 5}; have *one mode*, {5, 4, 6, 5, 8, 9}; or be *bimodal* (have two modes), {5, 2, 4, 5, 8, 9, 2}. In the case of a bimodal data set, there are two values that occur equally often. Data sets can even have three or more modes. If there is no mode, you should use the mean or median to describe your data set.

Memory Tip:
Mode begins with MO, which can stand for **M**ost **O**ften.

Sample Questions

1. Find the mean, median, and mode for the following set of data: {5, 0, 2, 9, 10, 3, 6,}.
2. The following data set represents the resting heart rates of eight students. Find the mean, median, and mode of these scores: {59, 73, 70, 71, 65, 70, 63, 68}.
3. Find the mean, median, and mode in the stem-and-leaf plot shown here.

Stem	**Leaf**
7	2 5 8 9
8	0 1 2 2 8 8
9	1 4 5 5 5 7 8 9
10	0 0

4. Which measure of central tendency would best represent the following game point scores for the local little league baseball team? {6, 3, 2, 1, 5, 20}
   A. mean
   B. median
   C. mode
   D. range

5. Mr. Lansford teaches six Algebra classes. There are 36 students in the first hour, 15 in the second hour, 33 in the third hour, 29 in the fourth hour, 36 in the fifth hour, and 28 in the sixth hour. If he wants to convince the principal that he needs smaller class sizes, should he use the mean, median, or mode to do this?

Answers

1. Mean $= \frac{5+0+2+9+10+3+6}{7} = 5$
   Median = 0 2 3 **5** 6 9 10
   There is no mode.
2. Mean $= \frac{59+73+70+71+65+70+63+68}{8} \approx 67.4$
   Median $= \frac{68+70}{2} = 69$
   Mode = 70
3. Mean = 88.45; median = 89.5(88 + 91/2); mode = 95
4. **B.** The median would be the best choice because the outlier, 20, would tend to pull the mean up and not give an accurate picture of the data. The term range is not a measure of central tendency.
5. He should use the mode because he wants to use the largest measure of central tendency possible to show that his class sizes are too large.

Range

The *range* of a data set describes how far the data are spread apart. For example, in the data set {22, 15, 28, 45, 33, 57, 39}, the data ranges from a low of 15 to a high of 57. We say the range is 57 – 15, or 42. It is the difference between the high and low values.

BOX-AND-WHISKERS PLOTS

A box-and-whiskers plot is a good way to display large sets of data. This method groups the data into four equal parts called *quartiles*. If you have 100 data values, 25 of the data values will be in the lower one-fourth of the plot, 25 in the upper one-fourth of the plot, and the remaining one-fourth will be in the "box."

Half the data in a box-and-whiskers plot is inside the box.

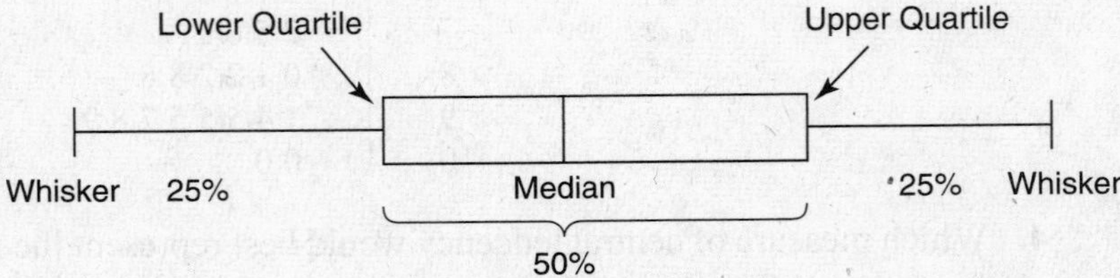

A box-and-whiskers plot is constructed using five numbers:

- The lowest number (the left whisker)
- The highest number (the right whisker)
- The median
- The lower quartile (the median of the lower half of the numbers)
- The upper quartile (the median of the upper half of the numbers)

The box is formed by joining the lower quartile, median, and upper quartile. The whiskers are drawn by joining the box to the high number on the right and the low number on the left.

Example 1:

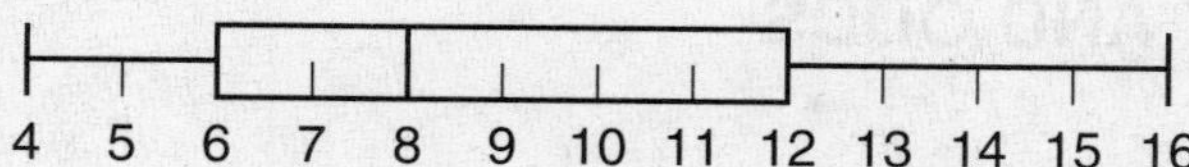

- Lowest number (also called the lower extreme) = 4
- Highest number (also called the higher extreme) = 16
- Median = 8
- Lower quartile = 6
- Upper quartile = 12

The range is 16 – 4 = 12. The *interquartile range* is 12 – 6 = 6.

Sample Questions

This box-and-whiskers plot shows the 2008 hourly compensation costs for 15 countries. Use it to answer the following questions.

1. Give the range, median, and quartiles of the data.
2. What fraction of the 15 compensation costs are between $12.25 and $21.75 per hour?
3. How many countries have compensation costs above $21.75?

Answers

1. The range is from approximately $5.00 to $27.40. Subtract: $27.40 – $5.00 = $22.40
   The median is about $17.00 per hour.
   The lower quartile is about $12.75 per hour.
   The upper quartile is about $21.75 per hour.
2. One-half of the data is contained in the box.
3. When you arrange 15 numbers in order, the eighth number is the median; therefore there are 7 numbers below the median and 7 numbers above the median. For the 7 numbers above the median, the fourth number represents the upper quartile. This means there are 3 numbers above that.

> ***Test Tip***: One excellent problem solving method is to try making up your own numbers. For example, let's make up 15 numbers and divide them up into a median, a lower quartile, and an upper quartile: {1, 1, 2, 4, 5, 6, 7, 8, 8, 9, 9, 11, 12, 13, 14}.
>
> 1 1 2 **4** 5 6 7 **8** 8 9 9 **11** 12 13 14
>
> The median (8) is in bold, the lower quartile (4) is in bold, and the upper quartile (11) is in bold. There are 3 numbers above the upper quartile. This method works for *any* 15 numbers.

SECTION 2: PROBABILITY, ODDS, AND STATISTICAL DATA

PROBABILITY AND ODDS

Probability

The term *probability* helps you to determine how often something is likely to happen. When the weather channel forecasts a 70% chance of rain, you know that the probability that it will rain is $\frac{70}{100}$, $\frac{7}{10}$, or 0.7. You cannot be 100% sure it will rain, or exactly when it will rain, but you can use this information to predict the likelihood of rain.

The probability that something will happen plus the probability it won't equal 100%.

The probability of an event is usually expressed as a number between 0 and 1. This number can be written as a fraction, decimal, or percent. When the probability is 0, this means the event will *never* happen, and when the probability is 1, it means the event will *always* happen.

Example 1: If the prediction of rain today is 90%, you should plan on taking your umbrella because 90% is much closer to 100% than 0%, making the event of rain very likely, although still not guaranteed.

Example 2: The probability of getting heads when you flip a fair coin is $\frac{1}{2}$. This means it is equally likely that you will get heads as tails because $\frac{1}{2}$ is just as close to 0 as it is to 1.

Example 3: The probability of drawing a black marble out of a jar is given as .28. This means that you are more likely to draw some other color because .28 is closer to 0 than it is to 1.

If you need to work out the probability that something will happen (an event), decide on the total number of things that could happen. Each of these is called an *outcome*. For example, if you throw a fair six-sided die (one of two dice), there are six possible outcomes: {1, 2, 3, 4, 5, 6}. The probability that you will roll any particular number is given as

$$\text{P(event)} = \frac{\text{number of favorable outcomes}}{\text{total possible outcomes}}$$

Example 1: The probability that you will roll a 3 is given as $P(3) = \frac{1}{6}$. Because there is only one 3, there is only 1 favorable outcome out of 6 possible outcomes.

Example 2: The probability that you will roll an even number is given as $P(\text{even}) = \frac{3}{6}\left(= \frac{1}{2}\right)$. Because there are 3 even numbers, there are 3 favorable outcomes out of 6 possible outcomes.

Example 3: The probability that you will roll a number divisible by 3 is given as $P(\text{divisible by }3)=\frac{2}{6}$ or $\frac{1}{3}$. There are 2 numbers that are divisible by 3 (3 and 6), so there are 2 possible favorable outcomes out of the 6 possible outcomes.

Example 4: If you roll an 8-sided die (numbered from 1 to 8), what is the probability of rolling a number divisible by 4? What is the probability of rolling a number not divisible by 4?

The possible outcomes are {1, 2, 3, 4, 5, 6, 7, 8}. Only two of these numbers are divisible by 4: {4, 8}. The other numbers are not divisible by 4: {1, 2, 3, 5, 6, 7}. The probability you will roll a 4 or an 8 is given by $P(\text{divisible by }4) = \frac{2}{8} = \frac{1}{4}$.

The probability you will roll a number *not* divisible by 4 {1, 2, 3, 5, 6, or 7} is given as $P(\text{not divisible by }4) = \frac{6}{8} = \frac{3}{4}$. Notice that the two fractions added together add to 1: $\frac{1}{4} + \frac{3}{4} = 1$. This means that the probability that something *will* happen plus the probability that it *will not* happen equals 1 (or is 100%).

Sample Questions

1. If you roll a 10-sided die numbered from 0 to 9, what is the probability that you will roll a zero?
2. In the spinner shown here, what is the probability that you will spin an A?

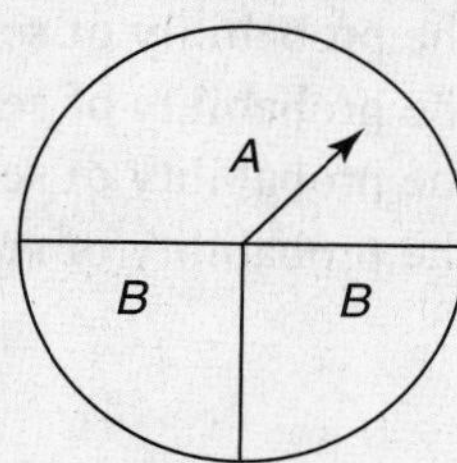

3. If you throw a fair 6-sided die, what is the probability that you will roll something other than a 4?
4. If you throw a fair 6-sided die, what is the probability that you will get a 7?
5. Using the spinner shown here, what is the probability that you will roll something other than a 5?

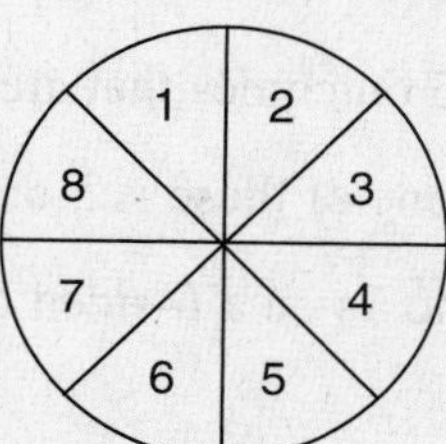

6. Using the spinner in question 5, what is the probability that you will spin a 3 or an even number?

7. Ann is looking in her mother's jewelry box. There are 3 diamond rings, 1 amethyst ring, 4 topaz rings and 2 citrine rings. If she reaches into the box without looking, what is the probability that the ring she selects is

(*a*) a diamond ring?
(*b*) an amethyst ring?
(*c*) a topaz ring?
(*d*) a citrine ring?

(i) What is the probability she will *not* select an amethyst ring?
(ii) What is the probability of selecting a diamond ring *or* a topaz ring?

8. If you select somebody at random from your class, what is the probability that he or she was born on the same day of the week that you were?

9. Using the spinner shown here, find the probability of spinning red if the probability of spinning blue, yellow, orange, or green is $\frac{1}{6}$.

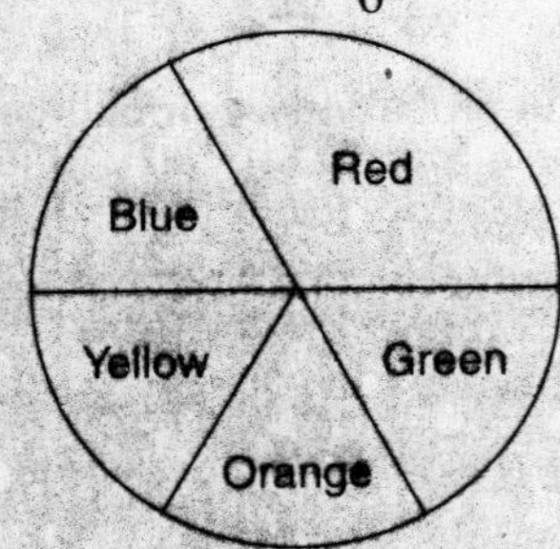

10. Given the group of numbers {2, 4, 6, 7, 9, 12, 14}, what is the probability that a number is even *and* less than 9?

11. A jar is filled with 15 marbles: 3 red, 5 yellow, 2 black, 1 green, and 4 blue. You reach into the jar (no peeking) and select one at random.

(a) What is the probability of selecting a yellow marble?
(b) What is the probability of selecting a black marble?
(c) What is the probability of selecting a marble other than a blue one?
(*d*) What is the probability of selecting a red or a green marble?

Answers

1. $\frac{1}{10}$. Out of 10 possible outcomes, only 1 is a zero.
2. $\frac{1}{2}$. When working with spinners, or any geometric figure, divide the spinner into equal portions. If the spinner is divided into 4 equal portions, you will find that the A takes up 2 of 4 portions or half the spinner.
3. $\frac{5}{6}$. There are 5 outcomes that are *not* a 4: {1, 2, 3, 4, 6}. Therefore, the probability that you will roll one of these is 5 out of 6.
4. 0. There are no 7s on a 6-sided die, so $P(7) = \frac{0}{6} = 0$.

5. $\frac{7}{8}$. The total possible outcomes on the spinner are 8. The probability of spinning a 5 is $P(5) = \frac{1}{8}$. The probability that you will not roll a 5 is $P(\text{not } 5) = 1 - \frac{1}{8} = \frac{7}{8}$.

6. $\frac{5}{8}$. This problem is tricky. The total number of outcomes on the spinner is 8. The favorable outcomes, 3 *or* an even number, are {2, 3, 4, 6, 8}. There are 5 favorable outcomes, and the total number of possible outcomes is 8. $P(3 \text{ or even}) = \frac{5}{8}$.

7. (*a*) $\frac{3}{10}$. 3 of 10 rings are diamond.

   (*b*) $\frac{1}{10}$. 1 of 10 rings is amethyst.

   (*c*) $\frac{4}{10} = \frac{2}{5}$. 4 of 10 rings are topaz.

   (*d*) $\frac{2}{10} = \frac{1}{5}$. 2 of 10 rings are citrine.

   i. $\frac{9}{10} \bullet 1 - \frac{1}{10} = \frac{9}{10}$

   ii. $\frac{3}{10} + \frac{4}{10} = \frac{7}{10}$

8. $\frac{1}{7}$. There are 7 days in the week: {Sunday, Monday, Tuesday, Wednesday, Thursday, Friday, Saturday}. The person you select can be born on only one of them.

9. $\frac{2}{6} = \frac{1}{3}$. The red portion of the spinner takes up twice as much space as each of the other colors, or $\frac{2}{6}$. Another way of looking at this problem is to decide that the colors blue, yellow, orange, and and green take up $\frac{4}{6}$ of the spinner. The remaining portion of the spinner (red) is $1 - \frac{4}{6}$ or $\frac{2}{6}$.

10. $\frac{3}{7}$. Of the 7 numbers given, only 3 are both less than 9 and even: {2, 4, 6}. Therefore, $P(\text{even and} <9) = \frac{3}{7}$.

11. (*a*) $\frac{5}{15} = \frac{1}{3}$. 5 out of 15 are yellow.

    (*b*) $\frac{2}{15}$. 2 out of 15 are black.

    (*c*) $\frac{11}{15}$. 11 marbles out of 15 are not blue.

    (*d*) $\frac{4}{15}$. There are 4 marbles that are either red or green.

Odds

The Robinson High School Knights are playing against the Plant High School Panthers. The odds of a Robinson win are given as 3 to 2, or $\frac{3}{2}$. By these odds, you know that the Knights are expected to win. Odds of 3 to 2 mean that if the teams play 5 games, the Knights should win 3 and the Panthers should win 2. The odds that the Knights will lose are given as 2 to 3.

Odds are another form of probability. The odds that you will throw heads on the toss of a coin is 1 to 1, or even, because only two things that can happen: One is heads and the other is tails.

Odds are most often expressed as a ratio or a fraction:

Odds and probability are related, but they are not the same.

$$\text{odds that something will happen} = \frac{\text{number of favorable outcomes}}{\text{number of unfavorable outcomes}}$$

$$\text{odds that something won't happen} = \frac{\text{number of unfavorable outcomes}}{\text{number of favorable outcomes}}$$

Example 1: On the toss of a fair 6-sided die, what are the odds that you *will* roll a 4?

$$\text{odds}(4) = \frac{\text{number of favorable outcomes}}{\text{number of unfavorable outcomes}} = \frac{1}{5}$$

There is only 1 favorable outcome, rolling the 4. There are 5 unfavorable outcomes, rolling a 1, 2, 3, 5, or 6.

Example 2: On the toss of a fair 6-sided die, what are the odds that you will *not* roll a 4?

$$\text{odds}(4) = \frac{\text{number of unfavorable outcomes}}{\text{number of favorable outcomes}} = \frac{5}{1}$$

Example 3: A basketball player makes a point 1 out of every 3 times he shoots. If he has to make the winning shot, what are the odds that his team will win the game?

Because he makes the shot 1 out of every 3 times, this means that out of every 3 shots, he makes the shot once and misses twice.

$$\text{odds}(\text{making the shot}) = \frac{\text{number of favorable outcomes}}{\text{number of unfavorable outcomes}} = \frac{1}{2}$$

Example 4: In a forest where deer are regularly caught and tagged, the odds that you will find a tagged deer are given as $\frac{1}{14}$. If you are a scientist tagging deer, predict how many deer out of 45 that are caught will already be tagged. If the odds are $\frac{1}{14}$, this means that 1 deer will be tagged for 14 that are not tagged (1 + 14 = 15). The probability that you will catch a tagged deer is $\frac{1}{15}$, or 1 tagged deer out of 15 deer. Multiply: $\frac{1}{15} \cdot 45 = 3$. Out of 45 deer, you can expect that 3 will already be tagged.

Sample Questions

1. A teacher is calling on students at random to read in a classroom of 24 students. If she has already called on 15 students, what are the odds that Bill will be the next person called on?
2. In a drawing, the probability that you will win is $\frac{1}{50}$. What are the odds against your winning?
3. A local nursery is giving away petunias and snapdragons as part of a promotion. If the odds of getting a petunia is $\frac{2}{3}$, predict how many snapdragons have been given away after 25 plants are distributed.
4. A can of mixed nuts is 40% peanuts. If you reach into the can at random, what are the odds that you will pick something besides a peanut?
5. If 4.3% of the students in the state of Florida drop out of high school. What are the odds that you will not drop out of high school?

Answers

1. $\frac{1}{8}$. There are only 9 people left to call on. Out of the 9 people left, Bill being selected to read counts as a favorable outcome. That leaves 8 unfavorable outcomes (someone other than Bill being selected to read). Odds (Bill is called on) = $\frac{1}{8}$.
2. $\frac{49}{1}$. There is a total of 50 outcomes; 49 of these outcomes are against your winning, and only 1 is in favor of your winning. Odds (not win) = $\frac{49}{1}$.
3. 15. Since the odds are $\frac{2}{3}$ and 2 + 3 = 5, this means that 2 of every 5 plants given away are petunias and 3 of every 5 plants are snapdragons. Multiply: $\frac{3}{5} \cdot 25 = 15$.
4. $\frac{3}{2}$. A 40% probability of picking a peanut means a 60% probability of picking something other than a peanut. $\frac{60}{40} = \frac{6}{4} = \frac{3}{2}$.
5. $\frac{957}{43}$. 4.3% means $\frac{4.3}{100} = \frac{43}{1000}$. If 43 of every 1000 students drop out, then 957 of every 1000 do not drop out.

THE COUNTING PRINCIPLE AND TREE DIAGRAMS

The *Counting Principle* can be useful when trying to find the probability of more than one event. For example, if one situation can occur in 2 ways and a second situation can occur in 3 ways, the total ways both things can occur is 2 × 3 ways.

A picture of the Counting Principle is called a *tree diagram*. For example, if you are making a Valentine card using a particular computer program, you are given 2 choices for the type of fold, a side-fold or a top-fold. Then, you can choose from 3 different designs for the front of the card. In this situation you have 2 separate events, the type of fold for the card, and the design. The number of different ways cards can be made is shown by the accompanying tree diagram.

The computer can make 6 different Valentine cards. You can make a tree diagram to show the total outcomes or multiply 2 types of fold by 3 types of designs to show that you have 2 × 3 or 6 possible outcomes.

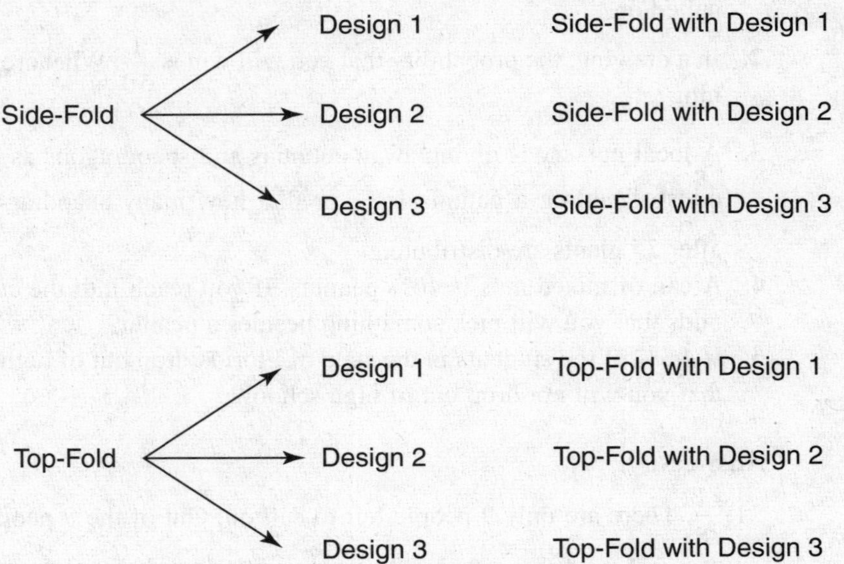

The Counting Principle can be useful for events that are repeated. For example, to find the possible outcomes (boy or girl) for a couple with three children, a tree diagram would look like this.

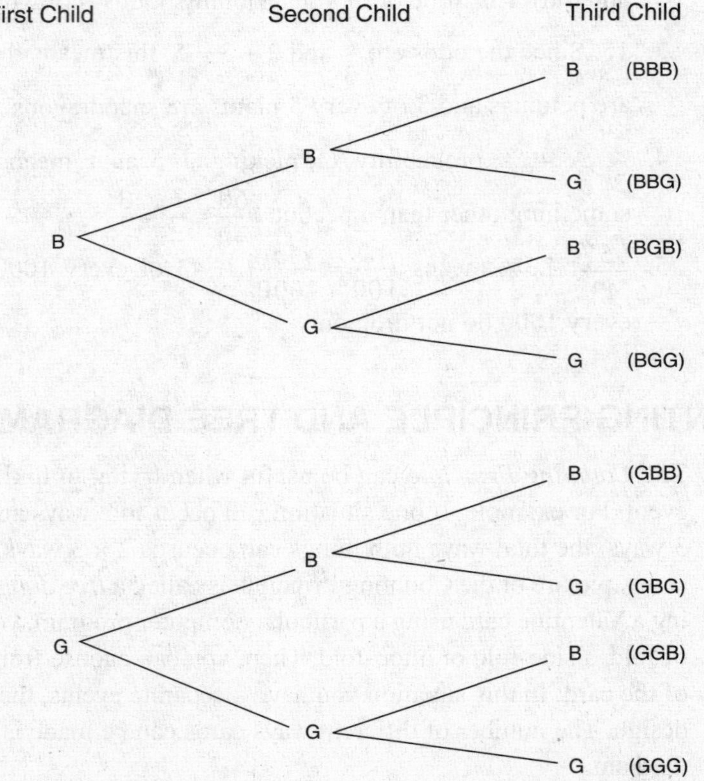

In this scenario, there are 8 possible outcomes: {BBB, BBG, BGB, BGG, GBB, GBG, GGB, GGG}. The Counting Principle could have been used to calculate the 8 possible outcomes by multiplying the number of outcomes for each event (2) three times (for three children). Therefore, $2 \times 2 \times 2 = 8$.

Example 1: Of the outcomes shown above, how many are possible for three boys?

There is only one outcome that shows three boys: {BBB}.

Example 2: What is the possible number of outcomes for rolling three 6-sided dice?

There are $6 \times 6 \times 6 = 216$ total outcomes possible. Making a tree diagram of this size would be tedious. You would have so many branches that it would be difficult to keep track of them.

Sample Questions

1. A deli stand at the Olympics has the menu shown in this table. How many sandwich choices are possible if every sandwich contains one kind of meat, one kind of bread, and one kind of cheese?

Bread	Meat	Cheese
White Wheat	Roast beef Turkey Ham Barbeque pork	American Swiss

2. Use the table above to find how many choices are available if the deli runs out of wheat bread.

3. Use the information from the table in question 1 to find the probability that the sandwich selected will have American cheese.

A. $\frac{1}{2}$

B. $\frac{1}{4}$

C. $\frac{1}{8}$

D. $\frac{1}{16}$

4. Scientists studying animals in the wild frequently tag them for identification. If the tag codes must be made up of two letters and three numbers, how many tag codes are possible?

A. 8
B. 20
C. 1000
D. 676,000

5. Use the spinners shown here to find the total possible outcomes from spinning both spinners.

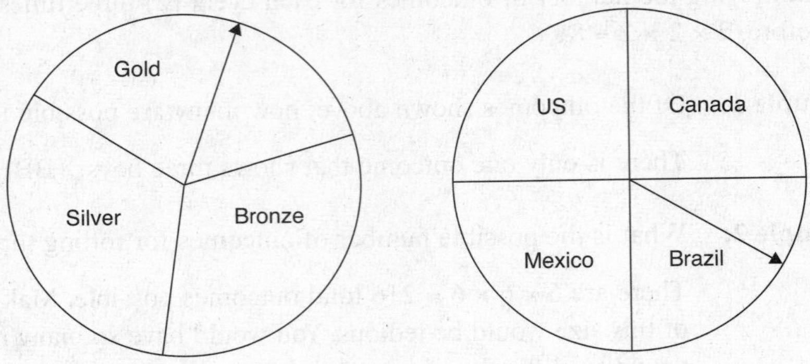

Answers

1. 16 possible outcomes. Use the Counting Principle and multiply the number of breads times the number of meats times the number of cheeses (2 × 4 × 2 = 16).
   Alternate solution: Make a tree diagram. Because it is time-consuming, you should make a tree diagram for a solution only if you can think of no other way to solve the problem.
2. 8 choices are available. Multiply 1 bread times 4 meats times 2 cheeses (1 × 4 × 2 = 8).
3. **A.** Sandwiches will have either American cheese or Swiss cheese; therefore there are only two possible outcomes.
4. **D.** There are 26 possible choices for one letter and 26 × 26 choices for two letters. There are 10 choices for one number (the digits 0 through 9) and 10 × 10 × 10 or 1000 choices for a three-digit number. So, there are 26 × 26 × 10 × 10 × 10 = 676,000 possible outcomes for a two-letter, three-digit tag code.
5. 12 outcomes. There are 3 medals and 4 countries. Using the Counting Principle, there are 3 × 4 = 12 possible outcomes.

INDEPENDENT AND DEPENDENT EVENTS

Independent Events

Sometimes you need to find the probability that two or more events will occur at the same time. First, you need to decide how the events are related to each other. When the outcome of one event has no effect on the results of the other event, the two events are said to be *independent*. Examples of this are rolling two dice, having four children, or spinning a spinner several times.

If you know the probability of each event, you can multiply to find the probability that both events will occur simultaneously (at the same time).

$$P(\text{A}) \cdot P(\text{B}) = P(\text{A and B})$$

To find the probability of two or more independent events, multiply.

Example 1: What is the probability that a couple will have four girls in a row?

Having a girl one time has absolutely no effect on whether a girl will be born the second, third, or fourth time. The probability of having a girl is given as

$P(\text{G}) = \frac{1}{2}$. The probability of having four girls is given as $P(\text{G then G then G then G}) = \frac{1}{2} \bullet \frac{1}{2} \bullet \frac{1}{2} \bullet \frac{1}{2} = \frac{1}{16}$.

Example 2: In a board game, to win the grand prize, you must roll a 2 three times in a row on a fair 6-sided die. What is the probability that this will happen?

The probability of rolling a 2 with a fair 6-sided die is given as $P(2) = \frac{1}{6}$. The probability of rolling a 2 three times in a row is calculated by multiplying: $\frac{1}{6} \bullet \frac{1}{6} \bullet \frac{1}{6} = \frac{1}{216}$. In other words, you can expect to roll a 2 three times in a row only one time out of 216 tries.

Example 3: Batting averages are calculated by dividing hits by times at bat. By using a person's batting average as a probability, you can calculate the probability that they will get a hit at any particular time at bat. If a baseball player has a batting average of .290, what is the probability he will get a hit two times in a row? What is the probability that he will get a hit the first time and miss the second time?

The probability of one hit is given as $P(\text{H}) = .290$, and the probability of two hits in a row is $P(\text{H then H}) = P(\text{H}) \bullet P(\text{H}) = .290 \bullet .290 = .0841$. If the probability of getting a hit $= .290$, the probability of missing is $P(\text{M}) = .710$. (Remember, both must add to 1.) The probability that he will get a hit on the first try and miss on the second try is given as $P(\text{H then M}) = .290 \bullet .710 = .2059$.

Using the Counting Principle for Independent Events

If you are not given the probability of each event, you may have to use the Counting Principle to find the probability. To do this, use the following steps:

Step 1: Find the total possible outcomes.

Step 2: Find which of the outcomes out of the total fit your problem (favorable outcomes).

Step 3: Set up a ratio : $\text{P}(\text{event}) = \frac{\text{favorable outcomes}}{\text{total possible outcomes}}$

Example 1: What is the probability that if you roll two fair dice you will get a sum of 7?

Step 1: There are 6 possible outcomes for rolling one die. For both dice the outcomes are 6 • 6 = 36. The outcomes can be confirmed by using the accompanying diagram.

Die 2 \ Die 1	**1**	**2**	**3**	**4**	**5**	**6**
1	2	3	4	5	6	7
2	3	4	5	6	7	8
3	4	5	6	7	8	9
4	5	6	7	8	9	10
5	6	7	8	9	10	11
6	7	8	9	10	11	12

Step 2: There are six ways to make a sum of 7 using two die (favorable outcomes):

Die 1	Die 2
1	6
2	5
3	4
4	3
5	2
6	1

Step 3: Set up a ratio:

$$\text{P}\left(\text{event}\right) = \frac{\text{favorable outcomes}}{\text{total possible outcomes}} = \frac{6}{36} = \frac{1}{6}.$$

Example 2: Using the table from Example 1, calculate the probability that the sum of the two dice will be at least 6.

Out of 36 possible outcomes, 26 outcomes show a sum of at least 6 (the sum must be 6 or higher). Therefore, the probability of rolling at least a 6 is given as $P(\text{sum} \geq 6) = \frac{26}{36} = \frac{13}{18}$.

Sample Questions

1. When you toss a pair of fair 6-sided dice, what is the probability that you will get a sum of 10?
2. When you toss a pair of fair 6-sided dice, what is the probability that you will get at most a 4?
3. If three coins are tossed at the same time, what is the probability that they will all land heads up?

Answers

1. $P(\text{sum of }10) = \frac{3}{36} = \frac{1}{12}$. Out of a total of 36 possible outcomes, there are three ways to make a sum of 10: {4 + 6, 5 + 5, and 6 + 4}.
2. $P(\text{sum} \le 4) = \frac{6}{36} = \frac{1}{6}$. The phrase "at most a 4" means that the sum must be no more than 4, in other words, 4 or less. Out of a total of 36 possible outcomes, there are six ways to make a sum of 4 or less: {1 + 1, 1 + 2, 1 + 3, 2 + 1, 2 + 2, 3 + 1}.
3. $P(\text{HHH}) = \frac{1}{2} \bullet \frac{1}{2} \bullet \frac{1}{2} = \frac{1}{8}$.

Dependent Events

Suppose you draw a black marble out of a jar containing 10 marbles, 4 of which are black. If you *do not* replace the black marble, this will affect your chances of drawing a black marble on the second try because there are only 9 marbles left, and only 3 of them are black.

When the outcome of one event affects the outcome of a second event, the two events are said to be *dependent.* The problem above involves two events in which the results of the first event have an affect on the results of the second event.

Example: To find the probability of drawing 2 black marbles from a jar containing 10 marbles (4 black, 3 green, and 3 red), *without replacement*, examine each event separately and then multiply the results of each:

$P(\text{black on first try}) = \frac{4}{10} = \frac{2}{5}$.

$P(\text{black on second try}) = \frac{3}{9} = \frac{1}{3}$ because after you draw 1 black marble out, there are 9 left, and only 3 of them are black.

$P(\text{black given black}) = \frac{2}{5} \bullet \frac{1}{3} = \frac{2}{15}$. Multiply the two probabilities together to find the probability of both events happening.

Suppose that the first marble *was* replaced; then the probability that 2 black marbles will be drawn in a row would be:

$P(\text{black then black}) = \frac{2}{5} \bullet \frac{2}{5} = \frac{5}{25} = \frac{1}{5}$

Sample Questions

1. Mike has 20 coins in his pocket. There are 3 quarters, 6 dimes, 5 nickels, and 6 pennies. If he takes one coin out at a time and does not put any back, what is the probability that the first 2 coins he draws from his pocket will both be pennies?
2. What is the probability (from question 1) that Mike will select a quarter first and then a dime? He does not replace the quarter.
3. Claire is looking in a box of 30 buttons for a matching button for her blouse. She is randomly taking buttons out of the box without putting them back in. What is the probability that she will select the correct button on the second try?

Answers

1. $\frac{6}{20} \bullet \frac{5}{19} = \frac{30}{380} = \frac{3}{38}$. Because Mike does not put the first coin back in his pocket, this is a problem in which the second outcome is affected by the first. After the first coin is drawn, only 19 remain, 5 of which are pennies.
2. $\frac{3}{20} \bullet \frac{6}{19} = \frac{18}{380} = \frac{9}{190}$. Once again, if Mike does not put the quarter back, the probability of the second event is affected. Now there are only 19 coins to choose from, and 6 of them are dimes.
3. $\frac{29}{30} \bullet \frac{1}{29} = \frac{1}{30}$. From the wording of the problem, you can assume that Claire picked the wrong button on the first try. If the probability of selecting the correct button on the first try is $\frac{1}{30}$, the probability of selecting the wrong one is $\frac{29}{30}$. If she does not replace the first button, 29 buttons are left in the box. The probability of selecting the correct button on the second try is then $\frac{1}{29}$. The probability of selecting the correct button on the second try is $P(\text{wrong then right}) = \frac{29}{30} \bullet \frac{1}{29} = \frac{1}{30}$.

Mutually Exclusive Events

When a coin is flipped, it can't come up both heads and tails at the same time. By the same token, a rolled die can't come up 1 and 6 at the same time. These are called *mutually exclusive events*. The probability of two mutually exclusive events (A and B) is $P(\text{A}) + P(\text{B})$. In other words, find the probability of each event and add.

Example 1: The probability that a rolled die will come up a 1 *or* a 6 is given as $P(1) + P(6) = \frac{1}{6} + \frac{1}{6} = \frac{2}{6} = \frac{1}{3}$.

Example 2: If all possible birthdays are written on slips of paper and put in a bag, what is the probability that you will draw a birthday in March *or* April?

Since you can't draw a birthday in both March and April, these are mutually exclusive events.

$P(\text{March}) = \frac{31}{366}$ and $P(\text{April}) = \frac{30}{366}$.

$P(\text{March } or \text{ April}) = \frac{31}{366} + \frac{30}{366} = \frac{61}{366}$.

When calculating the probability of mutually exclusive events, you need to calculate each fraction separately and then multiply.

Sample Questions

1. If one card is drawn from a deck of 52 cards, what is the probability that it will be either a heart or a king? (There are 13 hearts and 4 kings.)
2. If two dice are rolled, find the probability that either the first die *or* the second die will come up a 3.

Answers

1. $\frac{4}{13}$. P(heart or king) means you can draw a heart or a king but not a king of hearts. P(heart) $= \frac{13}{52}$, and P(king) $= \frac{4}{52}$. Add the probabilities of hearts and kings and subtract the probability of a king of hearts (so it isn't counted twice): $P(\text{heart or king}) = \frac{13}{52} + \frac{4}{52} - \frac{1}{52} = \frac{17}{52} - \frac{1}{52} = \frac{16}{52} = \frac{4}{13}$.
2. $\frac{1}{3}$. $P(3) = \frac{1}{6}$. $\frac{1}{6} + \frac{1}{6}$: $P(3 \text{ or } 3) = \frac{1}{6} + \frac{1}{6} = \frac{2}{6} = \frac{1}{3}$

ANALYZING STATISTICAL DATA

So far, we have looked at how data is displayed and measured. Now we will look at real-world data and analyze or interpret it. You may be asked on the FCAT to make predictions about some data based on what you find. You should also be able to recognize when a survey has been conducted improperly.

If the data is given in a chart or table, look it over closely by reading the title and locating the information that is relevant to your question. Frequently, you will be asked to find the *difference* between two sets of data. This means you should subtract.

Example: The double bar graph shown here compares the populations of several Florida counties in 1999 and 2008.

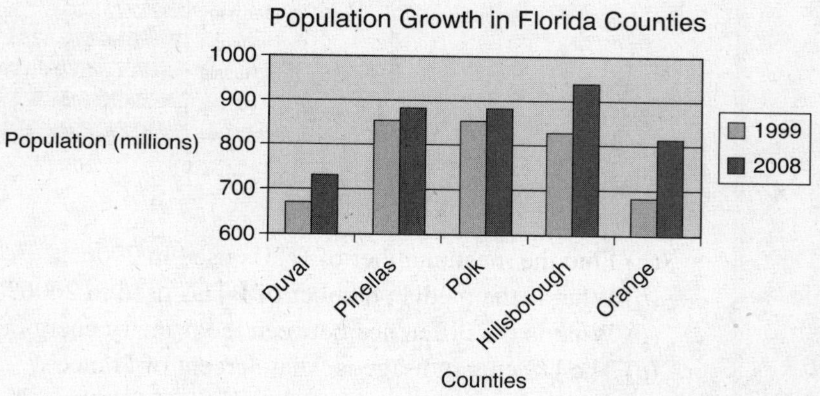

Which county had the highest percentage of growth between 1999 and 2008?

Method 1: Orange County stands out because there is a much larger distance between its two bars (1999 and 2008). You can see this by the relative sizes of the bars. This distance is much smaller for the other counties.

Method 2: You could actually calculate the percent of change for each county. Use the ratio $\frac{\text{amount of change}}{\text{previous amount}}$ and then do the division (amount of change ÷ prevous amount):

Duval = $\frac{65}{673} \approx 0.096$, which means growth was approximately 10%.

Percent increase or decrease = amount of change ÷ original amount

Pinellas = $\frac{26}{852} \approx 0.03$, which means growth was approximately 3%.

Polk = $\frac{23}{860} \approx 0.03$, which means growth was approximately 3%.

Hillsborough = $\frac{97}{843} \approx 0.12$, which means growth was approximately 12%.

Orange = $\frac{140}{677} \approx 0.21$, which means growth was approximately 21%.

This is a time-consuming process. Method 1 is preferable if it is at all possible.

Sample Questions

1. The world's consumption of primary energy—petroleum, natural gas, coal, nuclear, to name a few—is measured in British thermal units (BTUs). The accompanying graph shows the world's major consumers of primary energy as of 2006. Use this graph to answer the questions.

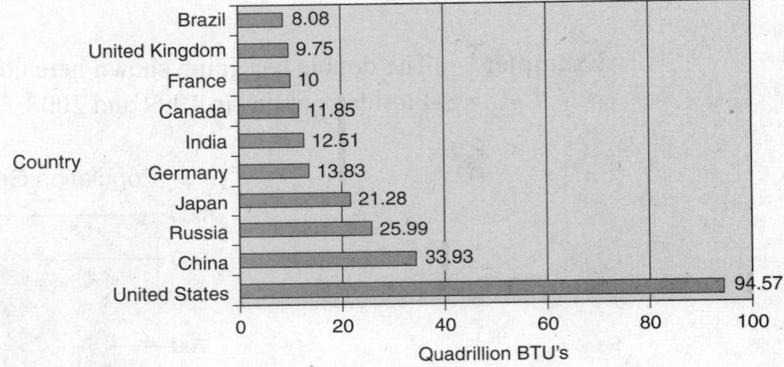

(*a*) Find the mean number of BTUs used in 2006.
(*b*) What is the median number of BTUs used in 2006?
(*c*) What is the difference between the primary energy usage of China and of India?
(*d*) The US energy usage is what percent of France's?
(*e*) Which country uses approximately one-fourth of the primary energy used by China?
(*f*) Which country uses approximately two-thirds of the primary energy used by Japan?

2. Four students ran for election at Washington High School. The results of the election are shown here.

Class	Doug Upp	Fanny Pack	Bill Dupp	May Knot	Total Votes by Class
Freshman	223	249	131	142	745
Sophomore	151	139	243	222	755
Junior	179	203	122	203	707
Senior	214	198	153	211	776
Total	767	789	649	778	2983

Based on the information in the table, which of the following statements are true?

A. Doug received more than one-fourth of the votes.

B. Fanny received more votes than Bill and May combined.

C. More seniors voted because there are more seniors than freshmen, sophomores, or juniors.

D. Bill received 30% of the votes.

Answers

1. (*a*) 24.179 quadrillion BTU. Add the BTU usage for each country and divide by 10.
   (*b*) 13.17 quadrillion BTU. Add 12.51 + 13.83 and divide by 2.
   (*c*) 21.42 quadrillion BTU. Subtract: 33.93 – 12.51.
   (*d*) 945.7%. Set up and solve the ratio $\frac{\text{United States}}{\text{France}} = \frac{x}{100} = \frac{94.57}{10} = \frac{x}{100}$.
   (*e*) Brazil. Multiply 33.93 by $\frac{1}{4}$ (or divide 33.93 by 4).
   (*f*) Germany. Multiply Japan's usage (approximately 21) by $\frac{2}{3}$: $21 \times \frac{2}{3} = 14$. Germany is the closest to that number.
2. **A.** Doug Upp received 767 out of 2983 votes. 767/2983 ≈ .2571 = 26.71%. This is more than $\frac{1}{4}$.

SECTION 3: NUMERICAL PATTERNS

Analyze the numerical pattern 2, 4, 6, 8, 10, 12, 14. . . . You can see that the next number in the pattern will be 16 because you observed that each number is 2 more than the previous number, or that the *difference* between each number is 2. This type of pattern is an *arithmetic sequence.* Sequences are of particular interest when known values can be used to predict future values.

ARITHMETIC SEQUENCES

An *arithmetic sequence* is a pattern of three or more numbers that share a common difference. Note that in this type of sequence each term is formed by adding a constant to a previous term. This constant is the *common difference*.

Examples: 1, 2, 3, 4, . . . has a common difference of 1.
5, 10, 15, . . . has a common difference of 5.
10, 20, 30, . . . has a common difference of 10.

Let's examine an arithmetic sequence more closely. We'll use the sequence 2, 6, 10, 14, ... In this sequence, the common difference is 4.

1st term	2
2nd term	6 (1st term plus 4)
3rd term	10 (1st term plus 2 × 4)
4th term	14 (1st term plus 3 × 4)
5th term	18 (1st term plus 4 × 4)
10th term	38 (1st term plus 9 × 4)
51st term	202 (1st term plus 50 × 4)

Examine the preceding pattern. To find the 51st term, we added the first term (2) to 50 times 4 (the common difference): 2 + (50 • 4). Using this same idea, then, the 101st term would be 2 + (100 • 4), or 402.

Example: Find the 48th term in the sequence 6, 9, 12, 15, ...

Step 1: First find the common difference: 9 – 6 = 3; 12 – 9 = 3; and 15 – 12 = 3. The common difference, then, is 3.

Step 2: Add the 1st term (6) to 47 times the common difference (3). 1st term + (47 • common difference) = 6 + (47 • 3) = 147.

Sample Questions

1. Find the common difference in the sequence 6, 13, 20, 27, . . .
2. Find the 33rd term in the sequence in question 1.
3. Find the common difference and the 45th term in the sequence 6, 1, –4, –9, . . .
4. Find the common difference and the 11th term in the sequence $\frac{1}{4}, \frac{3}{8}, \frac{1}{2}, \frac{5}{8}, \ldots$
5. If you are eligible to vote in your 1st presidential election at age 18, how old will you be in the 10th presidential election after that?

Answers

1. Find the common difference by subtracting: 13 – 6 = 7; 20 – 13 = 7; 27 – 20 = 7.
2. To find the 33rd term, add the 1st term to 32 times the common difference: 6 + 32 • 7 = 230.
3. The common difference is –5. The 45th term in the sequence is –214. Note that the sequence is *decreasing*, and so the common difference is negative. 1 – 6 = –5; –4 – 1 = –5; and –9 – (–4) = –5. To find the 45th term, add the 1st term to 44 times –5: 6 + 44(–5) = –214.
4. The common difference is $\frac{1}{8}$. The 11th term in the sequence is $\frac{3}{2}$. To find this term, add $\frac{1}{4}\left(\text{or } \frac{2}{8}\right)$ to 10 times $\frac{1}{8}$: $\frac{2}{8} + \frac{10}{8} = \frac{12}{8} = \frac{3}{2}$.
5. The common difference is 4 because presidential elections are held every 4 years. 18 + 9 • 4 = 54 years old.

GEOMETRIC SEQUENCES

Another important type of sequence is a *geometric sequence*. You can identify a geometric sequence because it is a pattern of three or more numbers in which each term after the first is formed by multiplying by a constant.

Example: 2, 4, 8, 16, ... Each term is found by multiplying the previous term by 2. In this example, the multiplier 2 is called the *common ratio* of the sequence.

Example: 6, 18, 54, ... Each term is found by multiplying the previous term by 3. The number 3 is the common ratio.

Finding the common ratio is fairly easy. Simply take any term, other than the 1st term, and divide it by the previous term: $18 \div 6 = 3$; also, $54 \div 18 = 3$.

Example: In the geometric sequence 100, 50, 25, 12.5, the common ratio is $\frac{1}{2}$ because $50 \div 100 = .5$; also, $25 \div 50 = .5$, and $12.5 \div 25 = .5$.

Let's examine the geometric sequence 5, 15, 45, . . . ; the common ratio is 3 because $15 \div 5 = 3$. Keep a close eye on how the common ratio is used.

1st term	5	
2nd term	15	(1st term times 3^1)
3rd term	45	(1st term times 3^2)
4th term	135	(1st term times 3^3)
5th term	405	(1st term times 3^4)
10th term	98,415	(1st term times 3^9)

> ***Test Tip***: Always check at least three numbers in the pattern for a common ratio to be certain the pattern is a sequence.

Did you notice that the exponent of 3 is always 1 less than the term number? For instance, in the 10th term, the power of 3 is 9. In the 5th term, the power of 3 is 4. Using this same logic, in the 7th term the power of 3 should be 6.

Example: Find the common ratio and the 8th term in the geometric sequence 3, −18, 108, −648.

Step 1: Find the common ratio by dividing: $-18 \div 3 = -6$; $108 \div -18 = -6$; and $-648 \div 108 = -6$. The common ratio must be −6.

Step 2: Find the 8th term by multiplying the 1st term by −6 to the seventh power:

$3(-6)^7 = -839{,}808$

$3 \bullet -279{,}396 = -839{,}808$

Example: To find the 11th term in the sequence 1, 2, 4, 8, . . . , first find the common ratio, 2. Then, begin at the first term (1) and multiply by the common ratio: 1×2.

		Calculator Keystrokes
1st term	1	1 × 2 =
2nd term	2	=
3rd term	4	=
4th term	8	=
5th term	16	=
6th term	32	=
7th term	64	=
8th term	128	=
9th term	256	=
10th term	512	=
11th term	1024	

> ***Test Tip***: Your FCAT calculator does not have a "power" key. You can find a term in a geometric series by using a feature on your calculator that performs repeated multiplications.

Notice that finding 11 terms requires hitting the = sign 10 times. Be sure to keep a careful count!

Sample Questions

1. Find the common ratio of the geometric sequence 2, 6, 18, 54, ...
2. Find the common ratio of the geometric sequence −1, 4, −16, 64, ...
3. Which of the following is a geometric sequence?
   A. 1, 3, 5, 7, 9, . . .
   B. $\frac{1}{2}$, 1, $\frac{3}{2}$, 2, . . .
   C. −4, 12, −18, 24, . . .
   D. −4, 8, −16, 32, . . .
4. The common ratio of a sequence is 4. The first term is −2. Find the 5th term in the sequence.
5. Jeff purchased a house in August of 1998 for $60,000. Which of the following represents the value of the house in August of 2006 if it appreciates (gains in value) at a rate of 5% per year?
   A. $60{,}000(.05)^7$
   B. $60{,}000(1.05)^8$
   C. $60{,}000(1.05)^7$
   D. $60{,}000(.5)^7$
6. Find the 6th term in the geometric sequence 2, −12, 72, −432, ...

Answers

1. 6 ÷ 2 = 3; 18 ÷ 6 = 3; and so on. The common ratio is 3.
2. 4 ÷ −1 = −4; 16 ÷ −4 = −4; and so on. The common ratio is −4.
3. **D.** The geometric ratio is −2. Both **A** and **B** are arithmetic sequences.
4. Multiply the 1st term, −2, by 4 to the fourth power: $-2 \times (4)^4 = -512$.
5. **B.** Consider the date of purchase (8/98). Then 8/99 would be term 1, continuing on to term 8, which is 8/2006.
6. The geometric ratio is −6. Multiply: $2 \times (-6)^5 = -15{,}552$.

VISUAL AND WORD PATTERNS

Visual patterns require you to recognize and analyze a pattern so that you can extend it into a logical progression. To analyze a pattern, look at where it begins and try to see what happens from step to step so that you can predict what will happen next.

Example 1: In this pattern, all the objects appear to be the same; therefore the difference must be in the way they are arranged or in how many there are in each group.

The pattern is numerical: 1, 1, 2, 3, 5, 8, ... The sequence is a famous one called the Fibonacci Sequence. Each term is found by adding the previous two terms: 1 + 1 = 2; 1 + 2 = 3; 2 + 3 = 5; 5 + 3 = 8. The next figure, then, should contain 13 (5 + 8) happy faces.

Example 2:

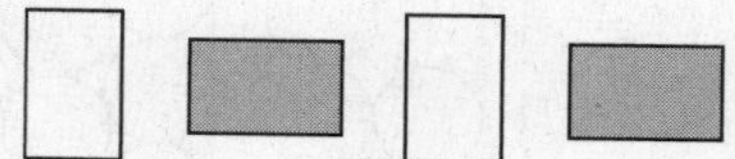

In this pattern of rectangles, two things are happening. The position of every other rectangle is different. Also, every other rectangle is shaded. From this information, you can predict that the next rectangle will be

Example 3: Use the pattern ALGEBRAALGEBRAALGEBRA to predict what letter will be in the 99th position.

There are 7 letters in the word ALGEBRA. You need to know the letter in the 99th position. Divide: $7\overline{)99}$. The quotient is 14 with a remainder of 1. The remainder of 1 is the important piece of information here because it tells you which letter in the word ALGEBRA will be in the 99th position.

Remainder	Letter
1	A
2	L
3	G
4	E
5	B
6	R
0	A

***Note:** No remainder (or a remainder of 0) means that you should select the *last* letter of the word, in this case, A.

If we need to find the letter in the 104th position, we will have to divide 7 into 104. The quotient will be 14, and the remainder (the important part) will be 6. The 6th letter of the word ALGEBRA, or R, is the letter in the 104th position of ALGEBRAALGEBRAALGEBRA. . . .

Sample Questions

1. Use the following pattern to predict the 10th term.

2. Use the word pattern FCATFCATFCAT to predict which letter will be in the 52nd position.

3. Predict the number of circles needed for the next group in the sequence.

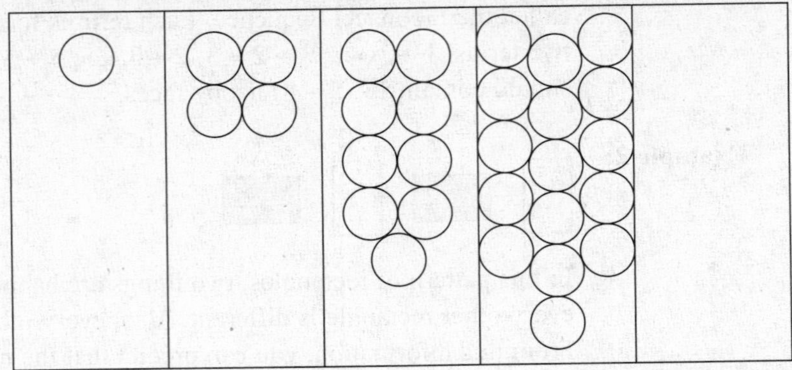

Being able to find patterns is an important part of understanding algebra.

4. In the following pattern, the squares represent seating arrangements. At the first table 4 people can be seated, and at the second table 6 people can be seated. Predict how many people can be seated in the 25th seating arrangement if the pattern continues.

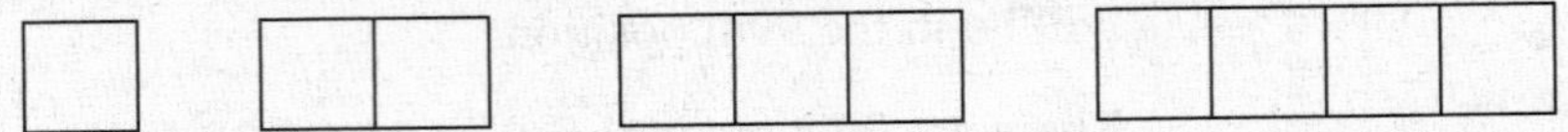

5. Find the missing number in the pattern 4, 10, ____, 46, 94, 190, 382.

Answers

1. The 10th term is . It takes 8 arrows to make a complete revolution and go back to the starting position. When you reach 8, go back to the beginning and keep counting. If you have to find a place that is such a high number that you cannot count the arrows, use the process shown in Example 3.

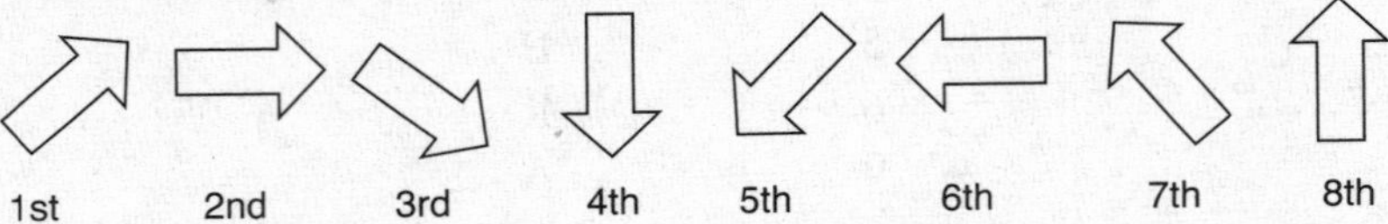

2. 52 divided by 4 = 13. Since there is no remainder, T must be the correct letter. (A remainder of 1 would yield F, a remainder of 2 would mean C, and a remainder of 3 would give A.)
3. The pattern is 1, 4, 9, 16. These numbers are perfect squares, and the next perfect square is 25.
4. The pattern is 4, 6, 8, 10, . . . , an arithmetic sequence with a common difference of 2. The number to be seated in the 25th seating arrangement is $4 + 24 \times 2 = 52$.
5. The missing number is 22. The pattern doubles and then adds 2 each time.

SUMMARY

As you complete Chapter 3, you should understand or be able to do most of the following:

- Determine the probability of complementary events, and calculate the odds for and against the occurrence of events.
- Determine probabilities of independent events.
- Determine the range, quartiles, mean, median and mode of a set of data.
- Read and interpret data in various formats.
- Read and solve problems based on Venn diagrams.

PRACTICE PROBLEMS

1. The data in the table represents Florida occupational wage estimates for some computer and mathematical occupations. Use the data to construct a bar graph in the space provided.

Occupation	Mean Annual Wage ($)
Information research	71,270
Computer programmers (applications)	55,240
Software engineers	61,240
Support specialists	31,270
Systems analysts	55,130
Actuaries	63,750
Mathematicians	59,210

1.

The accompanying table gives the nutritive value of one serving of several desserts. **Use this table to answer questions 2–6.**

Dessert	Calories	Protein (g)	Fat (g)	Carbohydrates (g)
Angel food cake	125	3	0	29
Cupcake	125	2	4	20
Cake with icing	445	4	14	77
Pound cake	110	2	5	15
Chocolate chip cookies	185	2	11	26

2. Which measure of central tendency best describes the calorie content of these desserts?

   A. mean
   B. median
   C. mode
   D. range

2. ________

3. Which measure of central tendency best describes the protein content?

   F. mean
   G. median
   H. mode
   I. range

3. ________

4. If the cake with icing weighs 121 grams, what percent of the total weight is carbohydrates?

4.

	⊘	⊘	⊘	
⊙	⊙	⊙	⊙	⊙
0	0	0	0	0
1	1	1	1	1
2	2	2	2	2
3	3	3	3	3
4	4	4	4	4
5	5	5	5	5
6	6	6	6	6
7	7	7	7	7
8	8	8	8	8
9	9	9	9	9

5. What is the mean number of carbohydrates contained in these desserts?

5.

6. By how much would the mean of the carbohydrates change if the outlier were removed?

6.

7. What type of graph would be appropriate to show the earnings of a book publisher from 1990 to 2005?

A. scatterplot
B. line graph
C. bar graph
D. pie chart

7. ______

8. What type of graph would best show the population of the ten largest cities in Florida?

F. double bar graph
G. bar graph
H. line graph
I. histogram

8. ______

9. The following ad was placed in the Help Wanted section of a local paper.

HELP WANTED
SALES!

Average Employee
Salary Is

$30,000 per year

The actual salaries are

Clerk	$10,000
Secretary	$13,500
Salesperson 1	$28,000
Salesperson 2	$27,000
Salesperson 3	$22,500
President	$79,000

Is the ad misleading? Why or why not?

THINK

SOLVE

EXPLAIN

9. ______

The Math Club's bake sale profits are shown here. **Use this table to answer questions 10–13.**

Month	December	January	February	March	April
Profit ($)	260	130	250	260	190

10. Calculate the mean.

10.

11. Find the median.

11.

12. Find the mode.

12.

13. Calculate the range.

13.

14. Four students were conducting a survey on how many hours of television the students at their school watched. Which of the students collected data that most accurately represented the school?

A. Mary went to first lunch and asked one person at each table.
B. Brett asked everyone in his science class.
C. Carl asked every member of the after-school sports teams.
D. Allison went to each first-period class on Monday and conducted the survey.

14. ________

15. The following data was collected by Algebra I classes in an experiment entitled "Do Tall People Run Faster Than Short People? The students went out to the track at Charlotte High School and tested their hypothesis.

Height (in.)	60	61	62	62	63	66	67
Speed (sec)	12.8	11.7	17	13.9	12.6	12.6	13.5

Height (in.)	67	68	68	70	71	72	74
Speed (sec)	13.7	13.1	14.9	13.2	13.4	12.1	12.4

(*a*) Plot the data on the accompanying graph.

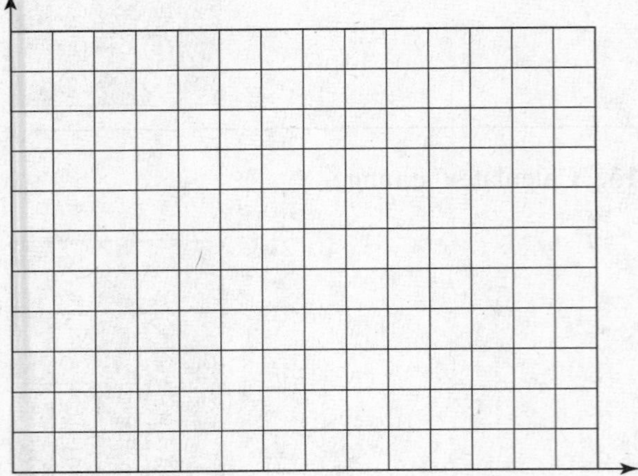

(*b*) What conclusion can you draw based on the data?

(*c*) What other factors might affect the outcome of this?

THINK

SOLVE

EXPLAIN

15. (*b*) ________

(*c*) ________

The stem-and-leaf plot shown here represents the population (in millions) of the top most populated 42 states. **Use it to answer questions 16–20.**

Stem	**Leaf**
1	1 2 2 2 6 7 8 8
2	1 5 6 7 8
3	2 3 3 8 9
4	0 3 3 7 7
5	1 2 4 4 7 9
6	1 8
7	6 7
8	1
9	8
11	2 9
12	1
15	1
18	1
20	1
33	1

Key: 33|1 = 33,100,000

16. What is the range of the data?

F. 33
G. 32
H. 33 million
I. 32 million

16. ________

17. What is the mode?

A. 12–33 million
B. 1.2 million
C. 9 million
D. 11 million

17. ________

18. What is the median?

F. 4,300,000
G. 4,500,000
H. 4,700,000
I. 6,000,000

18. ________

19. How many states have a population greater than 10 million?

A. 3
B. 5
C. 7
D. 9

19. ________

20. Only 42 states are represented in the stem-and-leaf plot. Why aren't the other 8 states represented?

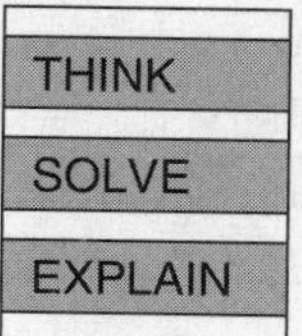

20. ________

The double line graph shown here represents marriage and divorce rates from 1920 to 1998. (*Source: 2000 Almanac and Book of Facts.*) **Use the graph to answer questions 21–25.**

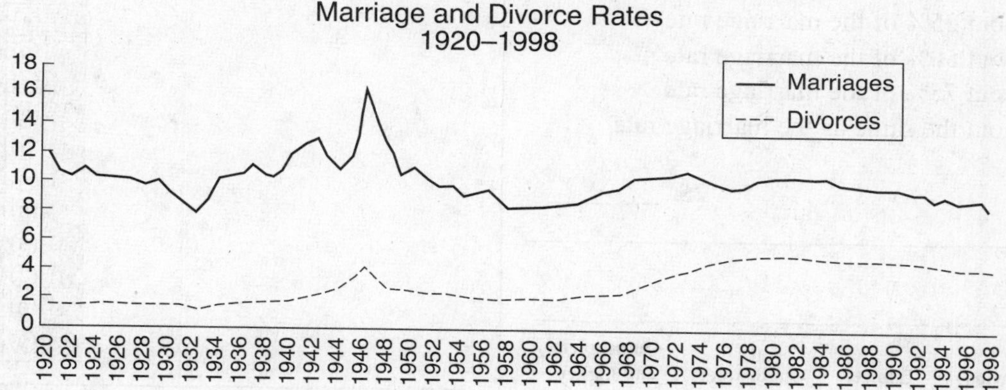

21. During what year was the divorce rate the lowest?

21.

22. During what year was the marriage rate the highest?

22.

23. Overall the divorce rate has

F. increased
G. decreased
H. stayed steady
I. mirrored the marriage rate

23. ________

24. If the graph shows marriage and divorce rates per 1000 of population, predict how many people got married in 2000.

A. 6
B. 8
C. 6000
D. 8000

24. ________

25. In 2000, the divorce rate was

F. about 25% of the marriage rate
G. about 50% of the marriage rate
H. about 75% of the marriage rate
I. about the same as the marriage rate

25. ________

26.

Year	Twins	Triplets	Quadruplets	Quintuplets and higher
2000	93,865	2,830	185	13
2001	94,779	3,121	203	22
2002	95,372	3,547	310	26
2003	96,445	3,834	277	57
2004	97,064	4,233	315	46
2005	96,736	4,551	365	57
2006	100,750	5,296	560	81
2007	104,137	6,148	510	79

The accompanying table shows the numbers of multiple births in the United States from 2000 to 2007. From 2000 to 2007, by what percent has the incidence of twin births in the United States increased?

A. 110%
B. 90%
C. 11.9%
D. 10.9%

26. ________

Mr. Harris' music class surveyed 40 students to determine their favorite music. The results are shown in this Venn diagram. **Use it to answer questions 27–30.**

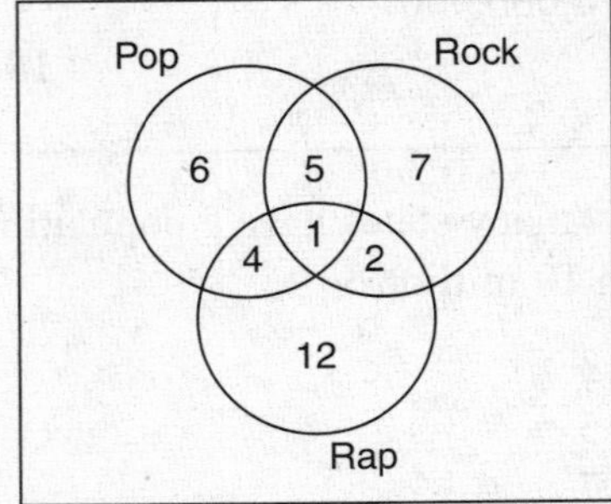

27. How many students like pop, rap, and rock?

27.

28. How many students like pop and rock but not rap?

28.

29. How many students do not like rap?

29.

30. How many students do not like pop, rap, or rock?

30.

31. The box-and-whiskers plot shown here represents the number of gold medals won at the 2002 Winter Olympics. What percent of the teams won between 2 and 6 gold medals?

1 2 3 5 6 10 12 15

31.

32. If you select a letter at random from the word OCCURRENCE, what is the probability that it will be a consonant?

32.

There are 100 letters in a Scrabble game as shown in the accompanying table. **Use this table to answer questions 33–36.**

Letters	Number of Tiles
E	12
A, I	9
O	8
N, R, T	6
D, L, S, U	4
G	3
B, C, F, H, M, P, V, W, Y, blank	2
J, K, Q, X, Z	1

33. What is the probability that you will select a T on the first draw?

33.

34. What is the probability that you will select an A if nine tiles have already been drawn and two were As?

34.

35. What is the probability that you will draw an E on the first draw and an A on the second draw if you do not replace the first tile?

F. 0.011
G. 0.019
H. 0.21
I. 0.75

35. ______

36. If you draw two tiles, what is the probability that both will be Es?

A. less than 1%
B. between 2% and 10%
C. between 1% and 2%
D. greater than 10%

36. ______

This table shows the total medals won during the 2002 Olympics by country. **Use it to answer questions 37–39.**

Country	Gold	Silver	Bronze	Total
Germany	12	16	7	35
USA	10	13	11	34
Norway	11	13	11	34
Canada	6	3	8	17
Austria	2	4	10	16
Russia	6	6	4	16
Italy	4	4	4	12
France	4	5	2	11
Switzerland	3	2	6	11
China	2	2	4	8
Netherlands	3	5	0	8
Finland	4	2	1	7
Sweden	0	2	4	6
Croatia	3	1	0	4
Korea	2	2	0	4
Bulgaria	0	1	2	3
Estonia	1	1	1	3
Great Britain	1	0	2	3
Australia	2	0	0	2
Spain	2	0	0	2
Czech Republic	1	0	1	2
Japan	0	1	1	2
Poland	0	1	1	2
Belarus	0	0	1	1
Slovenia	0	0	1	1

37. What percent of Russia's medals were gold?

37.

---

38. If 230 medals were distributed, which country received closest to 15% of the medals?

F. Norway
G. Canada
H. Austria
I. Italy

38. ________

---

39. Which of the following statements is true?

A. China won twice as many gold medals as Russia.
B. One-third of Italy's medals were gold.
C. Spain didn't win any silver or bronze medals because they went home early.
D. The United States won 34% of the medals.

39. ________

---

40. An advertisement reads: "We will not be undersold!!! If you can find a CD player of equal quality at a cheaper price, bring your receipt in and we will pay you 20% of the difference." Is this a good deal? Why or why not? Use mathematics to support your answer.

THINK
SOLVE
EXPLAIN

40. ________

---

41. What is the next number in the sequence 4, −2, 6, 0, 8, 2?

F. 6
G. 8
H. 10
I. 12

41. ________

---

42. If the dots continue in the same pattern, predict how many dots will be in the 10th grouping.

42.

---

43. Find the 13th number in the sequence 6, 10, 14, 18, . . .

43.

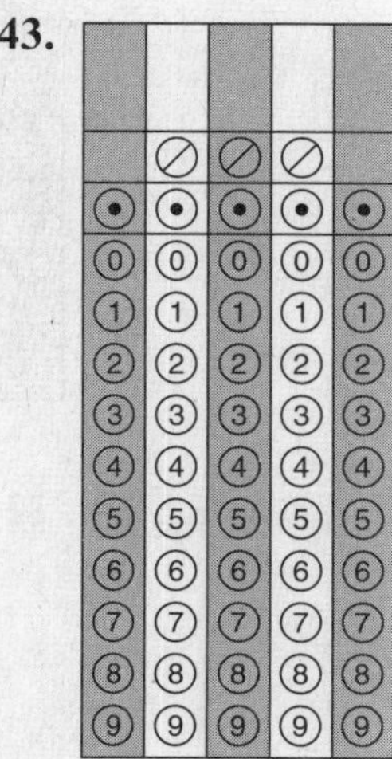

Answers Explained

1.

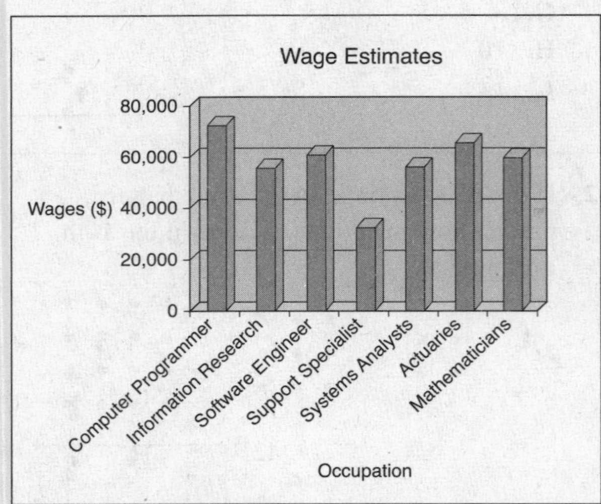

2. B. The median should be used when there are outliers that would cause the mean to stray from a central value or not be representative of the sample. 445 calories is an outlier.

3. H. There are three values of 2 under the heading Protein. This number best represents the central tendency.

4.

6 3 • 6

77 is 63.6% of 121. $\frac{77}{121} = .636 = 63.6\%$.

5.

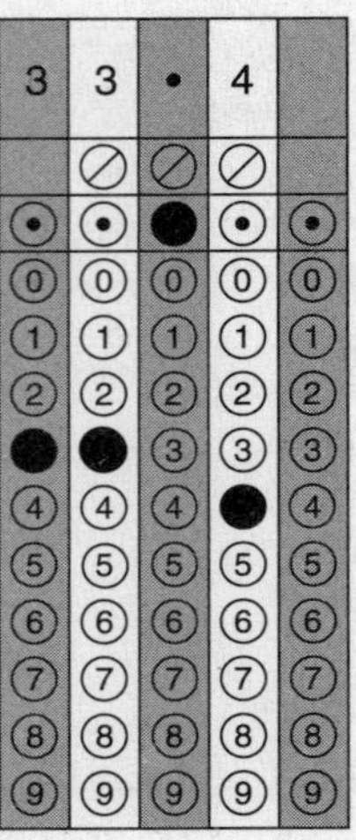

$$\frac{29+20+77+15+26}{5} = 33.4$$

6.

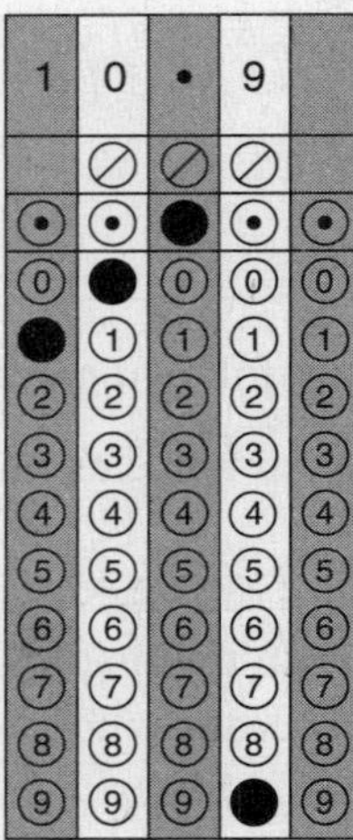

$$\frac{29+20+15+26}{4} = 22.5.$$

Subtract 33.4 − 22.5 = 10.9

7. B. A line graph is best suited to show change over time.

8. G. A bar graph is best suited to compare things.

9. The ad is misleading. Although the mean of the salaries is $30,000, none of the salespeople make $30,000 per year. The mean is not the best indicator of the measure of central tendency here because the president makes so much more than everybody else (outlier).

10.

2 1 8

$$\frac{260+130+250+260+190}{5}=218$$

11.

2 5 0

Put the numbers in order and select the number in the middle: 130, 190, **250**, 260, 260.

12.

2 6 0

Find the number that is repeated the most: 260.

13.

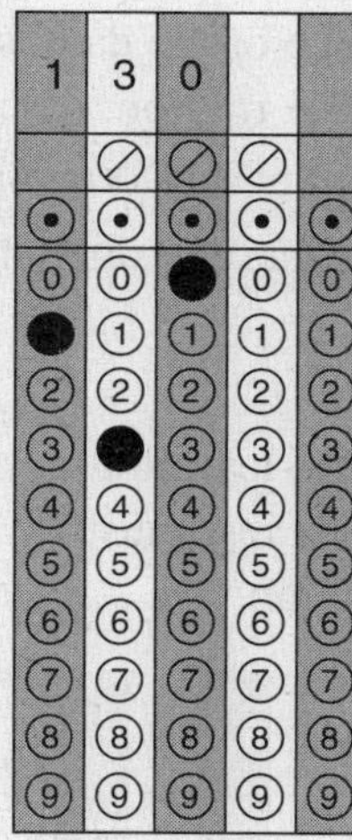

Highest – lowest = 260 – 130 = 130.

14. D. Allison's method allowed her to contact a larger number of students. In addition, she was able to contact a representative sample—everybody in school that day had a chance to participate.

15. (*a*)

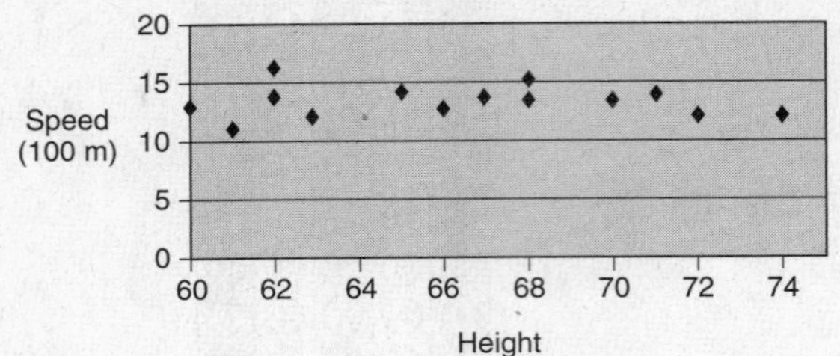

(*b*) There is no trend—tall people do not run faster than short people.

(*c*) The condition of the people running might be a factor. Also, some people just don't like to run or may not have worn the right kind of shoes for running on the day of the test.

16. I. The range of the data is 33 million – 1.1 million = 31.9 million.

17. B. The mode is 1,200,000 or 1.2 million.

18. F. The median is exactly halfway between 4|3 and 4|7.

19. C.

20. The stem-and-leaf plot represents only the top 42 states. The rest must have populations below 1 million.

21.

1	9	3	2

22.

1	9	4	6

(shortly after World War II ended)

23. F. Although the divorce rate leveled off around 1978, it has generally risen since the 1920s. The key word here is *overall*.

24. D.

25. G. The graph indicates that there were about 8000 marriages.

26. D. $\frac{\text{amount of change}}{\text{original amount}} = \frac{104{,}137 - 93{,}865}{93{,}865}$

$= 0.109 = 10.9\%.$

27.

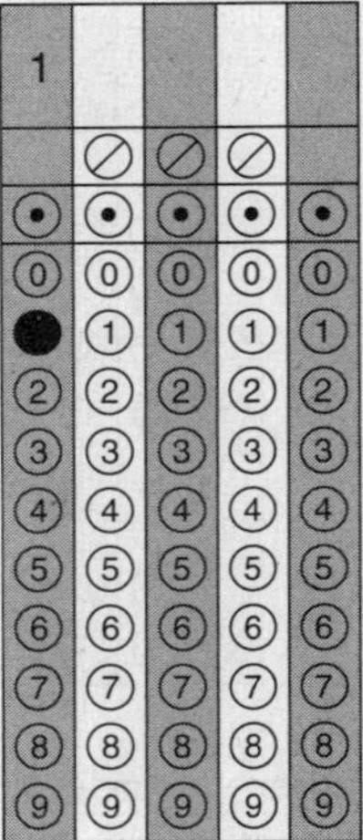

Find the place where all three circles intersect.

28.

5

Find the place where pop and rock intersect but do not count anything in the rap circle.

29.

21

Count all the students who like rap and subtract this number from 40.

30.

3

Add everything and subtract the sum from 40.

31.

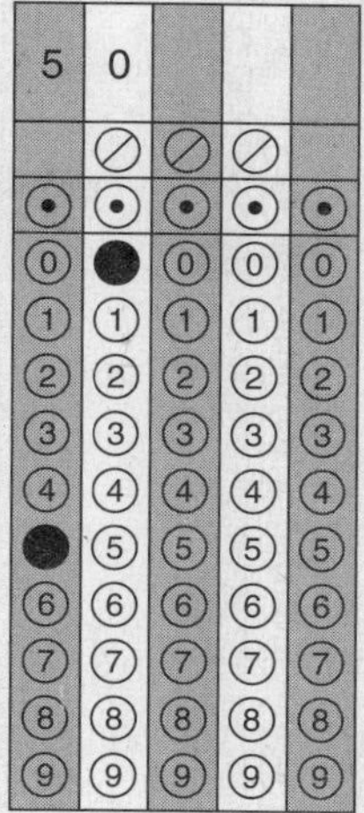

Half of the data always falls inside the box.

32.

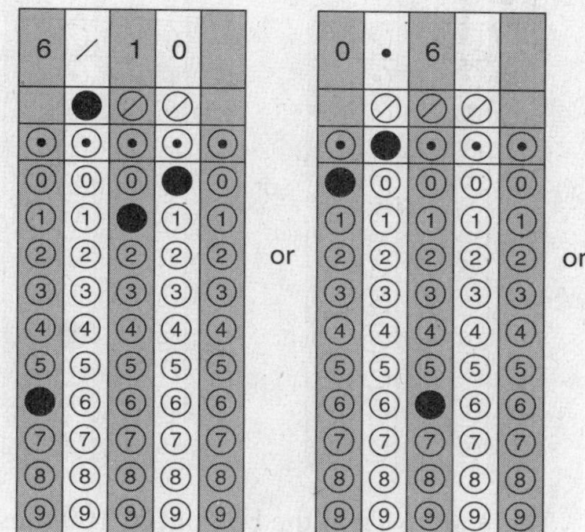

. 6

There are 6 consonants (C, R, and N) out of 10 letters. The probability can be expressed as a fraction, decimal, or percent.

33.

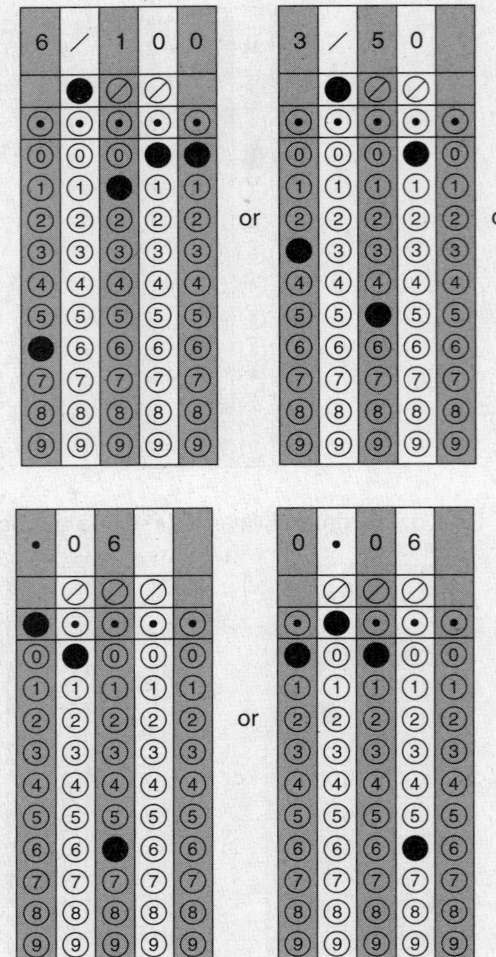

There are 6 Ts in the box out of 100 tiles. The probability can be expressed as a fraction, decimal, or percent.

34.

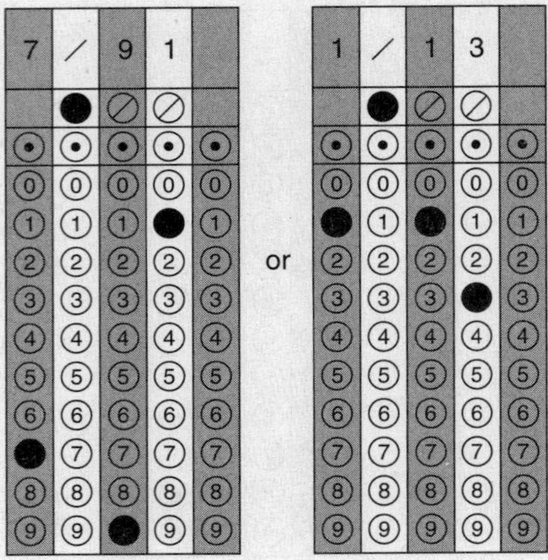

There are 7 As left after 2 have been drawn, and 91 total tiles after 9 have been drawn. The probability is 7 out of 91 or 1 out of 13. Since you were not asked for an approximate amount, it is best to leave your answer as a fraction.

35. F. Since tiles are not replaced in a Scrabble game, multiply the probability of getting an E the first time $\left(\frac{12}{100}\right)$ by the probability of drawing an A the second time $\left(\frac{9}{99}\right)$. Note that on the second draw only 99 tiles are left.

$$\frac{12}{100} \bullet \frac{9}{99} = \frac{3}{25} \bullet \frac{1}{11} = \frac{3}{275} = 0.0109 = 0.011.$$

36. C. Because tiles are not replaced, multiply the probability of selecting an E the first time by the probability of selecting an E after the first tile has been drawn:

$$\frac{12}{100} \bullet \frac{11}{99} = \frac{3}{25} \bullet \frac{1}{9} = \frac{1}{75} = 0.013 = 1.3\%$$

37.

3	7	•	5	

$\frac{6}{16} = 0.375 = 37.5\%.$

38. F. 15% of 230 = 34.5.

39. B. $\frac{4}{12} = \frac{1}{3}$.

40. If the difference is small, for example, $5.00, then 20% of $5.00 is $1.00. If you have to take the time to drive back to the store, it probably would not be worth your time. If the difference is large, for example, $50.00, then 20% of $50.00 is $10.00, which may turn out to be worth the time. Your explanation and sound mathematics are the key to earning points for solving this problem.

41. H. The pattern is: subtract 6, add 8.

42.

1	0	0		

Each pattern is a perfect square. The tenth pattern will have 10 × 10 dots.

43.

5	4			

This is an arithmetic sequence. Add the first term (6) to 12 times the common difference (4).

Practice Tests

PRACTICE TEST 1

Directions: Answer every question on this test. Place your answers in either the answer spaces or, where provided, in the "gridded-response" boxes. (Longer answers should be written on a separate sheet of paper.) Do all of your calculations on separate sheets of paper. Once you have answered every question, be sure to check your answers and to review the answer explanations at the end of the test.

1. The distance across a river is estimated to be $4\sqrt{59}$ yards. Which of the following numbers is closest to $4\sqrt{59}$ yards?

A. 118 yards

B. $30\frac{2}{3}$ yards

C. $15\frac{2}{5}$ yards

D. 11.7 yards

1. ________

2. Mason and three friends compared the distances from their houses to the mall. The distances were 5.4 miles, $5\frac{1}{3}$ miles, $5\frac{3}{8}$ miles, and $5\frac{3}{5}$ miles. Which distance is the shortest?

F. $5\frac{3}{5}$ miles

G. 5.4 miles

H. $5\frac{3}{8}$ miles

I. $5\frac{1}{3}$ miles

2. ________

3. Sharon is taking her children on a vacation to Disney World in Orlando. She is traveling at an average speed of 65 miles per hour. Which of the following represents the distance d that she has traveled after t hours?

A. $t = 65d$

B. $t = \frac{d}{65}$

C. $d = \frac{65}{t}$

D. $d = 65t$

3. ________

4. In 2000, approximately five-eighths of the US population was under 40 years old. If the population in 2000 was estimated at 264 million, what number should you multiply 264 million to find the population under 40?

F. 0.625

G. 0.63

H. 6.25

I. 6.

4. ________

GO ON →

5. Emelio is calculating the number of gallons of water in his swimming pool. He measured the pool and found the volume to be approximately $(9\text{ ft})^3$. He will need to multiply the volume in cubic feet by 7.5 to find the number of gallons the pool will hold. Which number should he multiply by 7.5?

A. 3
B. 27
C. 729
D. 6561

5. ________

6. Of 1290 hazardous waste sites in the United States, 53 are located in the state of Florida. To the nearest tenth, what percent of the total hazardous waste sites are located in Florida?

6.

7. Which of the following numbers will result in a smaller number when squared?

F. −1.5
G. −.5
H. .2
I. 2

7. ________

8. Find the value of $3(1.1)^2 - (-2)^0$.

8.

9. The average distance of the earth from the sun is 9.3×10^7 miles, and the speed of light is 186,000 miles per second. If the sun were suddenly to go out, approximately how many minutes would it be before people on earth would realize it? Show your work or explain how you arrived at your answer.

THINK
SOLVE
EXPLAIN

9. ________

10. Simplify the expression $-2^2 + (-2)^2$ completely. Show your work.

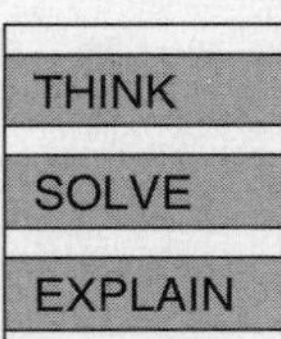

10. ________

11. In Yellowstone Park the temperature was 26 degrees at 10:00 A.M. By 4:30 P.M. the temperature had dropped to −14 degrees. What was the change in temperature between 10:00 A.M. and 4:30 P.M.?

A. −12 degrees
B. −14 degrees
C. −26 degrees
D. −40 degrees

11. ________

12. The formula for compound interest is $A = p(1 + r)^t$. Use this formula to find A to the nearest dollar if p = $1000, r = 0.029, and t = 2.

12.

13. Letisha worked 47.25 hours as a nursing assistant at a nursing home. She gets paid time-and-a-half for each hour she works over 40 hours. If her regular wage is $7.75 per hour, *estimate* how much she will be paid for the week. Show your work or explain how you arrived at your answer.

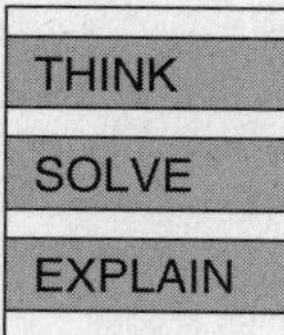

13. ________

14. In the first 6 days of his vacation, Jorge drives 470 miles per day. He spends the next 3 days at his destination and starts home, sightseeing along the way. After 4 days he realizes he has driven only a total of 500 miles toward home. If he drives at 65 miles per hour, *estimate* how many hours he will have to drive per day to get home in 3 days.

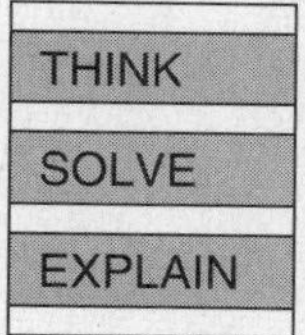

14. ________

15. The regular square pyramid shown here has a base measuring 8 units on each side and a height of 15 units. What is the volume, in cubic units, of the pyramid?

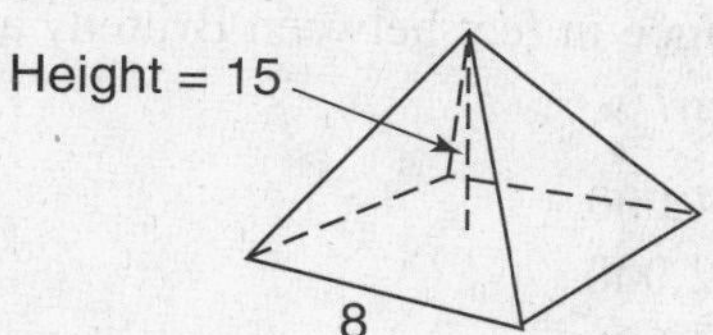

15.

GO ON

16. Plumbers' helpers earn an average of \$800 less per week than plumbers. AAA Plumbers employs 6 plumber's helpers and 12 plumbers. Let p represent the average monthly pay of a plumber. Which of the following equations correctly shows the relationship between the monthly payroll (P) and the wages of these employees?

F. $P = 6p + 12p$
G. $P = 12(p - 800) + 6p$
H. $P = 6(p - 800) + 12p$
I. $P = 12p - 800 + 6p$

16. ________

17. After Britteny sees lightning, it takes 9.87 seconds for her to hear the thunder. Sound travels at 1129 feet per second. Which of the following would be the best estimate of the distance in feet between Britteny and the storm?

A. 1,000
B. 2,000
C. 5,000
D. 10,000

17. ________

18. A pendulum in the physics building at a Florida university swings through an angle of 30 degrees as shown in the diagram.

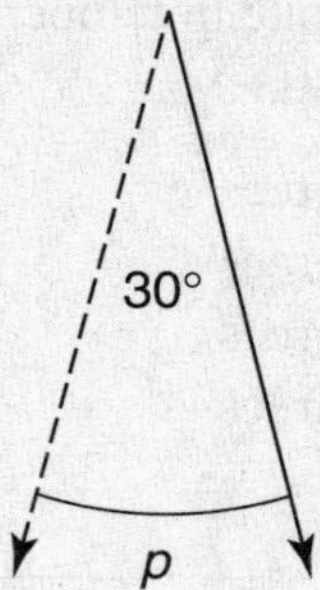

The pendulum forms an arc (p) that is what proportion of the full circle?

F. $\frac{1}{3}$
G. $\frac{1}{9}$
H. $\frac{1}{2}$
I. $\frac{1}{12}$

18. ________

19. A fishing boat travels due east out of Miami for 12 miles. From there it travels south for 6 miles to arrive at a favorite fishing spot over a shipwreck. How many miles would it have saved if it had traveled directly from Miami to the shipwreck?

A. 4.6 miles
B. 10.4 miles
C. 12 miles
D. 18 miles

19. ________

20. In the figure, what is the measure of $\angle QRS$?

F. 140 degrees
G. 120 degrees
H. 60 degrees
I. 50 degrees

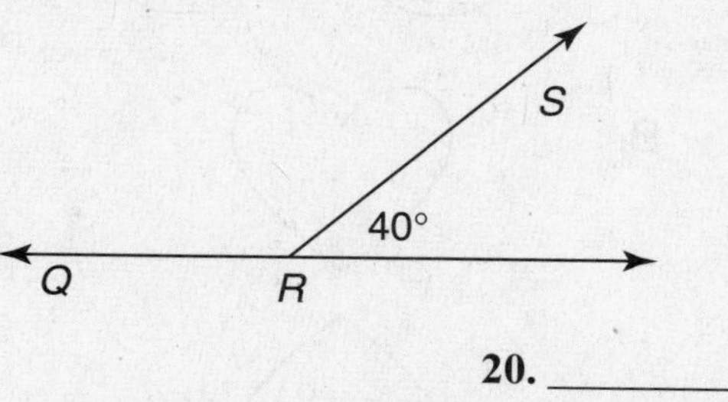

20. ________

21. Tiny Toy Company makes toys for toddlers. Suppose the income for Tiny Toy Company is represented by the equation $I = 6a$ and the expenses of production are represented by the equation $E = a + 4000$. In both equations, I is the income, E represents expenses, and a is the number of toys. Which equation shows the number of toys Tiny Toy Company has to make in order for their profit to be equal to their expenses (break-even point)?

A. $6a = a + 4000$
B. $6a + a = 4000$
C. $a = 6a + 4000$
D. $4000 - a = 6a$

21. ________

22. Albert needs to type a 100,000-word text. If he types 55 words per minute, to the nearest hour, how long will it take him to type the entire text?

22.

23. In the 45-45-90 triangle shown here, which of the following numbers represents the length of the hypotenuse?

F. $\sqrt{5}$
G. $2\sqrt{5}$
H. $5\sqrt{2}$
I. 5

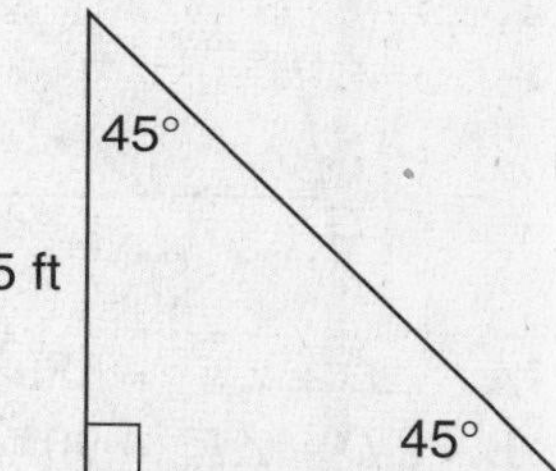

23. ________

GO ON

24. In the figure ∠1 is supplementary to ∠4. If m∠2 = 110°, find m∠3. Show your work, explain in words, or provide a proof that shows how you found m∠3.

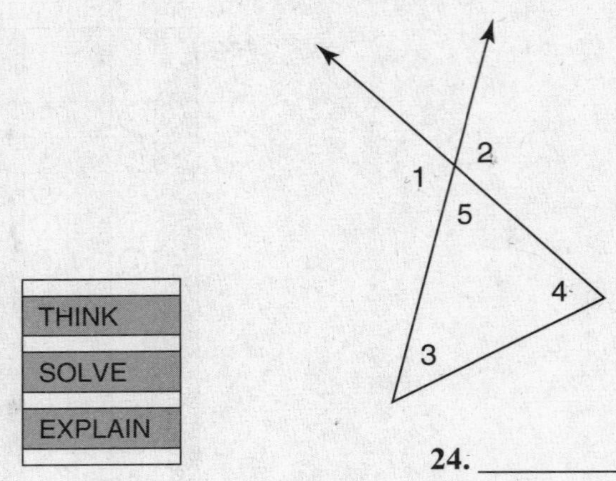

24. ________

25. In the figure, $\overline{DE}$ is parallel to $\overline{FG}$. Which of the following statements is true?

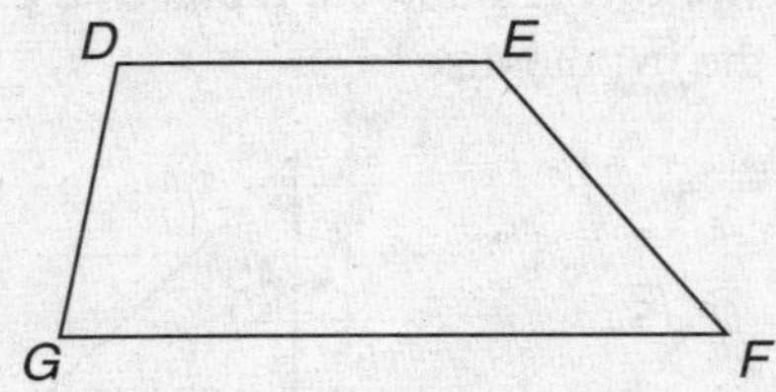

A. $\angle D + \angle E = 180°$
B. $\angle D + \angle F = 90°$
C. $\angle D + \angle G = 180°$
D. $\angle D + \angle F = 180°$

25. ________

26. Which of the following letters have more than one axis of symmetry?

F. B
G. Q
H. M
I. O

26. ________

27. Which of the following shows a translation of the original figure?

A.  

B. 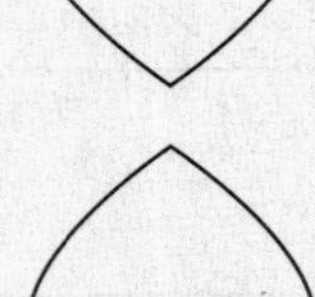

C. 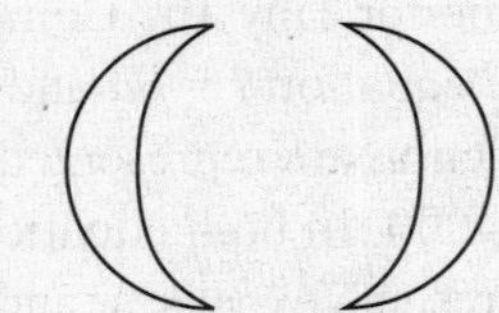

D. 

27. ________

28. If a grapefruit with a diameter of 5 in. is cut in half, what is the area (to the nearest square inch) of the cross section?

F. 79 in.^{2}
G. 78 in.^{2}
H. 26 in.^{2}
I. 20 in.^{2}

28. ________

29. Find the area of the shaded cross section of the pyramid shown here to the nearest tenth of a centimeter.

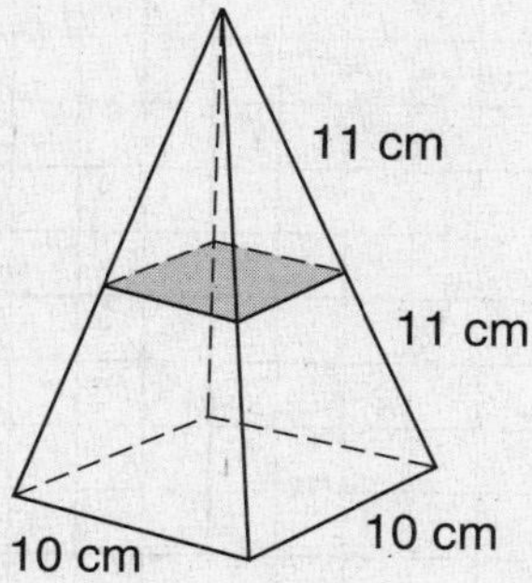

A. 100 cm^2
B. 25 cm^2
C. 49 cm^2
D. 20 cm^2

29. ________

30. The two cylinders shown here are similar. The ratio of their radii and heights is 2 : 3. Express the ratio of their volumes as a fraction.

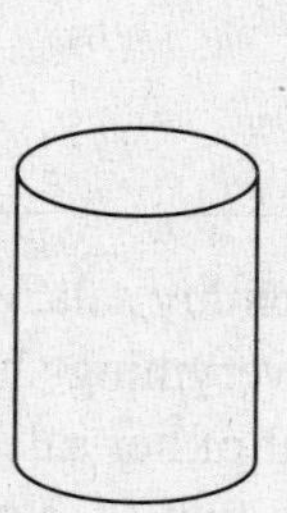

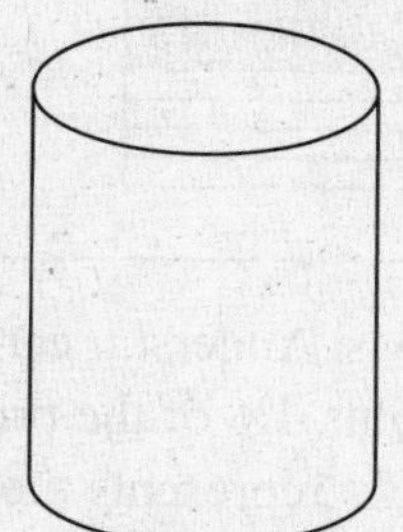

30.

31. A 35-year-old man's target heart rate t during exercise should be at least 70% of the difference of 220 and his age. The maximum heart rate during exercise should be 80% of the difference of 220 and his age. Which of the following inequalities expresses the effective heart rate for a person y years of age?

F. $70t < 220 - y < 80t$
G. $.7y < 220 - y < .8y$
H. $.7(220 - y) \leq t \leq .8(220 - y)$
I. $.7(y - 220) \leq t \geq .8(y - 220)$

31. ________

32. What is the distance between points A and B on the graph?

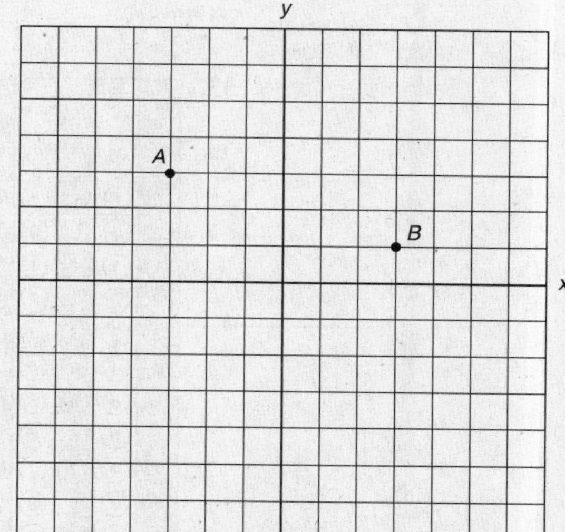

A. $4\sqrt{10}$
B. $2\sqrt{10}$
C. 20
D. 10

32. ________

GO ON

33. The graph shows the charges at an airport parking lot. What is the slope of the line containing the points on the graph?

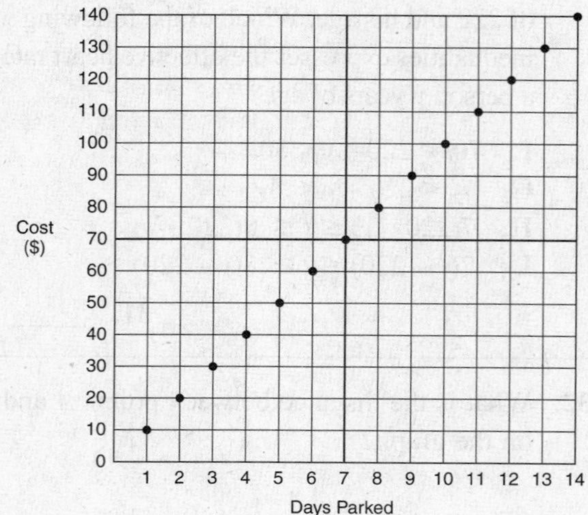

33.

34.

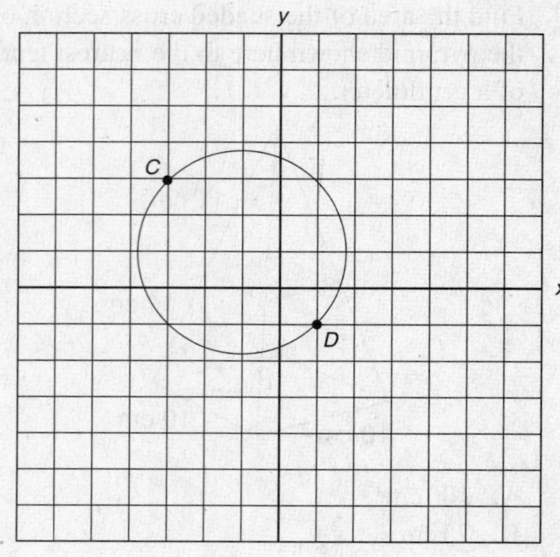

(*a*) Find the midpoint of diameter $\overline{CD}$.
(*b*) Find the radius of the circle.
(*c*) Find the area of the circle.

THINK
SOLVE
EXPLAIN

34. (*a*) ________
(*b*) ________
(*c*) ________

35. Ms. Anderson earns a monthly salary of \$500 plus 4% of the price of everything she sells. If *s* represents the amount of her sales, which equation could be used to find *M*, her monthly income?

F. $M = 500s$
G. $M = .04(s + 500)$
H. $M = 4s + 500$
I. $M = .04s + 500$

35. ________

36. What is the next term in the sequence –9, 1, 9, 15, . . . ?

36.

	/	/	/	
⊙	⊙	⊙	⊙	⊙
0	0	0	0	0
1	1	1	1	1
2	2	2	2	2
3	3	3	3	3
4	4	4	4	4
5	5	5	5	5
6	6	6	6	6
7	7	7	7	7
8	8	8	8	8
9	9	9	9	9

37. Ann has bought 56 feet of fencing to make a dog run. Initially, she constructed the dog run in her backyard and made it 8 feet wide and 20 feet long. After a month, however, she decided that this allowed very little area for her dog to move around. She needs to redesign the area to make the best use of her fencing and allow the most area for her dog. Design and label the dimensions of a new dog run in the space here that gives Ann's dog the most space.

THINK

SOLVE

EXPLAIN

37. ________

38. Chris has eight coins in his pocket. They are a combination of quarters and dimes. The value of the quarters and dimes is $1.55.

(*a*) Write a system of equations that represents the situation.

(*b*) Find how many quarters and dimes Chris has in his pocket. Show all your work or explain how you arrived at your answer.

THINK

SOLVE

EXPLAIN

38. (*a*). ________

(*b*) ________

39. Leigh Ann sold a total of 46 painted birdhouses and gourds at a local flea market. If she asked $25 for birdhouses and $15 for gourds, which system of equations would allow you to determine the number of birdhouses (b) and gourds (g) she sold if she made $1050?

A. $46b + g = 1050$
$25b + 15g = 1050$

B. $25b + 25g = 46$
$b + g = 1050$

C. $b + g = 46$
$25b + 15g = 1050$

D. $g + b = 46$
$25g + 15b = 1050$

39. ________

GO ON

40. Maurice has $250 to spend on school clothes. If he spends $45 each on three shirts, what is the most he can spend on the rest of the clothes he needs?

40.

41. Graph the system of inequalities on the grid provided. Shade their common solution.

$y < -1$ and $y \geq \frac{1}{2}x - 2$.

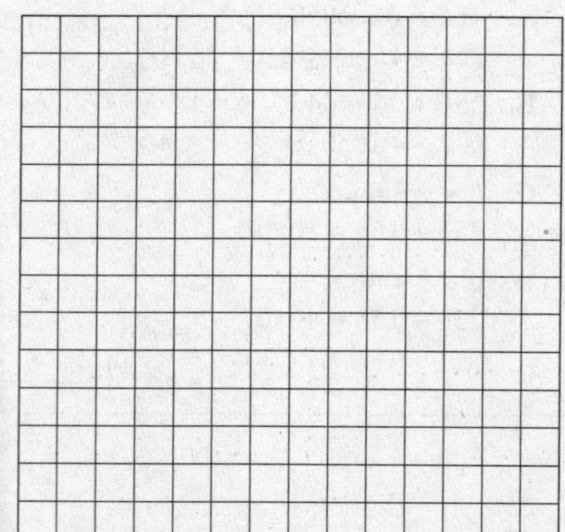

THINK

SOLVE

EXPLAIN

41. ________

42. The school menu offers the choices shown by the Venn diagram. A person making a choice from the shaded area would choose which of the following?

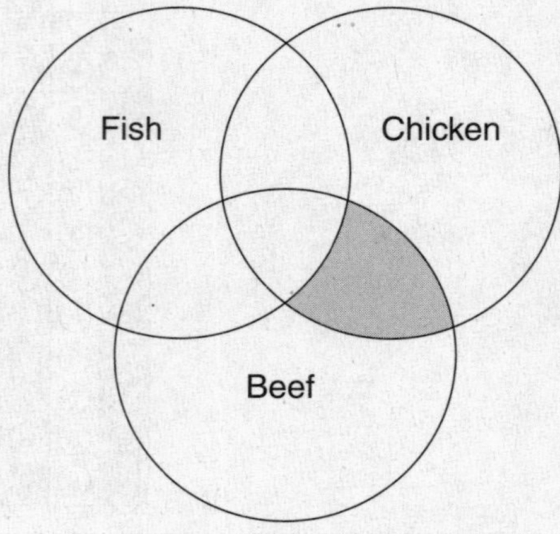

F. Beef but not fish
G. Chicken and beef but not fish
H. Beef but not chicken
I. Chicken but not beef

42. ________

43. Students in Mr. Bennett's class tested the theory that a person's height could be predicted from their arm span (distance from fingertip to fingertip when the arms are stretched out to the side). They gathered the following data:

Arm Span (in.)	Height (in.)
67	63
68	70
67	66
63	65
62	61
71	74
72	71
70	69
74	76
65	66

(*a*) Make a scatterplot that allows you to see the relationships between arm span and height.

(*b*) Draw a trend line.

(*c*) Use your trend line to predict the height of a person whose arm span is 76 inches.

THINK

SOLVE

EXPLAIN

43. (*a*) ________

(*b*) ________

(*c*) ________

44. The ACT average composite scores for several states are shown in the table.

State	ACT Score
California	21.4
Florida	20.6
Georgia	19.9
Hawaii	21.6
Illinois	21.5
Tennessee	20.0
Mississippi	18.7
Texas	20.3
Nevada	21.5
Utah	21.5

What is the mean ACT score for the states listed?

A. 20.7
B. 21.5
C. 21.4
D. 18.7

44. ________

45. Forrest's class recorded the following high temperature for 12 days: 85, 86, 86, 87, 88, 86, 87, 83, 82, 85, 84, 83. What is the median temperature?

45. ________

46. A fair 6-sided die is rolled. What is the probability that it will come up at most a 2?

F. $\frac{2}{3}$

G. $\frac{1}{3}$

H. $\frac{1}{2}$

I. $\frac{1}{6}$

46. ________

GO ON

47. Justin is working on the prom committee. He wants to assign jobs fairly, so he sets up two spinners for volunteer assignments. Cindy spins both spinners. What is the probability that she will put up decorations on Friday? Express your answer as a fraction.

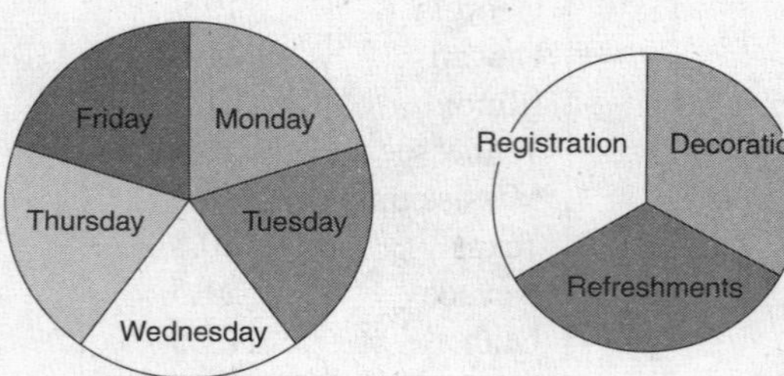

47.

49. Andrews High School conducted a survey of tenth graders each year from 2003 to 2007. Students were asked if they smoked and how often. The results are shown in the table.

Year	Frequently	Occasionally	Never
2003	103	26	191
2004	95	28	185
2005	118	20	216
2006	114	36	185
2007	101	23	209

To the nearest percent, what percent of students never smoked in 2005?

49.

48. The graph shows the percentage of the US population that was born in another country. If the most recent rate continues, predict the percentage of foreign-born people in the US population by the year 2010.

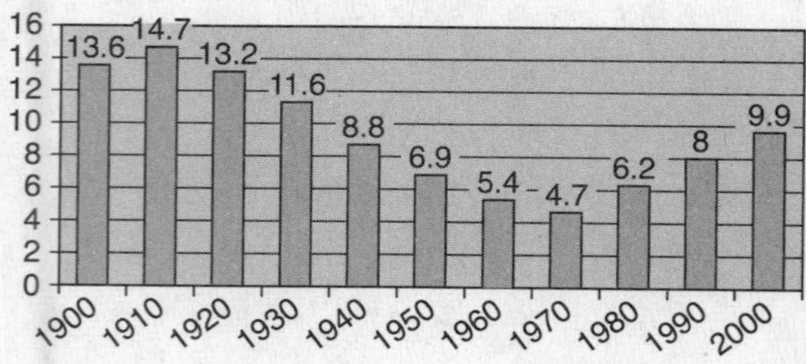

A. 11.5
B. 11.8
C. 13.2
D. 15

48. ________

50. Nam wanted to find out how students felt about rising gasoline prices. She asked the following question, "Do you think it is fair that gas prices keep rising?" She asked every student that came through the front door of the school on Monday. Was her survey valid? Write your answer and explain your reasoning.

THINK

SOLVE

EXPLAIN

50. ________

SOLUTIONS: PRACTICE TEST 1

Answer Key

1. **B**
2. **I**
3. **D**
4. **F**
5. **C**
6. **4.1%**
7. **H**
8. **2.63**
9. **8 minutes 20 seconds** or **$8\frac{1}{3}$ minutes**
10. **0**
11. **D**
12. **$1,059**
13. **$404**
14. **12**
15. **320**
16. **H**
17. **D**
18. **I**
19. **A**
20. **F**
21. **A**
22. **30**
23. **H**
24. **40°**
25. **C**
26. **I**
27. **D**
28. **I**
29. **B**
30. **8/27**
31. **H**
32. **B**
33. **$10**
34. (*a*): **(–1, 1)**
    (*b*): $\sqrt{8}$
    (*c*): **25.12**
35. **I**
36. **19**
37. See answer explanation
38. (*a*): See answer explanation
    (*b*): See answer explanation
39. **C**
40. **$115**
41. See answer explanation
42. **G**
43. (*a*): See answer explanation
    (*b*): See answer explanation
    (*c*): **77 in.**
44. **A**
45. **85.5**
46. **H**
47. **1/15**
48. **B**
49. **61°**
50. See answer explanation

Answers Explained

1. **B.** Change all the answers to decimal form and compare: $4\sqrt{59} = 30.72$. The closest answer is $30\frac{2}{3}$ yards. Find $\sqrt{59}$ and multiply by 4.
2. **I.** Change all the answers to decimal form and compare: $5\frac{1}{3} = 5.333$; $5\frac{3}{8} = 5.375$; $5\frac{3}{5} = 5.600$ (the largest).
3. **D.** The problem tells you that the answer must represent the distance (*d*). Therefore you should look for an answer written as *d* =. This, immediately eliminates **A** and **B**. To find distance, multiply the speed (or rate) by the time (use $d = rt$).
4. **F.** Change $\frac{5}{8}$ to decimal form ($5 \div 8 = 0.625$).
5. **C.** The problem is asking you to find the value of $(9\text{ ft})^3$. Multiply: 9 feet × 9 feet × 9 feet = 729.
6. 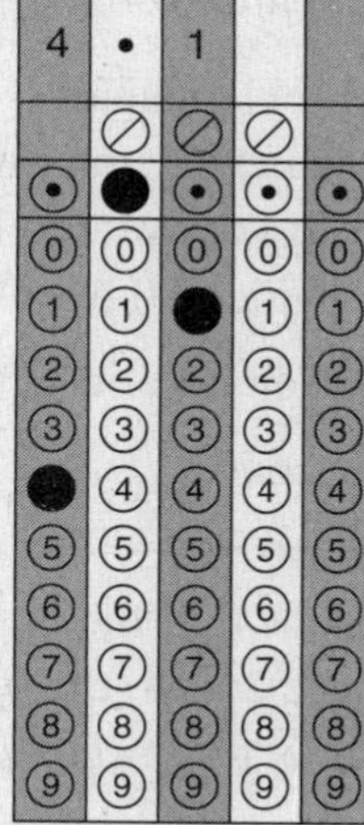

   Express as a fraction; then change to a decimal and a percent:

   $$\frac{part}{whole} = \frac{53}{1290} = .041 = 4.1\%.$$
7. **H.** $.2 \times .2 = .04$, which is smaller than .2. At first glance, –.5 appears to work, however, $-.5 \times -.5 = .25$. The positive answer is greater than –.5.

> ***Test Tip***: Multiply every answer you are given and compare if you are not certain.

8.

2 . 6 3

Work the problem step by step:

$3(1.1)^2 - (-2)^0 =$	Find powers first.
$3 \times (1.1 \times 1.1) - 1 =$	*Any* number to the zero power is 1. $(-2)^0 = 1$
$3 \times 1.21 - 1 =$	Work inside the parentheses.
$3.63 - 1 = 2.63$	Subtract 1.

9. Divide 9.3×10^7 by 186,000: $\frac{93{,}000{,}000}{186{,}000}$.

Make things easy on your calculator by canceling matching zeros: $\frac{93{,}000}{186} = 500$.

Check your problem again—light travels at 186,000 miles per second, so it would take 500 seconds. If you want to know the number of minutes, divide 500 by 60. It would take a little more than 8 minutes ($8\frac{1}{3}$ minutes = 8 minutes 20 seconds).

10. $-2^2 + (-2)^2 = -4 + 4 = 0$

Note: -2^2 stands for "the opposite of two squared"; since $2^2 = 4$, the *opposite* of 2^2 is -4. $(-2)^2$ stands for "negative 2 squared"; $-2 \times (-2) = 4$.

11. D. Find the difference between -14 and 26: $-14 - 26 = -40$.

12.

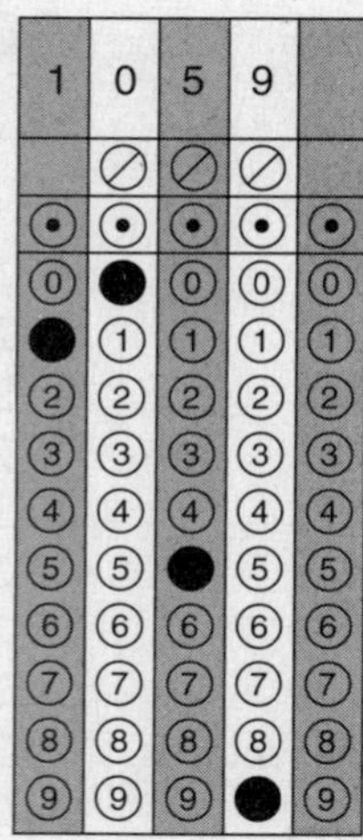

Begin with the formula $A = p(1 + r)^t$. Substitute the given values: $A = 1000(1 + 0.029)^2 = 1000(1.029)^2 = 1000(1.029 \cdot 1.029) = 1000 \cdot 1.058841 = \$1{,}058.84$, which is rounded to \$1059.

13. About \$404. Since the problem calls for you to *estimate*, before beginning round 47.25 hours to 47 hours and \$7.75 to \$8.00. Multiply 40 hours by \$8.00. Multiply 7 hours by \$8.00 and by 1.5 (for time-and-a-half). Then, $40 \cdot \$8 = \$320$ and $7 \cdot \$8 \cdot 1.5 = \$84$. Add: \$320 + \$84 = \$404.

14. About 12 hours per day. Jorge drove a total of 2820 miles ($470 \cdot 6$). If he drives 500 miles toward home, he will have 2320 miles left (or about 2300). Divide this by 3 to get the miles he must drive per day: $2320 \div 3 \approx 775$ miles per day. Divide 775 miles by 65 miles per hour to get the number of hours he must drive each day: $775 \div 65 \approx 12$.

15.

3	2	0		

Use the formula for the volume of a square pyramid: $V = \frac{1}{3}lwh$. Substitute the given values: $V = \frac{1}{3} \cdot 8 \cdot 8 \cdot 15 = \frac{960}{3} = 320$ cubic units.

16. H. $12p$ represents twelve plumbers, $(p - 800)$ is the expression for a helper's salary ($800 less than the plumber's salary). There are six helpers or $6(p - 800)$. The payroll is the sum of $12p$ and $6(p - 800)$.

17. D. Since this is an estimate, you should round 9.87 to 10 and round 1,129 to 1,000. Multiply.

18. I. There are 360 degrees in a circle. $\frac{30}{360} = \frac{3}{36} = \frac{1}{12}$.

19. A. Draw a picture:

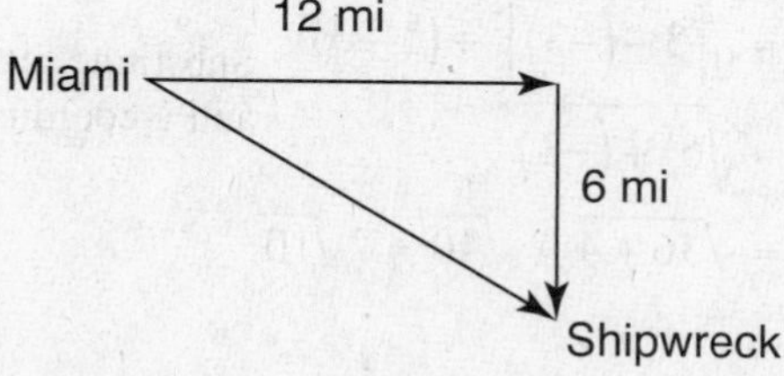

The drawing forms a right triangle The actual distance the boat went is 12 + 6 = 18 miles. If the boat had traveled directly from Miami to the shipwreck, it would have gone 13.4 miles. To find 13.4 miles, use the Pythagorean Theorem, $c^2 = a^2 + b^2$. Substitute the 12 and 6: $c^2 = 12^2 + 6^2$; $c^2 = 144 + 36 = 180$. If $c^2 = 180$, then $\sqrt{c^2} = \sqrt{180} \approx 13.4$ miles. They would have saved 18 – 13.4 or 4.6 miles.

20. F. The two angles are supplementary and must add to 180 degrees.

21. A. The break-even point is when income I equals expenses E or $6a = a + 4000$.

22.

3	0			

It will take Albert 100,000 ÷ 55 minutes to type the document. Divide this figure by 60 to find the number of hours: 100,000 ÷ 55 ÷ 60 = 30.30. Rounded to the nearest hour, it will take him about 30 hours to finish.

23. H. This is a 45-45-90 triangle or, in other words, an isosceles right triangle. The legs are congruent. To find the hypotenuse, use the Pythagorean Theorem: $c^2 = a^2 + b^2$. Substitute 5 for the legs: $c^2 = 5^2 + 5^2$; $c^2 = 25 + 25$. $c = \sqrt{50}$, which is equivalent to $5\sqrt{2}$.

Test Tip: The hypotenuse of a 45-45-90 right triangle is *always* the leg multiplied by $\sqrt{2}$.

24. $m\angle 3 = 40°$.
$\angle 1 + \angle 4 = 180$ (given)
$m\angle 2 = 110°$ (given)
$m\angle 1 = 110°$ (vertical angle to $\angle 2$)
$m\angle 4 = 70°$ (supplementary to $\angle 1$)
$m\angle 5 = 70°$ (forms a straight angle with $\angle 1$)
$m\angle 3 = 40°$ ($\angle 3 + \angle 4 + \angle 5 = 180°$).

25. C. If you think of $\overline{DG}$ as a transversal to the parallel lines, $\angle D$ and $\angle G$ are interior angles on the same side of the transversal and are therefore supplementary.

26. I. The letter O forms a circle with an infinite number of lines of symmetry.

27. D. A translation (or slide) involves moving the figure without changing its orientation or size. No flips or turns are involved.

28. I. A cross section of a grapefruit is a circle. If the diameter of the circle is 5 inches, the radius is 2.5 inches. Use the formula for the area of a circle:
$A = \pi r^2$. $A = 3.14 \times 2.5 \times 2.5 = 19.625$ square inches.

29. B. The plane cuts the pyramid exactly halfway up. The edge length of the square formed by the intersection of the plane and the pyramid is 5 cm, and the area is 5 cm × 5 cm = 25 cm².

30.

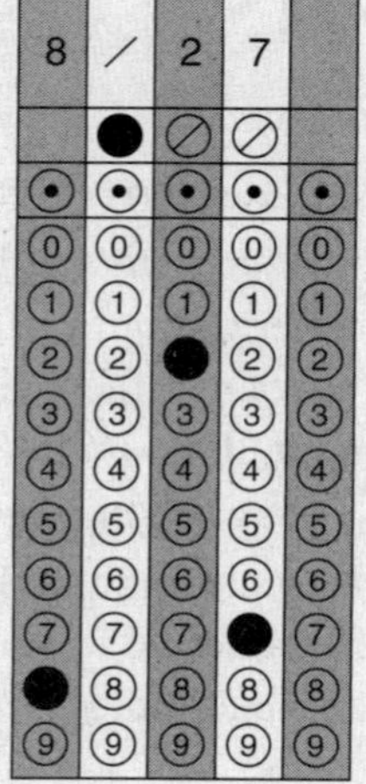

If the ratio of the radii is $\frac{2}{3}$, the ratio of their volumes will be $\frac{2^3}{3^3}$ or $\frac{8}{27}$.

31. H. The difference of 220 and a man's age y can be expressed as $(220 - y)$. 70% of this is expressed as $.7(220 - y)$ and 80% as $.8(220 - 7)$. Then, the target heart rate t is between them or greater than $.7(y - 220)$ and less than $.8(y - 220)$.

32. B.

Method 1: Form a right triangle as shown in the diagram and use the Pythagorean Theorem:

$c^2 = 6^2 + 2^2$

$c^2 = 36 + 4$ Simplify powers.

$c = \sqrt{40}$ Take the square root of both sides.

$c = 2\sqrt{10}$ Simplify $\sqrt{40}$ $(\sqrt{4 \cdot 10})$.

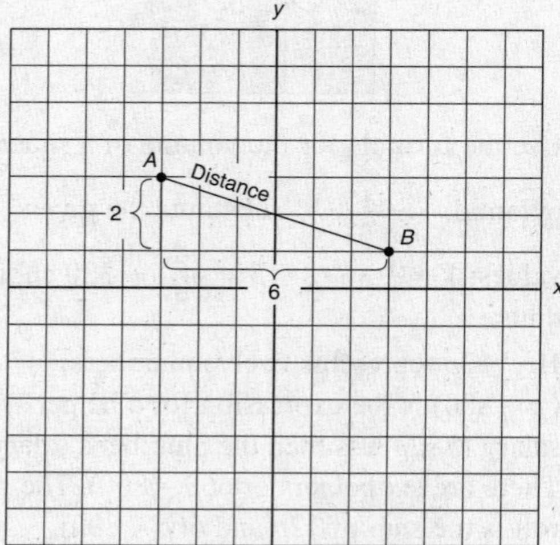

Method 2: Use the distance formula (which was derived from the Pythagorean Theorem). The formula for the distance formula will be given to you on the FCAT.

$$d = \sqrt{(x_2 - x_1)^2 + (y_2 - y_1)^2}$$

$$= \sqrt{[3 - (-3)]^2 + (1 - 3)^2}$$ Substitute the x- and y-coordinates.

$$= \sqrt{6^2 + (-2)^2}$$

$$= \sqrt{36 + 4} = \sqrt{40} = 2\sqrt{10}$$

33.

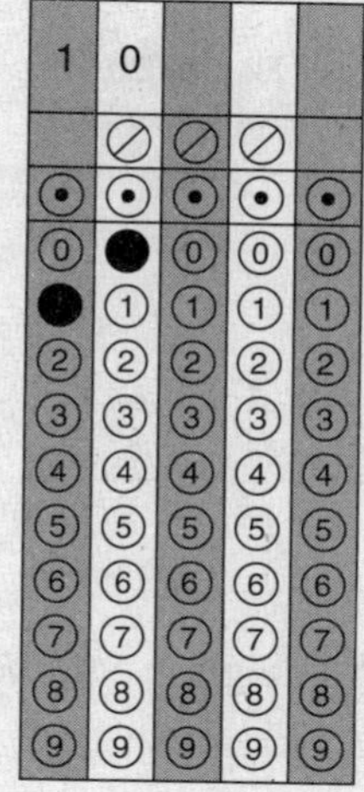

$$\text{slope} = \frac{\text{vertical change}}{\text{horizontal change}} = \frac{\$10}{1\text{ day}}$$

34. (*a*) Find the midpoint of diameter $\overline{CD}$. You can use the Midpoint Formula to find the midpoint of $\overline{CD}$. The Midpoint Formula is $\left(\frac{x_1 + x_2}{2}\right), \left(\frac{y_1 + y_2}{2}\right)$. Substitute the x- and y-coordinates into the formula:

$$\left(\frac{-3+1}{2}\right), \left(\frac{3+(-1)}{2}\right) = \left(\frac{-2}{2}\right), \left(\frac{2}{2}\right) = (-1, 1)$$

(*b*) To find the radius of the circle, you need to calculate the distance from the midpoint of the circle to a point on the edge of the circle. Use the Distance Formula $d = \sqrt{(x_2 - x_1)^2 + (y_2 - y_1)^2}$ and substitute the x- and y-coordinates for point $C(-3, 3)$ and the midpoint $(-1, 1)$:

$$d = \sqrt{[-3-(-1)]^2 + (3-1)^2}$$

$$= \sqrt{(-2)^2 + (2)^2} = \sqrt{8}.$$

(*c*) Substitute $\sqrt{8}$ for the radius in the formula for the area of a circle: $A = \pi r^2$. $A = 3.14 \bullet (\sqrt{8})^2 = 3.14 \bullet 8 = 25.12$. The area of the circle is 25.12 square units.

35. I. 4% of s should be written as $.04s$. This expression is increased by \$500.

36.

The pattern increases by 10, then 8, then 6. The next increase should be by 4.
$15 + 4 = 19$.

37. The labeled diagram should show a square with dimensions of 14 feet by 14 feet. A square is the shape that will afford the largest area.

38. (*a*) Let q stand for quarters and d stand for dimes. Then $q + d = 8$; $.25q + .10d = 1.55$.

(*b*) *Method 1*. Solve by substitution:
Let $q = 8 - d$ and substitute into the second equation:

$.25(8 - d) + .10d = 1.55$ Substitute $8 - d$ for q.

$2.00 - .25d + .10d = 1.55$ Distribute the 25 by multiplying.

$2.00 - .15d = 1.55$ Simplify.

$-.15d = -.45$ Subtract 200 from both sides.

$d = 3$ Divide by -15.

If there are 3 dimes, there must be 5 quarters because there were 8 coins.

Method 2: Set up a table. This method works very well if you don't have a *lot* of coins.

Quarters	Dimes	Value (\$)
0	8	0.80
1	7	0.95
2	6	1.10
3	5	1.25
4	4	1.40
5	3	1.55

39. C. The first equation identifies the numbers of items sold: $b + g = 46$. The second equation refers to the total cost: $25b$ (\$25 for birdhouses) plus $15g$ (\$15 for gourds) equals a total of \$1050.

40.

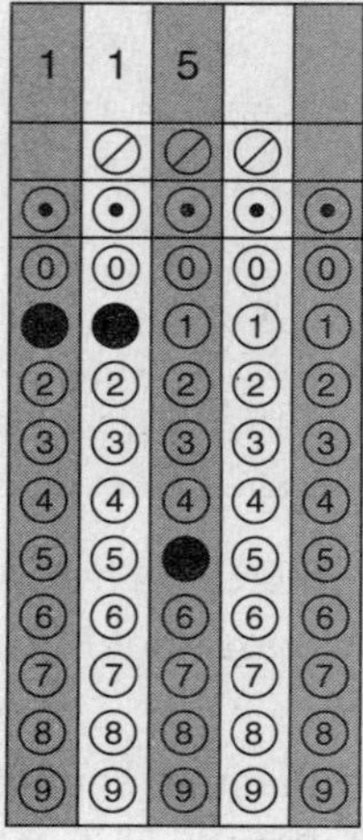

$\$250 - 3 \times 45 = \$115.$

41. Graph $y < -1$ by using a dotted line; shade the area below. Graph $y \geq \frac{1}{2}x - 2$. Graph with a solid line and shade above the line.

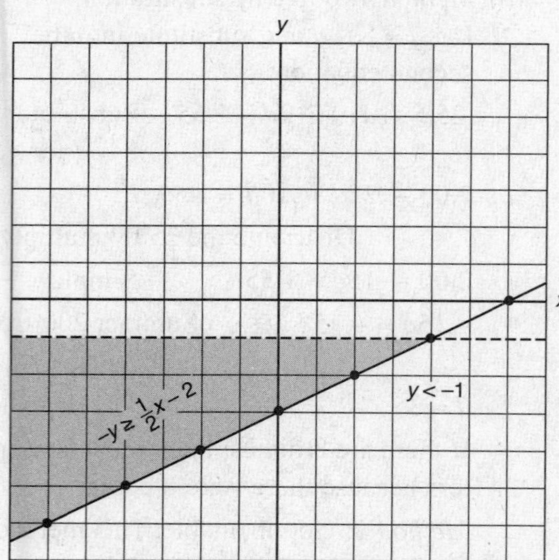

42. G. The shaded area is inside only the chicken and beef circles.

43. (*a*) and (*b*)

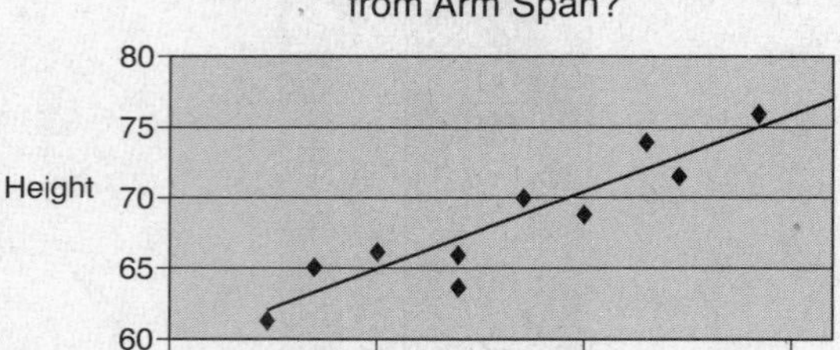

(*c*) The person's height would be about 77 inches.

44. A. The mean score is found by adding all the ACT scores and dividing by the number of scores given: $207 \div 10 = 20.7$.

45.

8 5 • 5

To find the median temperature, put all the temperatures in order: 82, 83, 83, 84, 85, **85**, **86**, 86, 86, 87, 87, 88. Since there is an even number of temperatures, take the average of the two temperatures in the center: $(85 + 86) \div 2 = 85.5$.

46. G. The expression "at most a 2" means the die could come up as either a 1 or a 2. This means there is a 2 out of 6 chance that the die will be at most a 2. $\frac{2}{6} = \frac{1}{3}$.

47.

1	/	1	5	

The probability she will work on Friday is $\frac{1}{5}$. The probability that she will work on decorations is $\frac{1}{3}$. The probability that both will happen is $\frac{1}{5} \times \frac{1}{3} = \frac{1}{15}$.

48. B. From 1990 to 2000, the foreign-born population increased by 1.9%. It would be reasonable to expect the percent to increase from 9.9% to 9.9 + 1.9 or 11.8%, which is **F**.

49.

6	1			

To find the percent, divide the number of students who did not smoke in 2005 by the total number of students in that year: $216 \div 354 \approx .61$. Expressed as a percent, $.61 = 61\%$.

50. The survey was not valid. Although Nam's survey was certainly large enough and represented the school well, her question was leading, making it more likely that those surveyed would say no.

PRACTICE TEST 2

Directions: Answer every question on this test. Place your answers in either the answer spaces or, where provided, in the "gridded-response" boxes. (Longer answers should be written on a separate sheet of paper.) Do all of your calculations on separate sheets of paper. Once you have answered every question, be sure to check your answers and to review the answer explanations at the end of the test.

1. Tim estimates he needs to buy a board at least $8\sqrt{2}$ feet long for the crossbeam of a gazebo. Which of the following could *not* be used as a board for the crossbeam?

A. 12 feet
B. $7\sqrt{3}$ feet
C. $11\frac{1}{3}$ feet
D. $11\frac{1}{4}$ feet

1. ________

2. A beekeeper is trying to increase the size of his honeybees by selecting and keeping only the largest. He measures four of the bees at 1.905 centimeters, 1.91 centimeters, 1.9 centimeters, and 1.92 centimeters. How large is the largest bee?

F. 1.92 centimeters
G. 1.91 centimeters
H. 1.9 centimeters
I. 1.905 centimeters

2. ________

3. The mean distance from Venus to the sun is given as approximately 67 million miles. What is this number expressed in scientific notation?

A. 6.7×10^5
B. 6.7×10^6
C. 6.7×10^7
D. 67×10^6

3. ________

4. Which of the following is arranged in descending order (from largest to smallest)?

F. $\pi, \frac{7}{2}, 2\sqrt{3}, \sqrt{10}$
G. $\frac{7}{2}, 2\sqrt{3}, \pi, \sqrt{10}$
H. $\frac{7}{2}, 2\sqrt{3}, \sqrt{10}, \pi$
I. $2\sqrt{3}, \frac{7}{2}, \sqrt{10}, \pi$

4. ________

5. Angelo wants to pack his umbrella in a suitcase. He measures the suitcase and finds the longest umbrella that will fit is $\sqrt{1476}$ inches. Which of the following numbers is equivalent to this measure?

A. 738
B. 246
C. $41\sqrt{6}$
D. $6\sqrt{41}$

5. ________

6. A Winter Olympics 50-kilometer men's Nordic skiing event was won by Bjoern Daehlie of Norway. His time was 2 hours 5 minutes and 8.2 seconds. To the nearest tenth of a minute, how many minutes did it take him to finish?

6.

7. Maria put $3\frac{3}{4}$ cups of flour in a bowl to make a cake. Then she realized she had put in $\frac{2}{3}$ cup too much. How much should she have put in the bowl?

F. $3\frac{1}{12}$ cups

G. $3\frac{1}{3}$ cups

H. $3\frac{2}{3}$ cups

I. $4\frac{1}{2}$ cups

7. ________

8. Express the value of $\frac{4^2 - 2}{2 + 5 \times 8}$ as a fraction in lowest terms.

8.

9. Jason was asked to find $\frac{1}{4}$ of $12\frac{4}{9}$. Since he had forgotten how to change the mixed number into an improper fraction, he tried a different method. He figured that $\frac{1}{4}$ of 12 was 3 and $\frac{1}{4}$ of $\frac{4}{9}$ was $\frac{1}{9}$. He then stated that his final answer was $3\frac{1}{9}$. Use properties of numbers to explain how this worked.

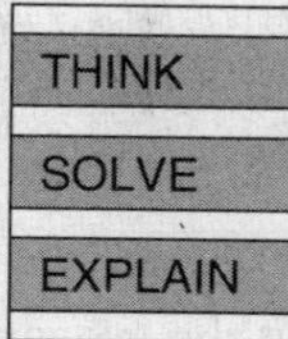

9. ________

10. Two hundred fifty tenth graders from Mason High School were scheduled to visit Epcot Center for a field trip. Two-thirds of them were able to travel on charter buses. Show and explain how many students were able to travel on these buses.

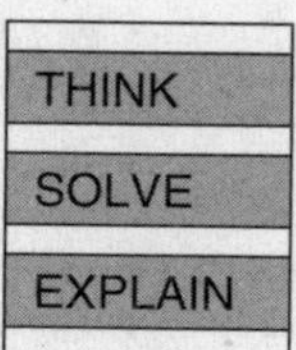

10. ________

11. Carbon monoxide emissions went down by 5% from 2007 to 2008. If these emissions were 94,410 thousand short tons in 2007, what were they in 2008?

A. 99,130 thousand short tons
B. 94,405 thousand short tons
C. 89,690 thousand short tons
D. 89,410 thousand short tons

11. ________

12. Antonio jogs across the park every morning from point *A* to point *B* and back. He has calculated the distance from point *A* to point *B* to be $3\sqrt{2}$ miles. What is the total distance he jogs? Express your answer as a decimal.

12.

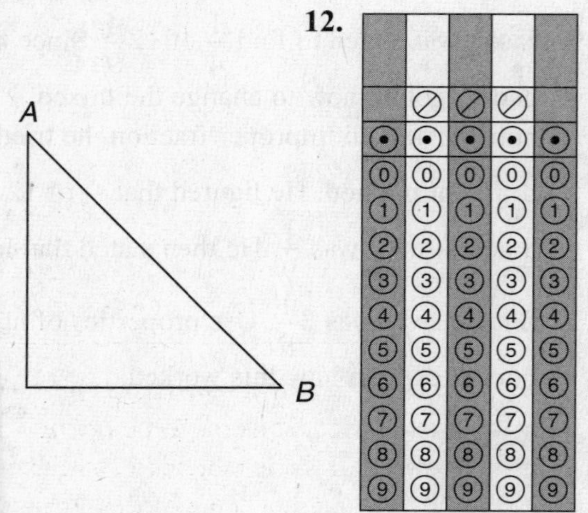

13. You buy the following items at the grocery store: 2 cans of soup at $1.29 each, 3 bags of chips at $2.79 each, and 4 two-liter bottles of soda at $.99 each. What is your approximate bill? Show your work or explain how you arrived at your answer.

THINK
SOLVE
EXPLAIN

13. ________

14. Sid is filling his 20,000-gallon swimming pool with two hoses. One hose has a diameter of $\frac{3}{4}$ inch and fills the pool at a rate of 5.6 gallons per minute. The second hose has a diameter of $\frac{5}{8}$ inch and fills the pool at a rate of 4.2 gallons per minute. Sid is concerned that if he turns the hoses on and goes to work at 7:00 A.M. the pool will run over before he returns home at 6 P.M. Will the pool run over before he returns? *Estimate* how many hours it will take to fill the pool using both hoses. Show your work or explain how you arrived at your answer.

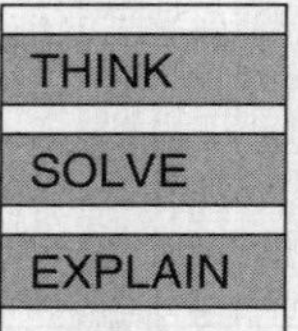

14. ________

15. Congratulations! A soup has been named after you and you get to design the label. What are the dimensions of the label if you have to add $\frac{1}{4}$ inch to the width to allow for glue and subtract $\frac{1}{4}$ inch from the height to allow for the rims? Draw a picture of the label and identify the dimensions.

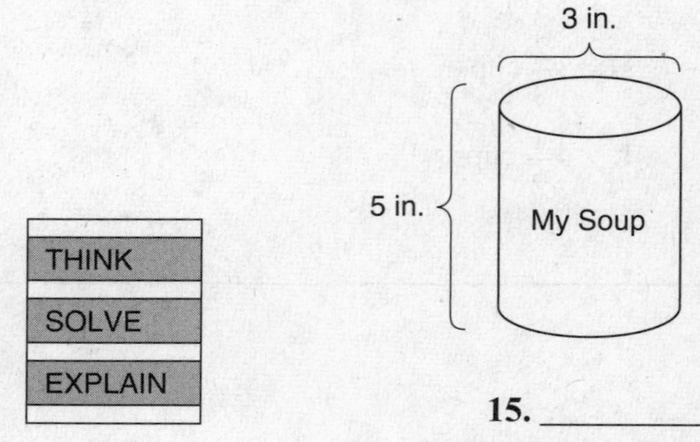

15. ________

16. The right circular cone has a base with a radius measuring 6 centimeters and a height of 10 centimeters. What is the volume in cubic centimeters?

10 cm

$r = 6$ cm

16.

17. If your heart beats an average of 70 beats per minute, how many times does it beat in a day?

F. 1680
G. 50,400
H. 4200
I. 100,800

17. ________

18. If a car averages 60 miles per hour, how long will it take it to travel 775 miles?

A. 46 hours
B. 12 hours, 55 minutes
C. 12 hours, 9 minutes
D. 8 hours

18. ________

19. Nancy jogs home from school every day. Usually the gates to the practice field are locked. On the days the gates are open she can cut across the field. How much shorter is the distance when she cuts across the field?

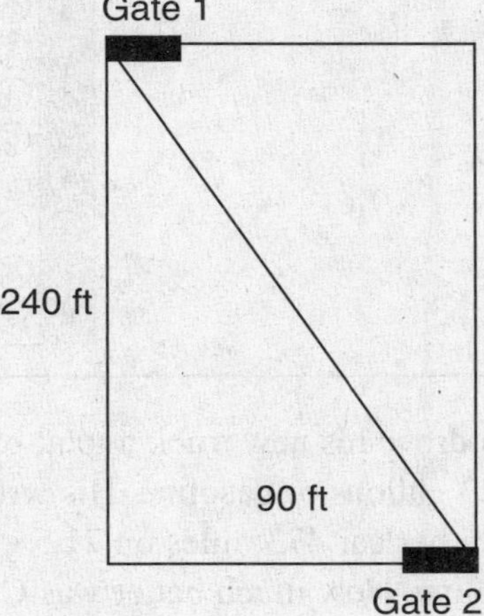

F. 256 feet
G. 222 feet
H. 182 feet
I. 74 feet

19. ________

20. An oak tree in the town square was hit by lightning and died. It needs to be cut down, but the town council is worried it will hit one of the local businesses. One of the businessmen came up with an idea to measure the height of the tree indirectly. This is the drawing he made. How tall is the tree?

(Answer grid is on page 324.)

GO ON

20.

21. Don drove his new truck a total of 368 miles on 23 gallons of gasoline. His wife, Carolyn, drove her car 473 miles on 21.5 gallons of gasoline. How much better was Carolyn's gas mileage per gallon than Don's?

A. 7 miles per gallon
C. 5 miles per gallon
B. 6 miles per gallon
D. 3.5 miles per gallon

21. ________

22. A piece of copper wire is $2\frac{1}{2}$ feet long. If the wire is cut into eight equal pieces, how many inches long will each piece be?

22.

23. In the isosceles triangle shown here, $m\angle X = 25°$. What is the measure of the complement to $\angle Z$?

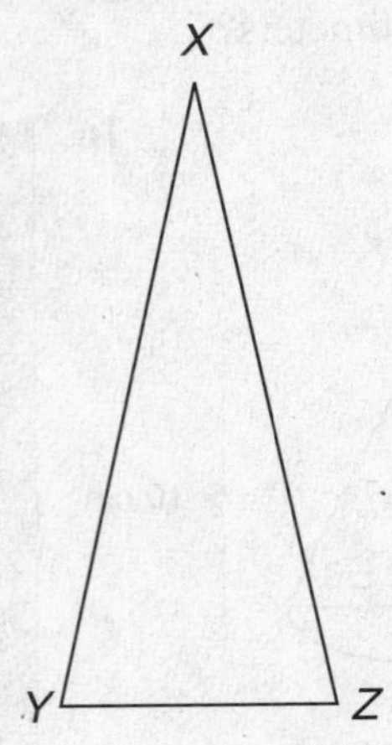

F. 155°
G. 102.5
H. 77.5
I. 12.5°

23. ________

24. In trapezoid *ABCD* shown here, $\overline{AB} = 20$ cm, and $\overline{CD} = 12$ cm. If point *M* is the midpoint of $\overline{BC}$ and point *N* is the midpoint of $\overline{DA}$, find the length of $\overline{MN}$. Show your work, explain in words, or show a proof how you found the length of $\overline{MN}$.

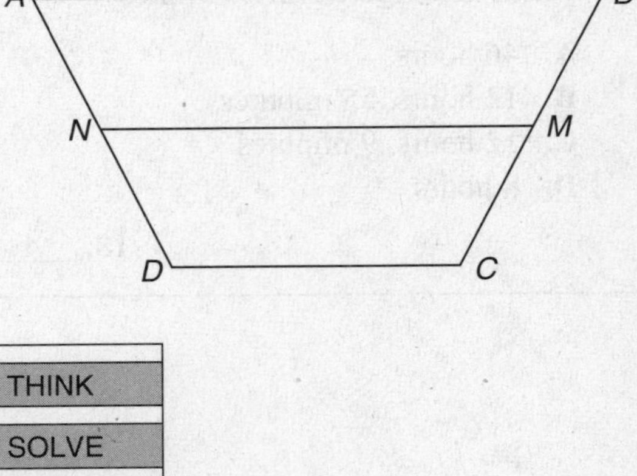

THINK
SOLVE
EXPLAIN

24. ________

25. $\overline{AB} \perp \overline{BC}$ and $\overline{EF} \perp \overline{DE}$. What further information would be required to make these two figures similar?

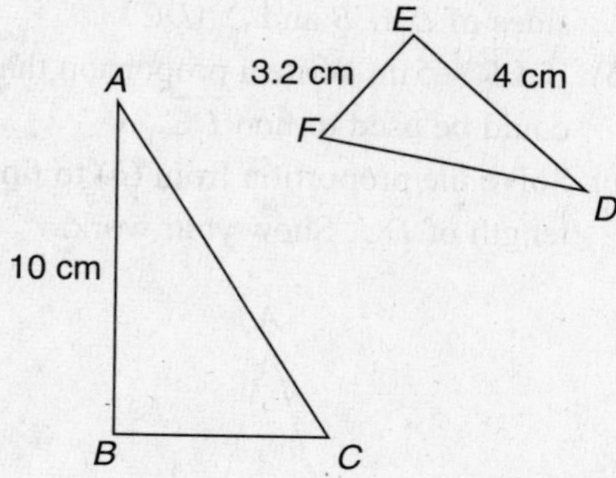

A. $\overline{BC} = 8\,\text{cm}$
B. $\angle A = 60°$ and $\angle C = 30°$
C. $\overline{AB} = 12\,\text{cm}$
D. None

25. ________

26. Which of the following proves these triangles are congruent?

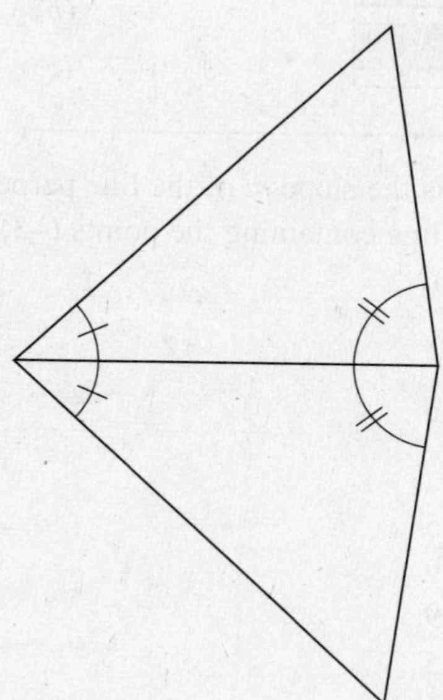

F. side-side-side (SSS)
G. angle-side-angle (ASA)
H. side-angle-side (SAS)
I. hypotenuse-leg (HL)

26. ________

27. John bought a painting by a local artist at an art gallery on Little Gasparilla Island. He wanted to put it in his suitcase to take home. A diagram of the suitcase is shown here. What are the dimensions of the largest picture that will fit into his suitcase?

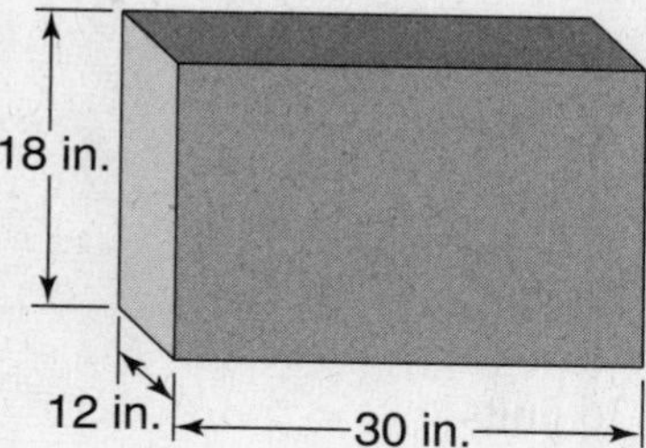

A. 30 in. × 32 in.
B. 30 in. × 21 in.
C. 30 in. × 24 in.
D. 30 in. × 12 in.

27. ________

28. A spider forms a triangle-shaped web beginning at point A and traveling to point B. From there, it draws the web to point C and then back up to point A. To the nearest centimeter, find the perimeter of the triangle formed by the web in the cube if the edge length of the cube is 12 cm.

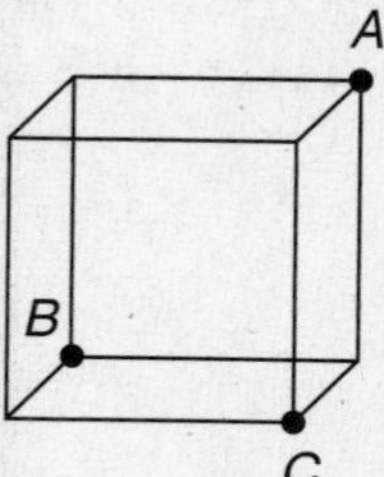

F. 51 cm
G. 36 cm
H. 24 cm
I. 17 cm

28. ________

GO ON →

29. The sides of the two regular hexagons shown are in the ratio of 1 : 3. If the area of the smaller hexagon is 12 units, what is the area of the larger hexagon?

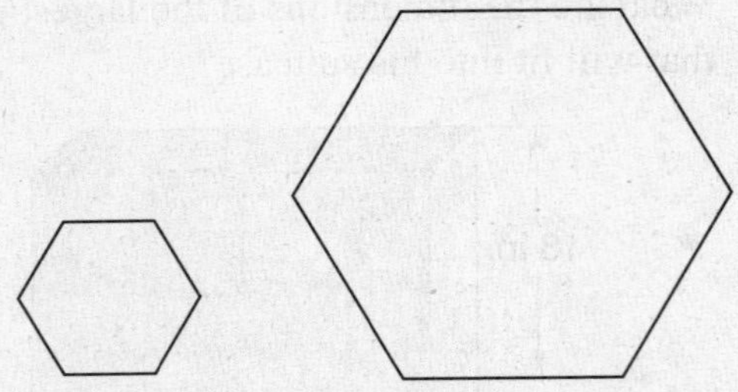

A. 36 units
B. 72 units
C. 84 units
D. 108 units

29. ______

30. A blueprint of a home shows a scale of 1 inch = 12 feet. If the actual opening for a set of patio doors is 18 feet, how big is this opening on the blueprint?

30.

31. In the diagram shown, $\overline{EB}$ is parallel to $\overline{DC}$, $\overline{AB}$ = 7 meters, and $\overline{BC}$ = 2 meters.

(*a*) What is the ratio of the corresponding sides of $\triangle AEB$ and $\triangle ADC$?

(*b*) If $\overline{EB}$ = 5 m, Write a proportion that could be used to find $\overline{DC}$.

(*c*) Solve the proportion from (*b*) to find the length of $\overline{DC}$. Show your work.

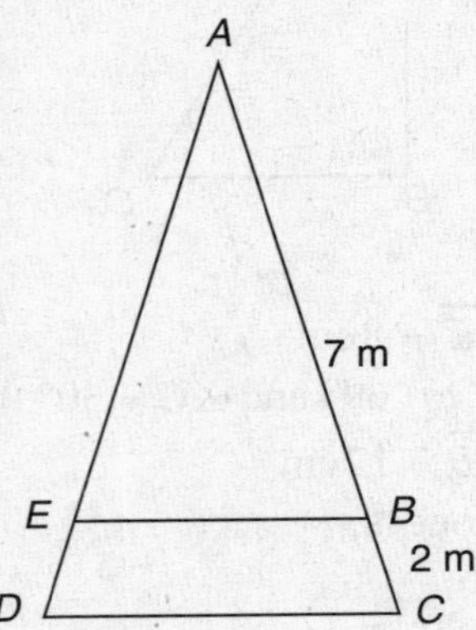

THINK
SOLVE
EXPLAIN

31. (*a*) ______
(*b*) ______
(*c*) ______

32. What is the slope *m* of the line perpendicular to the line containing the points (–3, 4) and (2, –5)?

F. $\frac{5}{9}$

G. $\frac{9}{5}$

H. $-\frac{5}{9}$

I. $-\frac{9}{5}$

32. ______

33. Polygon *ABCD* is a rectangle positioned on a grid. If point *A* is located at (–1, 6) and point *B* is located at (4, –6), how long is side $\overline{AB}$?

33.

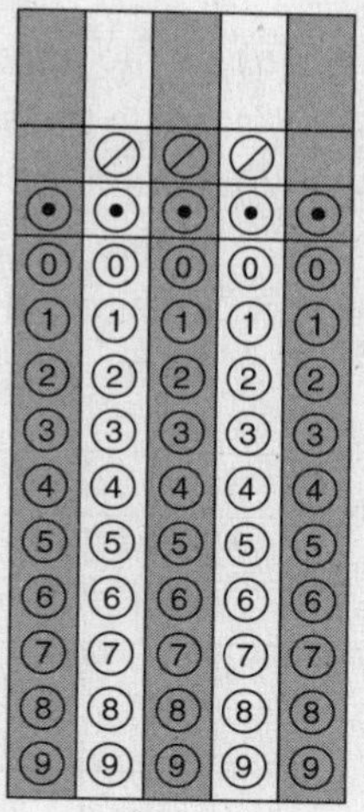

34. The graph represents the distance traveled by a boat during a trip to Key West.

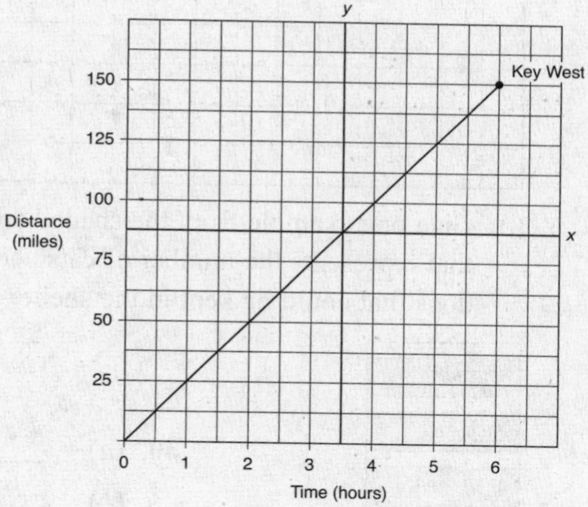

(*a*) Find the slope of the line.
(*b*) Explain what the slope represents.

THINK
SOLVE
EXPLAIN

34. (*a*) ________
(*b*) ________

35. Which expression comes next in this pattern? $2a, 4a^2, 8a^3, 16a^4, \ldots$

A. $25a^5$
B. $32a^5$
C. $32a^6$
D. $24a^5$

35. ________

36. If one more layer is added at the bottom of the stack of cubes shown here, how many cubes will there be in all?

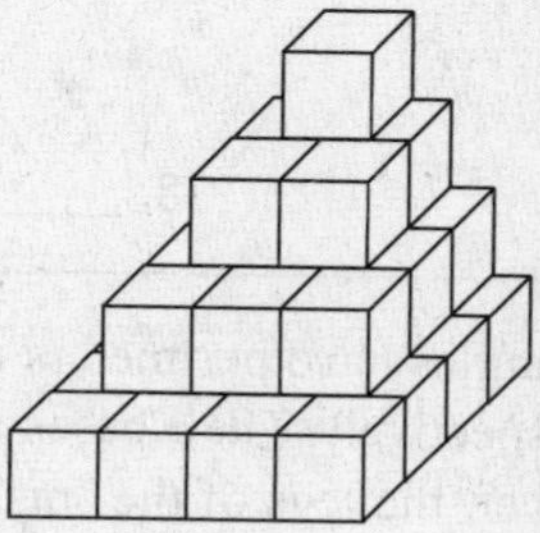

36.

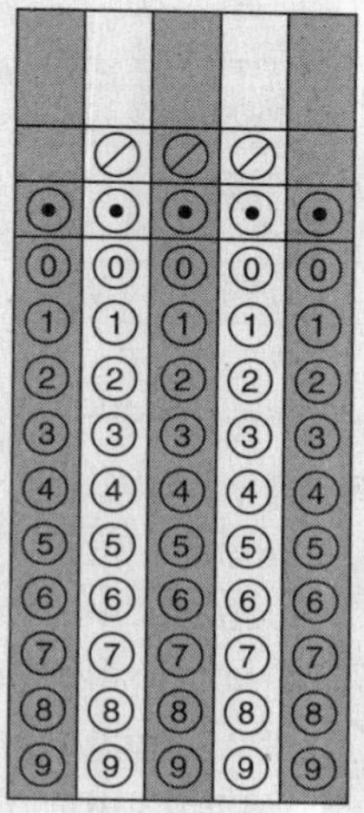

37. For which value of x is $\sqrt{x}$ undefined under the set of real numbers?

F. –4
G. 2
H. 4
I. 0

37. ________

GO ON

38. This picture shows a semicircular solarium which is to be added to a home. Once the addition is completed, what will the total square footage of the home be?

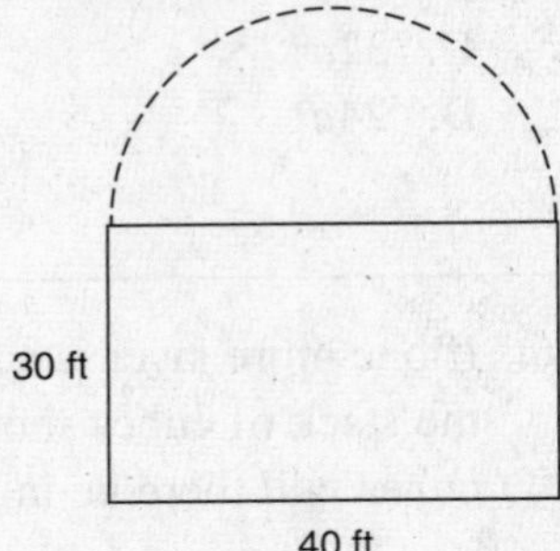

THINK
SOLVE
EXPLAIN

38. ________

39. Two isosceles triangles have perimeters of 17 and 41 inches, respectively. Their bases are congruent. However, the legs of the smaller triangle are one-third as long as the legs of the larger triangle. Which set of equations would allow you to find the length of the legs of the larger triangle?

A. $2a + b = 41$
$\frac{1}{3}a + b = 17$

B. $2(a + b) = 41$
$\frac{1}{3}(a + b) = 17$

C. $2a + b = 41$
$\frac{2}{3}a + b = 17$

D. $a + b = 41$
$\frac{1}{3}a + b = 17$

39. ________

40. The local humane society is running out of space for cats (c) and dogs (d). They can take no more than 100 animals, and no more than 40 cats.

(a) Write a system of inequalities that represents the situation in terms of c and d.

(b) Graph the system of inequalities and shade the area that shows the number of cats and dogs the humane society can keep.

(c) Give one example from the shaded area that represents the number of cats and dogs that could be kept in the shelter.

THINK
SOLVE
EXPLAIN

40. (a) ________
(b) ________
(c) ________

41. This graph shows the number of bachelor's degrees conferred during the time period 1889–2000. What major change took place between the time period 1979–1980 and 1989–1990? (*Source: 2001 World Almanac and Book of Facts.*)

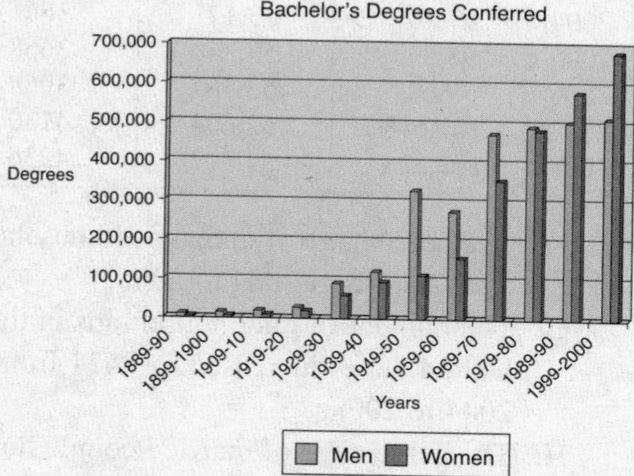

F. More than 100,000 bachelor's degrees were conferred on men between the ages of 24 and 30.
G. Women began to outnumber men in the number of bachelor's degrees received.
H. No degrees were conferred between the years 1980 and 1988.
I. The number of degrees given to women between 1980 and 1988 went up by 20%.

41. ________

42. The test scores for Ms. Kenneally's Geometry class are shown in the histogram. What is the range of the scores?

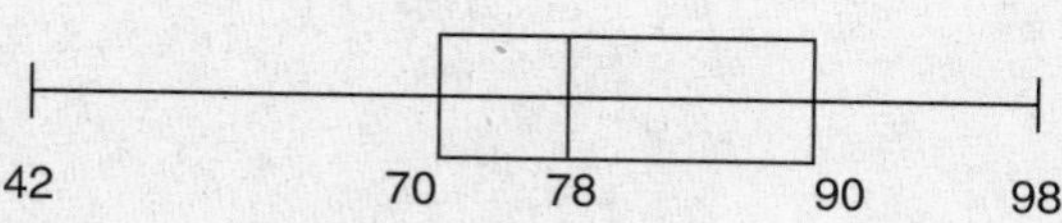

A. 20
B. 56
C. 78
D. 98

42. ________

43. The Rain Man Roof Service gave nine estimates during a week for cleaning tile roofs. The estimates were: \$150, \$175, \$180, \$160, \$175, \$200, \$175, \$150, \$185, and \$190. What was the mode of these estimates?

43.

44. Two brothers decide that the winner of two out of three coin flips will get to drive their father's new convertible. What is the probability that the contest will be over after only two flips?

F. $\frac{2}{3}$
G. $\frac{1}{2}$
H. $\frac{1}{3}$
I. $\frac{1}{4}$

44. ________

GO ON →

45. The batting average of a baseball player is really a probability that the hitter will hit the ball at any given time based on past hits. Batting averages are expressed as decimals instead of as fractions or percents. If Greg has a batting average of .300, what are the odds he will hit the next ball?

A. 3 to 10
B. 3 to 7
C. 7 to 10
D. 7 to 3

45. ________

46. The following menu was offered in a quick-serve restaurant.

Entrée	Vegetable	Dessert
Fried chicken	Broccoli	Brownie
Baked fish	Carrots	Fruit
Barbecue pork	Spinach	

Each person receives one entrée, one vegetable, and one dessert. What is the probability that the next person in line will select fried chicken, spinach, and fruit?

46.

	/	/	/	
.	.	.	.	.
0	0	0	0	0
1	1	1	1	1
2	2	2	2	2
3	3	3	3	3
4	4	4	4	4
5	5	5	5	5
6	6	6	6	6
7	7	7	7	7
8	8	8	8	8
9	9	9	9	9

47. Based on this table comparing US households by type from 2004 to 2008, which of the following is *not* true?

Year	Total US Households	Married Couple Households	Unmarried Couple Households
2004	97,107	53,171	3661
2005	98,990	53,858	3668
2006	99,627	53,567	3958
2007	101,018	53,604	4130
2008	102,528	54,317	4236

F. The percentage of unmarried households increased from 2004 to 2005.
G. The number of total households in the United States increased by 5421 from 2004 to 2008.
H. The percentage of married couple households decreased from 2004 to 2008.
I. The percentage of unmarried households increased from 2004 to 2008.

47. ________

48. The US coastline and shoreline lengths (in statute miles) are shown in this table.

Area	Coastline	Shoreline*
Atlantic Ocean	2069	28,673
Gulf of Mexico	1631	17,141
Pacific Ocean	7623	40,298
Arctic Ocean	1060	25,210

* shoreline includes offshore islands, sounds, bays, rivers, and creeks.

To the nearest whole percent, the Pacific and Arctic coastlines are what percent of the US coastline?

48.

	/	/	/	
.	.	.	.	.
0	0	0	0	0
1	1	1	1	1
2	2	2	2	2
3	3	3	3	3
4	4	4	4	4
5	5	5	5	5
6	6	6	6	6
7	7	7	7	7
8	8	8	8	8
9	9	9	9	9

49. The Murdock family is going on vacation. It will take them 6 days to reach their destination.

(*a*) Use the grid to draw a line graph showing that they travel 500 miles per day over a period of 6 days.

(*b*) Indicate on the line where the family will be after $3\frac{1}{2}$ days.

(*c*) How many miles are left on the trip after $3\frac{1}{2}$ days?

THINK

SOLVE

EXPLAIN

49. (*a*) ________

(*b*) ________

(*c*) ________

50. A local restaurant claimed that it lost 30% of its business as a result of the ban on smoking in public places. Their statement was based on a survey sponsored by the tobacco industry's public relations firm. Is this claim valid? Why or why not?

50. ________

STOP

SOLUTIONS: PRACTICE TEST 2

Answer Key

1. **D**
2. **F**
3. **C**
4. **H**
5. **D**
6. **125.1 minutes**
7. **F**
8. **1/3**
9. See answer explanation
10. **166** or **167**
11. **C**
12. **8.484**
13. **$15**
14. **33**
15. See answer explanation
16. **376.8**
17. **I**
18. **B**
19. **I**
20. **30**
21. **B**
22. **3.75**
23. **I**
24. **16 cm**
25. **A**
26. **G**
27. **B**
28. **F**
29. **D**
30. **1.5in.**
31. (*a*): **7/9**
    (*b*): $\frac{7}{9} = \frac{5}{x}$ or $\frac{7}{5} = \frac{9}{x}$
    (*c*): **6.4**
32. **F**
33. **13**
34. (*a*): **25**
    (*b*): **25 mph**
35. **B**
36. **55**
37. **F**
38. **1828 square feet**
39. **C**
40. (*a*): $c + d \leq 100$ and $c \leq 40$
    (*b*): See answer explanation
    (*c*): **20 cats, 40 dogs**
41. **G**
42. **B**
43. **175**
44. **G**
45. **B**
46. **1/18**
47. **F**
48. **70**
49. (*a*): See answer explanation
    (*b*): See answer explanation
    (*c*): **1250 miles**
50. See answer explanation

Answers Explained

1. D. Any board used has to be $8\sqrt{2}$ or 11.31 feet long or longer. **D** is the only length shorter than that.

2. F. Line the numbers up vertically and fill in with zeros:

1.920
1.905
1.910
1.900

Ignore the columns where all numbers are the same and compare those where the numbers are different.

3. C. First, write 67 million as a number in standard form: 67,000,000. Change it to scientific notation by moving the decimal point to the right of the first nonzero digit: 6.7000000. Multiply by 10 raised to the power of the number of places you moved the decimal point from its original position (7).

4. H. Change all the numbers to the same form. In this case, changing to decimal form would be easiest: $\pi \approx 3.14$; $\frac{7}{2} = 3.5$; $2\sqrt{3} \approx 3.46$; $\sqrt{10} \approx 3.16$.

5. D. $\sqrt{1476} \approx 38.42$. $41\sqrt{6} \approx 100.41$ and $6\sqrt{41} \approx 38.42$.

6. Convert hours and seconds to minutes: 2 hours × 60 = 120 minutes. Convert 8.2 seconds to minutes by dividing by 60: $8.2 \div 60 = 0.13\overline{6}$. Add: 120 minutes + 5 minutes + $0.13\overline{6}$ = 125.1 minutes.

1	2	5	.	1

Also can be gridded as 125.

7. F. Subtract: $3\frac{3}{4} - \frac{2}{3} = 3\frac{9}{12} - \frac{8}{12} = 3\frac{1}{12}$.

8.

1	/	3		

Since $\frac{4^2 - 2}{2 + 5 \times 8}$ is to be expressed as a fraction, leave the fraction bar and simplify the numerator and denominator: $\frac{4^2 - 2}{2 + 5 \times 8} = \frac{16 - 2}{2 + 40} = \frac{14}{42} = \frac{1}{3}$.

9. $\frac{1}{4}$ of $12\frac{4}{9}$ can be written algebraically as $\frac{1}{4}\left(12\frac{4}{9}\right)$ or $\frac{1}{4}\left(12 + \frac{4}{9}\right)$. He can then use the distributive property by multiplying each term inside the parentheses by $\frac{1}{4}$.

10. $\frac{2}{3}$ of $250 = \frac{2}{3} \times 250 = 166.\overline{6}$ Either 166 or 167 students could travel on the charter bus.

11. C. 94,410 × .05 = 4720.5. Subtract: 94,410 − 4720.5 = 89,689.5. thousand short tons.

12.

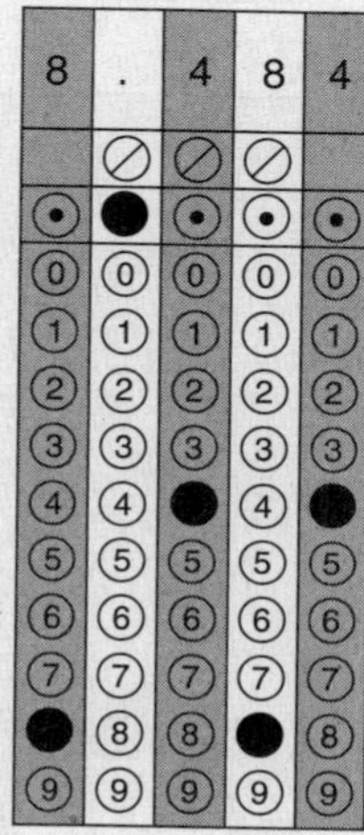

Multiply: $3\sqrt{2} \times 2 = 6\sqrt{2} \approx 6 \times 1.414 \approx 8.484$.

13. About \$15. Since you are approximating your bill, you should round the cost of items. Round the soup to \$1, the chips to \$3, and the soda to \$1; 2 cans of soup would be \$2, 3 bags of chips would be \$9, and 4 bottles of soda would be \$4, for a total bill of about \$15.

14. Sid is safe. It will take about 33 hours to fill the pool. Running both hoses adds 5.6 + 4.2 = 9.8 gallons per minute to the pool. Rounded, this is about 10 gallons per hour. 20,000 divided by 10 = 2000 minutes. 2000 minutes divided by 60 (minutes per hour) is about 33 hours.

15. The circumference of the can represents the length of the label. $C = \pi d = 3.14 \times 3 = 9.42$ inches. Add .25 inch for a total length of 9.67 inches. The height of the can is 5 inches. Subtract: $5 - .25 = 4.75$ inches for the height of the label.

9.67 in.

4.75 in.

16.

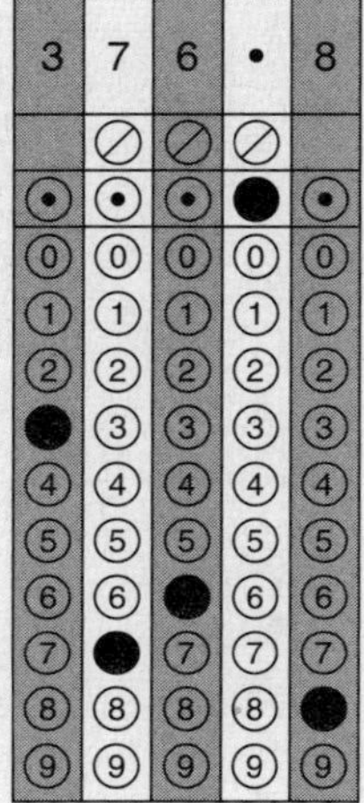

Use the formula for the volume of a cone: $V = \frac{1}{3}\pi r^2$. $V = \frac{1}{3} \times \pi \times 6^2 \times 10 = \frac{1}{3} \times \pi \times 36 \times 10 = 120\pi = 376.8$.

17. I. Multiply 70 by 60 to get beats per hour and then by 24 to get beats per day.

18. B. Divide 775 by 60. This gives you 12.91666 hours or nearly 13 hours. The closest answer is **C**. You can convert .91666 to minutes by subtracting the hours (12), leaving .91666 in your calculator display. Convert this to minutes by multiplying by 60. You should get 55 minutes.

19. I. Use the Pythagorean Theorem: $c^2 = a^2 + b^2$. Substitute the values you were given: $c^2 = 90^2 + 240^2$. $c^2 = 8100 + 57{,}600 = 8100 + 65{,}700$; $c = \sqrt{65{,}700} = 256.3$ feet. Going around the long way, she walks 330 feet, so the distance is about 74 feet shorter when she cuts across.

20.

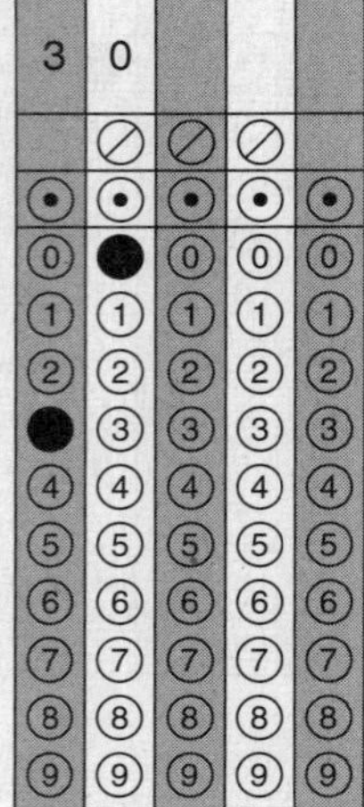

The businessman wants to use a proportion to solve the problem:

$$\frac{\text{man's height}}{\text{man's shadow}} = \frac{\text{tree's height}}{\text{tree's shadow}}.$$

Then, $\frac{6}{7} = \frac{t}{35}$, where t stands for the tree's height. Solve by cross-multiplying: $7t = 210$. Divide 210 by 7 to find $t = 30$.

21. B. The miles per gallon number was $368 \div 23 = 16$ for Don, and $473 \div 21.5 = 22$ for Carolyn. The difference was $22 - 16 = 6$ miles.

22.

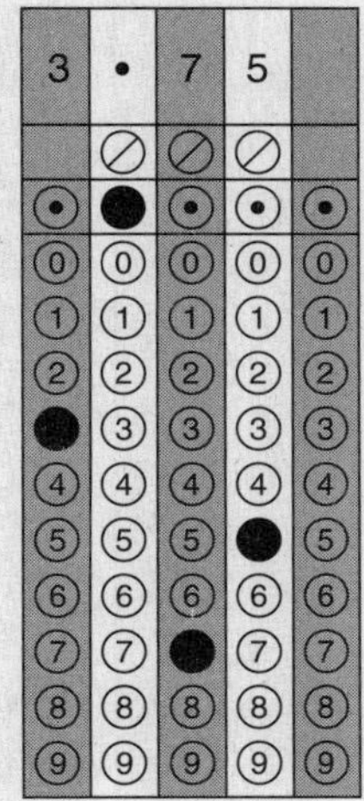

Convert $2\frac{1}{2}$ to inches (24 inches + 6 inches). Divide by 8.

23. I. If $m\angle X = 25°$, then $m\angle Y$ and $\angle Z$ must add to $180 - 25 = 155$ degrees. Because the triangle is isosceles, $\angle Z = 155 \div 2 = 77.5$ degrees. The complement to 77.5 is $90 - 77.5 = 12.5$. Remember that the sum of complementary angles is 90 degrees.

24. 16 cm. The length of the line connecting the midpoints of the legs of a trapezoid can be found by averaging the bases. $(20 + 12) \div 2 = 16$.

25. A. Figures are similar when their corresponding sides are proportional: $\frac{4}{3.2} = \frac{10}{8}$.

26. G. Two angles and the side between them.

27. B. Use the Pythagorean Theorem to find the height of the picture: $c^2 = 18^2 + 12^2 = 468$; $c = \sqrt{468} \approx 21$.

28. F. The distance from point A to point B is $\sqrt{12^2 + 12^2} = \sqrt{288} = 16.97$. The distance is the same from point B to point C and from point C back to point A. $16.97 \times 3 = 50.91$.

29. D. The ratio of the sides can be written as 1/3. The ratio of their areas is written as $1^2/3^2 = 1/9$. Use this ratio to make a proportion, substituting 12 for the area of the smaller hexagon and x for the area of the larger hexagon. $1/9 = 12/x$, $12 \times 9 = x$, or $x = 108$.

30.

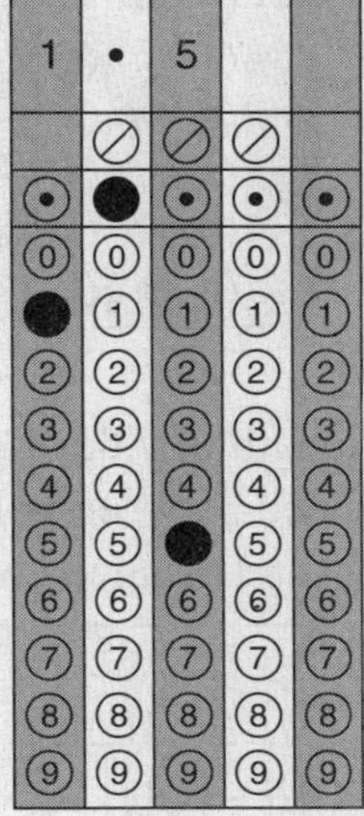

Set up a proportion: $\frac{1 \text{ in.}}{12 \text{ ft}} = \frac{x}{18 \text{ ft}} = 12x = 18$, or $x = 18 \div 12 = 1.5$ in.

31. (*a*) $\frac{7}{9}$.

(*b*) $\frac{7}{9} = \frac{5}{x}$ or $\frac{7}{5} = \frac{9}{x}$

(*c*) Cross-multiply to solve: $7x = 45$, $x = 45 \div 7 \approx 6.4$.

32. F. The slope of the line containing the points $(-3, 4)$ and $(2, -5)$ is $\frac{-5 - 4}{2 - (-3)} = \frac{-9}{5}$.

The line perpendicular to that is the opposite reciprocal or $\frac{5}{9}$.

33. 1 3

To find the length of side $\overline{AB}$ you need to find the distance between point A and point B. Use the Distance Formula:

$$\begin{aligned} d &= \sqrt{(x_2 - x_1)^2 + (y_2 - y_1)^2} \\ &= \sqrt{[4-(-1)]^2 + [-6-6]^2} \\ &= \sqrt{5^2 + (-12)^2} \\ &= \sqrt{169} \\ &= 13. \end{aligned}$$

34. (*a*) The slope of the line is $\frac{25}{1} = 25$.

(*b*) A slope of 25 represents the speed of the boat in miles per hour. The boat is traveling 25 miles for every hour.

35. B. The numbers are increasing by a multiple of 2 each time. The variable's power is increasing once with each term.

36. 5 5

There is 1 (1 × 1) in the top layer, 4 (2 × 2) in the second layer, 9 (3 × 3) in the third layer, 16 (4 × 4) in the fourth layer, and 25 (5 × 5) in the fifth layer.

37. F. The square root of any negative number is undefined under the set of real numbers.

38. 1828 square feet. Since the solarium is semicircular, find its area using the formula for the area of a circle and then divide by 2. Use $A = \pi r^2$; then the area of the solarium is $A = 3.14 \times 20 \times 20 = 1256$ divided by $2 = 628$. The area of the original home is $30 \times 40 = 1200$. Add both areas.

39. C. Let the length of the bases of the two triangles be represented by b. Let the length of the sides of the larger triangle be represented by a and the length of those of the smaller triangle by $\frac{1}{3}a$. Then, $\frac{1}{3}a + \frac{1}{3}a + b = 17$ and $a + a + b = 41$ or, simplified, $2a + b = 41$ and $\frac{2}{3}a + b = 17$.

40. (*a*) If c stands for cats and d stands for dogs, $c \leq 40$ and $c + d \leq 100$.

(*b*)

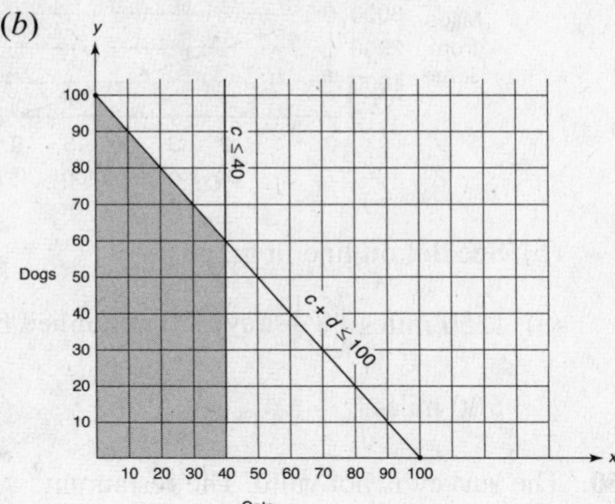

(*c*) 20 cats and 40 dogs is one possible answer.

41. G.

42. B. The range of scores represents the difference between the highest (98) and the lowest (42) scores: $98 - 42 = 56$.

43.

1	7	5		

The mode represents the data entry seen most often.

44. G. There are four possibilities with two throws of the dice: heads-heads, heads-tails, tails-heads, or tails-tails. For the winner to win after only two tosses, he will need to get either heads-heads or tails-tails. The probability of this happening is 2 out of 4, or $\frac{1}{2}$.

45. C. .300 is a probability of $\frac{300}{1000}$ or $\frac{3}{10}$. The *probability* that he will hit a ball is $\frac{3}{10}$. Three out of ten times he will hit the ball, and seven out of ten times he will not. Therefore, the *odds* are 3 to 7.

46.

1	/	1	8	

There are $3 \times 3 \times 2 = 18$ different ways to select a meal. Only one of these is fried chicken, spinach, and fruit. The probability of making this selection is $\frac{1}{18}$.

47. F. To find the percent of married couple households, divide the number for that year by the total US households for that year. For example, the percent of married couple households for 2004 is 53,171 ÷ 97,107. Change to a percent by moving the decimal point two spaces to the right.

Year	Total US Households	Married-couple Households	Unmarried-couple Households
2004	97,107	53,171 (55.8%)	3661 (3.8%)
2005	98,990	53,858 (54.4%)	3668 (3.7%)
2006	99,627	53,567 (53.8%)	3958 (4.0%)
2007	101,018	53,604 (53.1%)	4130 (4.1%)
2008	102,528	54,317 (53.0%)	4236 (4.1%)

48.

7 0

Add Pacific and Arctic coastline numbers (7623 + 1060) and divide by the length of the total coastline (2069 + 1631 + 7623 + 1060). 8683 ÷ 12,383 = 70.1%.

49. *(a)*

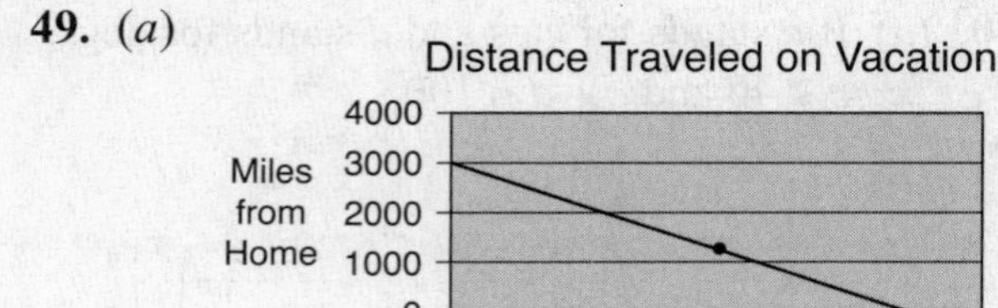

(b) See dot on line in graph.

(c) 1250 miles $\left(2\frac{1}{2}\right.$ days left multiplied by 500 miles$\left.\right)$.

50. The survey is not valid. The restaurant claimed it lost 30% of its business, but the survey was conducted by a party that was not impartial.

Index